Advances in
APPLIED PHYCOLOGY

Advances in APPLIED PHYCOLOGY

— *Editors* —
Rajan Kumar Gupta & Vidya Dhar Pandey
Department of Botany
Pt. L.M.S. Government Post Graduate College
(NAAC Accredited Grade "A" College)
(Affiliated to H.N.B. Garhwal University, Srinagar)
Rishikesh, (Dehradun) Uttaranchal, India

2007
DAYA PUBLISHING HOUSE
Delhi - 110 035

ISBN 81-7035-476-5

Published by : **Daya Publishing House**
1123/74, Deva Ram Park
Tri Nagar, Delhi - 110 035
Phone: 27383999
Fax: (011) 23244987
e-mail : dayabooks@vsnl.com
website : www.dayabooks.com

Showroom : 4760-61/23, Ansari Road, Darya Ganj,
New Delhi - 110 002
Phone: 23245578, 23244987

Laser Typesetting : **Classic Computer Services**
Delhi - 110 035

Printed at : **Chawla Offset Printers**
Delhi - 110 052

PRINTED IN INDIA

Preface

Algae, a large and diverse group of photosynthetic organisms, have been the subject of both basic and applied research over the years. This group of organisms has begun to emerge from the early descriptive stage into an experimental one, largely because of the current interest in their phototrophic metabolism, biological nitrogen fixation, environmental and ecological implications, source of innumerable products of commercial importance and prokaryotic genetic organization in blue-green algae (cyanobacteria). Their peculiar features and potential applications in agriculture, aquaculture, bioremediation, bioenergy, human nutrition and pharmaceutical industries have attracted the attention of workers from diverse fields. Cyanobacterial and algal biomass production has assumed industrial dimension. In recent years, we are witnessing interdisciplinary research in phycology.

Different aspects of algae and their potential biotechnological applications are being revealed by workers world over. India is richly endowed with the algal flora, and has long tradition of phycological research. Algae form a significant part of the subjects like Botany, Microbiology, Agriculture and Biotechnology being taught in the Indian Universities.

The research activities in Phycology have expanded in several directions. During last two decades; there has been a natural explosion of information on various aspects of these organisms. The rapid advances and developments in phycology prompted us to assemble the up-to-date information in the form of present book targeted to both post graduate students and researchers.

This volume consist of wide ranging 21 chapters which encompasses topics on ecology, morphology, physiology, stress responses, bioremediation, heavy metal toxicity, bio-fuel production, biotechnology and molecular biology. The contributing authors are nationally and internationally acknowledged experts in the field of phycology. The book with the intention of providing a sufficient depth of the subject to satisfy the needs at a level which will be comprehensive and interesting. We have tried to synthesise all the information which will be useful and hope that this book would be informative to the students, teachers, scientists and researchers in the field of basic and applied phycology.

The editors wish to thank and give appreciations to all the authors whose contributions have enriched this volume. We also express our deep sense of gratitude to our parents whose blessings have always prompted us to pursue academic activities.

It is possible that in a work of this nature, some mistakes might have crept in text inadvertently and for these we own undiluted responsibility.

The editors profoundly thank M/s Daya Publishing House, New Delhi, which is known for its reputation in quality scientific publication, for kindly accepting to publish this book on applied phycology.

The first editor is thankful to University Grants Commission, New Delhi for the financial support and wish to place on record his special thanks to his wife Mrs Alka for her cooperation in all his academic and scientific endeavours. Finally, we will be always remain debtor to all our well wishers for their blessings without which this book would not have come to light.

Rajan Kumar Gupta

Vidya Dhar Pandey

Contents

Preface *v*

1. Micro-algal Diazotrophs and their Mutants in Natural and Artificial Symbioses 1
A. Vaishampayan, R.P. Singh, R.K. Singh, H.K. Jaiswal, J.P. Shahi, T. Dey, R.P.N. Mishra, S. Sharma, H.K. Sharma and A.B. Prasad

2. Bioprospecting of Marine Algae 42
J. Rath and S.P. Adhikary

3. Fueling the Future: By Microalgae as a Source of Renewable Energy 56
C. Dayananda, R. Sarada and G.A. Ravishankar

4. Factors Influencing Algicide Production by Microcystis sp and its Effect on Selected Cyanobacteria 75
P. Jaiswal, R. Prasanna and P.K. Singh

5. Seaweeds as a Human Diet: An Emerging Trend in the New Millennium 85
P.V. Subba Rao, Vaibhav A. Mantri, K. Ganesan and K. Suresh Kumar

6. Potential Biotechnological Applications of Cyanobacteria 97
G. Abraham

7. Chlorophyll Fluorescence Analysis: A Potential Tool for Rapid Measurement of Photosynthesis 104
Jagar Narayan Maurya

8. *Botryococcus*: A Hydrocarbon Producing Microalga 115
Yashveer Singh

9. **Food, Feed and Nutraceutical Applications of Algae** 131
Vishwanath Prasad and R.K. Gupta

10. **Nitrogen Fixing Cyanobacteria and their Potential Applications** 142
K.K. Choudhary, S.S. Singh and A.K. Mishra

11. **Cyanobacterial Hydrogen Metabolism with Special Reference to Transcriptional Regulation of Hydrogenases** 155
D. Vyas and R.K. Gupta

12. **Algae and Cyanobacteria as the Living Resources for Sustainable Environmental Management** 169
Chandra Bhushan Singh

13. **Cyanobacterial Remediation of Heavy Metal Pollution** 179
M.R. Suseela and Kiran Toppo

14. **Heavy Metal Toxicity to Rice Field Cyanobacteria** 186
Mihir Kumar Das

15. **Cyanobacterial Toxins: Risks and Impacts on Human Health** 196
B.V. Singh and Shalini Singh

16. **Algae and Disease Prevention** 209
U.K. Chaturvedi and Iqbal Habib

17. **Molecular Approaches to Cyanobacteria** 213
S. Singh, B.V. Singh and D.W. Dhar

18. **Organic Farming and Environmental Biodiversity** 229
Y.V. Singh

19. **Cyanobacteria as a Source of Pharmaceutical Compounds** 250
V.D. Pandey, R.K. Gupta and S.K. Singh

20. **Cellular Water in Desiccation-Tolerant Cyanobacteria** 261
B.S. Tiwari and S.N. Tripathi

21. **Nutraceutical, Pharmaceutical and Bioactive Potential of Cyanobacteria** 275
N.K. Rathore, R. Raghuvanshi and S. Singh

Author Index 295

Subject Index 297

© R.K. Gupta & V.D. Pandey (*Editors*)
Advances in Applied Phycology, 2007
Daya Publishing House, New Delhi

Chapter 1

Micro-algal Diazotrophs and their Mutants in Natural and Artificial Symbioses

A. Vaishampayan*[1], R.P. Singh[1], R.K. Singh[1], H.K. Jaiswal[1], J.P. Shahi[1], T. Dey[1], R.P.N. Mishra[1], S. Sharma[1], H.K. Sharma[1] and A.B. Prasad[2]

[1]*Department of Genetics and Plant Breeding, Institute of Agricultural Sciences, Banaras Hindu University, Varanasi – 221 005 (India)*
[2]*Emeritus Fellow, All-India Council for Technical Education, Centre for Bioinformatics, Ranchi – 834 002 (India)*

ABSTRACT

The established role of N_2-fixing cyanobacteria (both free-living and *Azola-Anabaena* symbiosis) in improving the fertility of rice fields has prompted isolation of natural and mutant strains of these photobio-N fertilizers resistant against various abiotic stresses, promising enough to add to the quatlitative and quantitative rice grain yield along with the physico-chemical properties of the soil. This review focuses on such genetic improvements in the cyanobacterial biofertilizers and their prospective advantages in symbiosis with plants worked out during the last couple of decades.

Introduction

The great majority of cyanophycean algae are known to possess a unique combination of "higher plant type" O_2-evolving photosynthesis and "bacteria type" oxyphobic N_2 fixation (Stewart 1980).

Corresponding Author: *Tel: 0542-2575358; 09415201138, E-mail: vaishampayan_genetist@yahoo.co.in

The significant role of these diazotrphs (generally known as dinitrogen-fixing cyanobacteria) in improving the soil fertility of paddy fields at the sole expense of photosynthetic energy produced by their own, is well documented (Singh 1961, Roger and Kulasooriya 1980, Venkataraman 1981, Watanabe 1984, Prasad and Vaishampayan 1987, Vaishampayan 1994b, 1998a, 2001). Though, the viability and N_2-fixing potential of this agriculturally useful solar energy driven low cost microflora is affected by a variety of abiotic stress factors (Häder *et al.*, 1995, Sinha and Häder 1997, Häder and Worrest 1997, Sinha *et al.*, 1998), including UV-B radiations (Sinha *et al.*, 1999), herbicides (Vaishampayan 1993f, 1995b, 1998, 2002a, Vaishampayan and Mishra 1990, Prasad and Vaishampayan 1994b, Vaishampayan *et al.*, 1998e) and synthetic nitrogenous fertilizers (Singh *et al.*, 1978a,b, Vaishampayan 1982d, 1983a,b, 1984b, 1995a, 1996, Vaishampayan and Hemantaranjan 1984). While UV-B causes irreversible genetic, physiological and biochemical damages (Sinha and Häder 1996a,b), the herbicides kill these natural nitrogen resources (as non-target species) owing to their being an uncoupler and the inhibitor of photosynthetic electron transport for the herbs (Dodge 1975). On the other hand, the synthetic nitrogenous fertilizers repress N_2 fixation above a threshold level as these organisms switch off their N_2-fixing machinery as soon as the concentration of NH_4^+ in the external environment exceeds its biological consumption, thereby converting the N_2 fixer into rather an NH_4^+ consumer (Vaishampayan and Singh 1981a,b, Vaishampayan 1982c, 1983c,d,g,h). The net result, therefore, is the fixation of not more than 27-30 kg N $ha^{-1}year^{-1}$ by the diverse free-living N_2-fixing cyanobacterial flora (Singh and Singh 1989), while the value is 300-400 kg N $ha^{-1}year^{-1}$ with respect to the *Azolla-Anabaena* symbiotic N_2 fixing complex. It is important to mention at this juncture that the vegetative mass culture of the *Azolla-Anabaena* system is often limited by its critical thermo-sensitivity, higher P dependence, and scanty spore culture (Vaishampayan 1992b, 1994a), and hence the need for systematic and meticulous genetic improvements.

Favourably, the cyanophytes are fast growing, maintain a simple genome organization, respond to DNA transformation, and display oxygenic photosynthesis, chlorophyll/amino acid biosynthesis, and an efficient homologous recombination system (Vaishampayan, 1983d 1989, Vaishampayan and Sahay 1984, Haselkorn 1992, 1995, Vaishampayan *et al.*, 2003). These properties have enabled the development of various classes of cyanobacterial mutants, genetically improved for the aforementioned agronomic traits, but these face competition with the predominant native species after field inoculation. There has been a pressing need, thus to (*i*) establish artificial symbiotic associations of the improved cyanobacterial strains with fruit, vegetable, cereal and oilseed plants in order to enable them to be away from the competition problem and fix dinitrogen in close proximity to the user for its rapid uptake and maximum beneficial utilization; and (*ii*) genetically improve *Azolla-Anabaena* mutualistic N_2-fixing system to suit to the higher thermal regime and P limited conditions, preferably through spore culture. The present review is centred around the recent developments in these areas.

Cyanobacterial Mutants of Useful Agronomic Traits

The morphologically diverse groups of asymbiotic cyanobacterial species known for their N_2-fixing properties have been listed in Table 1.1, and their supportive effects on average rice grain yield are shown in Table 1.2. Researches on cyanobacteria under laboratory conditions have been in progress parallel to the field experiments. Refinement of technology for mass multiplication of cyanobacteria for their effective use as biofertilizers has always been under consideration. Choice of suitable strains that may not only survive in the adverse and extreme ecological conditions in rice field but also be a good nitrogen fixer under such conditions has been one of the concerns (Vaishampayan 1981,1982a,b,1984a). In this context, apart from some of the important natural N_2-fixing cyanobacterial isolates, we have their improved mutant strains developed to suit to the modern agronomic fields.

Table 1.1: A List of Known Nitrogen-fixing Cyanobacterial Genera

Unicellular	*Filamentous Non-heterocystous*	*Filamentous Heterocystous*		
Aphanothece	*Lyngbya*	*Anabaena**	*Anabaenopsis*	*Aulosira*
Chroococcidiopsis	LPP group	*Calothrix**	*Camptylonema*	*Chlorogloea*
Dermocarpa	*Microcoleus chthonoplastes*	*Chlorogloeopsis*	*Cylindrospermum*	*Fischerella**
Gloeocapsa (*Gloeothece*)*	*Myxosarcina*	*Gloeotrichia*	*Haplosiphon*	*Mastigocladus*
Myxosarcina	*Oscillatoria*	*Nodularia*	*Nostoc**	*Nostochopsis*
Pleurocapsa group*	*Plectonema boryanum*	*Rivularia*	*Scytonema**	*Scytonematopsis*
Synechococcus	*Pseudoanabaena*	*Stigonema*	*Tolypothrix*	*Westiella*
Xenococcus	*Schizothrix*	*Westiellopsis*		
Trichodesmium				

* Some strains of these genera live symbiotically with other plants (after Venkataraman, 1993; Prasad and Vaishampayan, 1994a; Vaishampayan, 1994c; Sinha and Hader, 1996a; Sinha and Vaishampayan, 1997; Vaishampayan *et al.*, 2001)

Table 1.2: Average Rice Grain Yield with 10 kg ha^{-1} Cyanobacterial Biofertilizer Applied in Experimental Fields Unsupplemented or Supplemented with Inorganic N Fertilizer

*N Level**	*Rice grain yield (kg ha^{-1})*			*Reference*
	With Biofertilizer	*Without Biofertilizer*	*Per cent Increase*	
0	2,541	2,079	22.2 per cent	Venkataraman (1981)
50 N	4834	4,176	15.75 per cent	Vaishampayan *et al.* (1996)
100 N	5,485	5,112	7.29 per cent	Vaishampayan (2002b,2005a)
100 N	5,654	5,079	11.32 per cent	Vaishampayan *et al.* (2000a)

* 50 and 100 N represent the 50 per cent and 100 per cent of the recommended N levels.

Most earlier studies on the response of cyanobacteria to pesticides, fungicides, insecticides and herbicides have been confined to toxicity screening (Anand 1990), but later the possible modes of action of many of these chemicals were studied in the different cyanobacterial systems (Vaishampayan *et al.*, 1998e). While some of the pesticides at doses recommended for field application were not found to affect the cyanobacteria (Goyal, 1989), the others proved to be extremely injurious, causing inhibition of either N_2 fixation/fixed N assimilation directly, or indirectly through inhibiting chlorophyll biosynthesis, photochemical generation of reducing power, photophosphorylation, fixed N assimilation (references cited in Table 1.3).

In attempts to genetically improve the N_2-fixing cyanobacteria for developing pesticide-resistance, substantial work has been carried out during the last two decades. Astier *et al.* (1980) were the first to show that *Aphanocapsa* sp. could resist up to 10 ppm concentration of the herbicide diuron. Following this, Tiwari *et al.* (1981) reported a 100 ppm 2-4,D-resistant mutant of *Anacystis nidulans*. But, in both the cases, the cyanobacteria used were non-N_2-fixing. *Nostoc muscorum* has been a throughly worked out filamentous and heterocystous N_2-fixing cyanobacterium in which mutagenesis for resistance to a number of pesticides has been successfully induced, *e.g.*, dithane (Vaishampayan and Prasad 1981b), blitox (Vaishampayan and Prasad 1982a), monuron (Vaishampayan 1984e,f, Vaishampayan *et al.*, 1992a), diuron (Vaishampayan 1984g, 1985a), mercury-fungicide (Prasad *et al.*, 1986), atrazine (Mishra *et al.*, 1991a), sandoz6706 (Vaishampayan *et al.*, 2000b). and uracil (Vaishampayan *et al.*, 2004). In another unicllular N_2-fixing cyanobacterium, *Gloeocapsa* sp., mutants resistant to 10-100 ppm concentrations of the herbicides, glyphosate, machete, PCP and swep were reported (Singh *et al.*, 1987). Genetic recombination for many characters are now established in N_2-fixing cyanobacteria (Vaishampayan *et al.*, 2001), which includes both inter-specific (Prasad and Vaishampayan 1984, Vaishampayan 1984c, 1988a,b, Vaishampayan and Prasad 1984) as well as inter-generic (Singh *et al.*, 1987, Vaishampayan *et al.*, 2000c) transfer of herbicide and fungicide resistance markers in the N_2-fixing cyanobacterium, *Nostoc muscorum* (involving *Gloeocapsa* sp. as donor in case of the latter), suggesting that herbicide-resistance may be a plasmid-born characteristic (Vaishampayan and Mishra 1990, Vaishampayan *et al.*, 1998e), similar to the plasmid-born resistance to the herbicide paraquat reported in the bacterium, *Pseudomonas* sp. (Saleh *et al.*, 1989).

Table 1.3: Biological Effects of Various Pesticides Studied in Cyanobacteria

Pesticide	*Organism Level (ppm)*	*Effect*	*Sensitivity*	*Reference*
Atrazine	*Nostoc muscorum*	Photosynthetic inhibitor	10.00 (R)	Mishra *et al.* (1991a)
BHC	*Anabaena* sp.	Inhibitory to N_2 fixation	50.00	Das (1977)
Blitox	*Nostoc muscorum*	do	10.00 (R)	Vaishampayan and Prasad (1980,1981a,1982a)
Carbaryl	do	Mutagenic	do	Vaishampayan (1982f,,1985b)
2,4-D	do	Inhibitory to N_2 fixation	11.00	Lundkvist (1970)
do	*Cylindrospermum* sp.	do	do	do
do	*Nostoc punctiforme*	do	do	do
do	*Tolypothrix tenuis*	do	4.50	Hamdi *et al.* (1970)
Diquat	*Nostoc muscorum*	Mutagenic	10.0	Vaishampayan (1984i,j, 1985c,d)
Dithane	do	Photosynthetic inhibitor	10.00	Vaishampayan and Prasad (1981b,1982b,c)
Diuron	do	Inhibitory to CO_2/N_2 fixation	2.34 (R)	Vaishampayan (1982e,1984g,1985a)
Eptam	*Calothrix brevissima*	Inhibitory to N_2 fixation	0.10	Ibrahim (1972)
Hg-fungicide	*Nostoc muscorum*	Inhibitory to CO_2/N_2 fixation	10.00 (R)	Prasad *et al.* (1986)
Isocil	do	do	do (R)	Vaishampayan and Mishra (2003); Vaishampayan *et al.* (2004)
Lasso	do	Mutagenic	do (R)	Singh *et al.* (1979)
Linuron	*Anabaena* sp.	Inhibitory to N_2 fixation	10.00	Das (1977)
Machete	*Nostoc muscorum*	Mutagenic	do (R)	Singh and Vaishampayan (1978)
Malathion	Many species	Inhibitory to N_2 fixation	100.0	Da Silva *et al.* (1975) MCPA
do	*Nostoc muscorum*	do	19.00	Lundkvist (1970)
do	*Nostoc punctiforme*	do	do	do

Contd...

Table 1.3–Contd...

Pesticide	Organism Level (ppm)	Effect	Sensitivity	Reference
Malathion	*Cylindrospermum* sp.	do	do	do
Molinate	*Tolypothrix tenuis*	do	25.00	Hamdi *et al.* (1970)
Monuron	*Nostoc muscorum*	Inhibitory to CO_2/N_2 fixation	0.34 (R)	Vaishampayan (1984e,f); Vaishampayan and Mishra (1989); Vaishampayan *et al.*(1992a)
do	*Anabaena* sp.	do	do	Das (1977)
Paraquat	*Nostoc muscorum*	Mutagenic	10.0	Vaishampayan (1982g, 1983e,1984h,1985c)
Propanil	*Tolypothrix tenuis*	Growth stimulatory	0.010	Ibrahim (1972)
do	*Calothrix brevissima*	do	do	do
do	*Anabaena cylindrica*	do	0.030	Wright *et al.* (1977)
do	*Nostoc entophytum*	do	do	do
do	*Gloeocapsa alpicola*	Growth inhibitory	0.005	Wright *et al.* (1977)
do	*Tolypothrix tenuis*	do	0.100	Hamdi *et al.* (1970)
do	*Calothrix brevissima*	do	do	Ibrahim (1972)
do	*Nostoc entophytum*	do	0.170	Wright *et al.* (1977)
do	*Anabaena variabilis*	do	5.000	do
do	*Tolypothrix tenuis*	Chlorophyll biosynthesis inhibitor	1.800	Hamdi *et al.* (1970)
do	*Anabaena cylindrica*	Photosynthetic inhibitor	8.000	Wright *et al.* (1977)
do	*Tolypothrix tenuis*	do	do	do
do	*Nostoc entophytum*	do	do	do
do	*Nostoc muscorum*	do	10.00 (R)	Vaishampayan *et al.* (1978)
Sandoz 6706	do	Phosphorylation inhibitor	10.00 (R)	Vaishampayan *et al.* (2000b)
Trifluralin	*Tolypothrix tenuis*	Inhibitory to N_2 fixation	25.00	Hamdi *et al.* (1970)

"R" represents the successfully isolated mutants resistant to 10-100 times higher concentrations of the respective herbicides

Equally important scientific concern has been towards evaluating the effect of nitrogen and non-nitrogen commercial fertilizers on cyanobacterial N_2 fixation. Higher concentrations of the fertilizers were found detrimental to the growth of 12 cyanobacterial species (Singh 1975). Later, Rodgers (1982) found that ammonium N fertilizers at concentrations as low as 0.2 μM markedly repressed cyanobacterial N_2 fixation in fields. The non-nitrogen fertilizers did not affect heterocyst differentiation. Anand and Karuppuswamy (1987) found that lower concentration of fertilizers supported all morphological types of cyanobacteria. However, Fernandez-Valentine *et al.* (1997) found that the highest level of ammonium fertilizer (140 kg N ha^{-1}) deep-placed in rice fields near Valencia, Spain, did lead to a significant reduction in cyanobacterial nitrogenase activity, and that the partial reduction in activity increased over the cultivation cycle, being highest at the end. With the increasing practice of the cultivation of high yielding rice varieties, the input of inorganic N fertilizer is also increasing recurrently which is injurious to the N_2-fixing machinery of cyanobacteria, and hence the need to develop derepressed cyanobacterial mutants with the capacity of uninterrupted N_2 fixation even in the presence of high dose of inorganic N fertilizer (Singh and Singh 1978, Mishra *et al.*, 1991b). Anand (1992) reported that certain cyanophycean members could continue to fix nitrogen even in the presence of commercial nitrogen fertilizers. Suseela and Goyal (1995) screened several strains for their capacity to fix nitrogen in the presence of ammonium nitrogen and found that nitrogenase was active even at 100 mg ammonium ml^{-1}. Certain strains of *Anabaena* were found to excrete the fixed ammonium (Subramanian and Shanmugasundaram 1986a,b). Anand and Parmeswaran (1992) reported that there have been strains that released ammonium as these grew in normal medium.

Although some cyanobacterial strains, which inhabit the rice fields (Venkataraman 1975) release small quantities of the major fertilizer product, *i.e.*, ammonia, during active growth phase, yet most of the fixed products are made available mainly through autolysis and microbial decomposition (Martinez 1984). Under these circumstances, it is difficult to control flow of nitrogen compounds needed for the development of rice plants (Vaishampayan 1988a). A possible solution to this problem is to develop strains of cyanobacteria which release ammonium continuously. However, in this effort, some classes of cyanobacterial mutants either had a totally defunct GS, leading to exogenous glutamine requirement (Kerby *et al.*, 1985, Verma *et al.*, 1990) or had too low an expression of nitrogenase with the exogenous NH_4^+ to serve any practical purpose (Prasad *et al.*, 1991). The most common features of these derepressed mutants are reduced glutamine synthetase (GS) activity together with increased nitrogenase (Boussiba *et al.*, 1984, Lattore *et al.*, 1986, Spiller *et al.*, 1986, Subrahmanian and Shanmugasundaram 1986). The main strategy to induce extracellular release of ammonia is to supply a specific inhibitor of GS activity, *e.g.*, L-methionine-DL-sulfoximine (MSX), a glutamine/glutamate and methionine analogue (Stewart and Rowell 1975). The continuous presence of the inhibitor was a limitation for long-term ammonia production and thus for practical applications; this is because MSX leads rapidly to a deficiency of glutamine and other nitrogenous compounds which are necessary for metabolism of the cell with cessation of ammonia production and finally lysis of the cells. It was shown possible to overcome at least partly this limitation and to lengthen production period by adding small amounts of glutamine (Ramos *et al.*, 1984) or by allowing the cells to recover in the absence of inhibitor (Brouers 1986). Another approach for increasing the long-term productivity is the use of immobilized resting cells (Musgrave *et al.*, 1982, Hall *et al.*, 1985). Substantial ammonia production from dinitrogen has also been observed in the presence of MSX (Musgrave *et al.*, 1982, Ramos *et al.*, 1987, Hall *et al.*, 1985, Newton and Cavins 1985).

In contrast to the mutants of enteric bacteria that lack the *amt* system (Jayakumar *et al.*, 1989, Castorph and Kleiner 1984) and release ammonia, the nitrogen-fixing cyanobacterial mutant (Boussiba 1997) releases ammonia continuously while fixing nitrogen, but still possesses an ammonium uptake system as evidenced by the ability to accumulate $^{14}CH_3NH_3^+$. Mutants of *Anabaena variabilis* resistant to the ammonium analogue, ethylene diamine (EDA), and to the glutamane analogue, MSX, have been reported to release ammonium (Spiller *et al.*, 1986, Kerby *et al.*, 1986, 1987). These exhibit, however, a slower growth as compared to their parents. Another major problem of using cyanobacteria as a biofertilizer is the competition between indigenous and introduced strains, the former generally dominating. It was assumed that ammonium-excreting mutants, isolated from strains indigenous to the rice-field, would overcome the constraints of the rice-field environment better than strains derived from other habitats. The reinoculation and establishment of these mutants in rice-fields would thus be comparatively more successful.

In this respect, *Anabaena siamensis*, a rice-field isolate from Thailand proved promising, and is already marketed as algal biofertilizer for rice-fields. Its efficiency in increasing the growth and yield of rice plants is apparently due to its high nitrogen-fixing capacity (Antariakanonda 1982a,b). It was reported to release a variety of amino acids during active growth phase (Antariakanonda 1984) but not ammonium as observed for other nitrogen fixers found in rice-fields. Selection of mutants resistant to MSX was found to lead to reduction in GS activity with consequent release of unassimilated ammonium into the medium without the induction of MSX (Spiller *et al.*, 1986). It is assumed that a strain releasing ammonium continuously would be a better biofertilizer (Lattore *et al.*, 1986). The MSX-resistant mutant of *A. siamensis*, SS_1 (Boussiba 1997) seems to conform to this expectation (Thomas *et al.*, 1990) as it releases ammonium due to the high activity of nitrogenase, both being controlled by cell density of the culture. The direct effect is apparently that of light availability to each cell that becomes progressively limited as cell density increases. The rate of ammonium release is consequently maximum only during the early log phase of growth in batch cultures. A similar pattern is also seen under immobilized conditions. Based on the above observations, SS_1 growing in continuous culture at low cell density (Chlorophyll value of 5-7 µg ml^{-1}) seems to be an ideal system for sustained ammonium release. The rate of ammonium release by SS_1 is lower than the rates obtained for other mutants of *A. variabilis*, *i.e.*, 35-50 µ mol mg Chl-$a^{-1}h^{-1}$ in batch and immobilized cultures (Kerby *et al.*, 1986, Spiller *et al.*, 1986). Although both the parent and SS_1 strains have a similar pattern of nitrogenase activity, the activity of SS_1 enzyme is 30 per cent higher than that of the parent during steady-state of growth and 50 per cent higher under immobilized conditions. In the presence of ammonium, SS_1 nitrogenase activity is about 5-fold higher than that of the parent due to a stronger repression by the end product in the parent. GS also seems defective in SS_1, exhibiting less than 50 per cent of the parent's enzyme activity. Hence it appears that the decreased susceptibility of nitrogenase activity to ammonia repression and the defective GS are the main factors controlling the ammonium excretion in SS_1 mutant. The other ammonium-excreting mutants of *A. variabilis*, *i.e.*, SA_1, ED_{81} and ED_{92}, were found to have derepressed nitrogenase activity and lower GS activity (Hien *et al.*, 1988). Analysis of GS and its mRNA into EDA-resistant mutants suggested that ED92 is a regulatory mutant containing less GS mRNA and consequently less GS protein, as found for *Anabaena azollae* growing in symbiosis (Nierzwicki-Bauer and Haselkorn 1986). On the other hand, ED_{81} is a structural mutant with a catalytically deficient GS, resulting in reduced activity of GS, which synthesizes protein at equal amount to its parent as found for *Nostoc* sp. 7801 growing in symbiosis (Kerby *et al.*, 1987).

Fertilization of rice plants under laboratory conditions by application of another MSX-resistant mutant of *A. variabilis*, SA_1, was found successful (Lattore *et al.*, 1986). The shorter doubling time of SS_1 and the lack of a lag period at the beginning of the growth cycle as compared to SA_1 (Spiller *et al.*, 1986), which exhibits a lag period, are obvious assets for mass cultivation of the former. These characteristics make possible the production of considerable amounts of inoculum material within a short period. The usefulness of SS_1 as a biofertilizer to rice plants in the actual isolate location, Thailand, should be studied in order to scale up the technology transfer from laboratory to the rice-field.

Of special mention in this regard are the two mutant strains of *Nostoc muscorum*, resistant to the herbicide amitrole (3-amino-1,3,4-triazole) and the fungicide carbendazine (2-methoxycarbamoyl-benzamidazole), isolated by Mishra *et al.* (1991b) and Prasad *et al.* (1991), respectively, tested with rice (Vaishampayan *et al.*, 1996) which were found to support an increase in grain yield by 28-32 per cent in three traditional rice varieties, *i.e.*, "Saket-4", "Lanjhee" and "Adamchini", and 15-20 per cent in three high yielding rice varieties, *i.e.*, "Pant-4", "HUR-36" and 'Saryu-52". These two cyanobacterial mutants were found important not only with respect to their stably high resistance to the respective agro-chemicals, but also in view of their nitrogenase-derepressed nature by virtue of which they had a higher heterocyst frequency and uninterrupted nitrogen fixation even in the presence of combined nitrogen (see Vaishampayan *et al.*, 1998e). A mutation for carbendazine or amitrole resistance is suspected to have adversely affected the NH_4^+-metabolizing enzyme system as much as to inactivate the control factor for the repression of heterocysts (the N_2-fixing compartments) and N_2 fixation (Mishra *et al.*, 1991b, Prasad *et al.*, 1991). These mutants have shown higher heterocyst frequency and nitrogenase activity as compared to the wild type parent *Nostoc muscorum* (Table 1.4). Further, compared to the wild type parent, the mutants have shown release of unassimilated ammonium in the exogenous medium at a nearly doubly enhanced rate. The latter activity has been shown to be due to a combination of (i) ammonium produced as a result of N_2 fixation and (ii) metabolic conversion of nitrate and nitrite to ammonia and accumulation of the latter due to the weak ammonia-assimilating system. Biochemically, apart from N_2 fixation, the nitrogen metabolic pathways leading to the reduction of nitrate and nitrite (through nitrate reductase) is operating well in the derepressed mutants, but due to the weakening of its ammonia-assimilating pathway, the excess of ammonium is being discharged into the exogenous medium, and hence their significant supportive effects on increased grain yield in rice. These results emphasize the acceptability and efficient use of cyanobacterial mutants for an economic rice cultication. At this stage, resistance of at least one representative member from all major chemical groups of pesticides needs to be introduced one after the other to the efficient N_2 fixing cyanobacterial species. A systematic approach in this direction shall have to be made before this phycotechnology is practically and more vigorously transferred to the rice fields, as some of such reconstructed N_2-fixing cyanobacterial strains are expected to develop cross-resistance to a large number of related agro-chemicals, that would have wider applicability as viable bio-N fertilizer in wet agro-chemicalized fields.

Table 1.4: Heterocyst Frequency and Nitrogense Activity of a Nitrogen-fixing Cyanobacterium *Nostoc muscorum* and its Derepressed Mutants in Molecular Nitrogen and 1 mM Ammonium Chloride supplemented Medium*

Strain	*Resistant to*	*Heterocyst Frequency*		*Nitrogenase Activity*		*Reference*
		-N	*+N*	*-N*	*+N*	
Parent	-	6.05±0.08	0.00±0.00	42.823±0.856	0.00±0.00	Vaishampayan (1983i)
Mutant-1	Sodium azide (0.2 mg/ml)	6.82±0.16	5.02±0.34	30.530±1.412	26.388±1.747	Singh and Singh (1978)
Mutant-2	Carbendazine (0.8 mg/ml)	7.39±0.22	5.22±0.12	38.583±0.072	31.146±0.182	Prasad *et al.* (1991)
Mutant-3	Amitrole (0.5-1.0 mg/ml)	6.77±0.03	4.31±0.27	35.745±1.023	31.016±0.768	Mishra *et al.* (1991b)

* The values represent means + SE. In comparison to wild type (parent), paddy yield with mutants was enhanced by 25-30 per cent, 28-32 per cent and 28-30 per cent, respectively, in the traditional paddy varieties, *i.e.*, "Saket-4", "Lamnjhee" and "Adamchini" and 12-15 per cent, 15-22 per cent and 15-20 per cent in the high yielding paddy varieties, *i.e.*, "Pant-4", "HUR-36" and "Saryu-52" (Vaishampayan *et al.*, 1996, 2001).

Cyanobacterial Symbioses in Nature

A variety of symbiotic associations, involving cyanobacteria as the microsymbiotic partner, are on record inhabiting the macrosymbionts, *i.e.*, plants; animals (echuroid works and marine sponges), hollow shafts of the hairs of polar bears, non-photosynthetic protests (belonging to the group Glaucophyta); and bacteria (Smith and Douglas 1987). Plants exhibiting cyanobacterial symbioses in nature are (i) algae: diatom (Rai 1990); (ii) fungi: lichens (Rai *et al.*, 1983); (iii) bryophytes: liverworts, hornworts, and mosses (Duckett *et al.*, 1977, Dalton and Chatfield 1985, Meeks *et al.*, 1985; Henriksson *et al.*, 1987); (iv) pteridophyte: *Azolla* (Canini *et al.*, 1990, Tang *et al.*, 1990, Peters 1991); (v) gymnosperm: *Macrozamia* (Lindblad *et al.*, 1991) and (vi) angiosperm: *Gunnera* (Söderbäck *et al.*, 1990, Jäger *et al.*, 1997). The diatom, *Rhizosolenia*, is the one alga which associates the cyanobiont, *Richelia intracellularis*, while lichens (the fungal macrosymbionts) display a wide array of cyanobionts ranging from the unicellular forms of cyanobacteria (*Gloeocapsa*, *Gloeothece*, *Synechocystis* and *Hyella*) to the unbranched filamentous, heterocystous forms (*Calothrix*, *Nostoc* and *Anabaena*) and branched filamentous, heterocystous forms, *i.e.*, *Fischerella* (Rippka *et al.*, 1979). *Nostoc*, however, is the only genus that is found in symbiotic association with bryophytes (*Anthoceros*), gymnosperms (cycads) and angiosperm (*Gunnera*).

The symbiotic association of the cyanobiont *Anabaena azollae* with the water fern *Azolla* has been most thoroughly studied, as nitrogen fixed by this complex is very efficient in water-logged paddy fields (Peters and Meeks 1989). Obviously, the range of cyanobacterial symbioses is wider than that of other symbiotic N_2 fixers, *e.g.*, *Frankia* and *Rhizobium*. In view of the N_2-fixing nature of the majority of cyanobionts, their role in providing fixed nitrogen (ammonia) to the host plant is well studied, whereby variations have been noted in the form in which fixed N moves from cyanobiont to the plants (Gusev and Korzhenevskaya 1990). The frequency of heterocysts (the N_2-fixing compartments) in the filamentous cyanobionts is significantly higher than that in free-living cyanobacteria, and this is one of the reasons why cyanobionts fix nitrogen at remarkably higher rates than the free-living cyanobacteria (Wolk 1996), though some of the cyanobionts have low levels of glutamine synthetase (GS) as compared to the free-living cyanobacteria due to an altered gene regulation at the transcription or activation level. As a consequence, the newly fixed ammonia is mostly liberated by the cyanobiont to the host system. The other remarkable feature of the naturally occurring cyanobacterial symbiosis is the homogeneous distribution of GS in their heterocysts and vegetative cells, in sharp contrast to the free-living cyanobacteria where GS in the vegetative cells is half that of the heterocysts (Rai 1990). The cyanobiont may transfer both fixed C and fixed N to the plant (in the case of bipartite lichens), or only fixed N, while the plant fixes C to meet its own requirement (in the case of tripartite lichens and *Azolla*). However, in the case of liverworts, cycads and *Gunnera* the cyanobionts have a non-functional photosynthesis and thus while supplying fixed N they receive fixed C in return from the plants.

Despite the large variety of cyanobacterial symbioses prevalent in nature, nitrogen fixation by *Azolla-Anabaena* symbiosis is most significant, fixing approximately 2-7 kg N ha^{-1} d^{-1} in paddy fields. This is equivalent to 10-20 kg of $(NH_4)_2SO_4$ and also comparable to the amount of assimilable N fixed by the forage leguminous crops, a value which is much higher than the agronomically significant value (5 kg N ha^{-1} $crop^{-1}$ of rice) of nitrogen fixation, supporting 15-40 per cent increase in rice yields recorded at the International Rice Research Institute, Philippines, Central Rice Research Institute Cuttack, as well as at the Banaras Hindu University, Varanasi (Dey 1999, Bhan 2000). Moreover, being rich in proteins, fats and amino acids, this water fern-cyanobacterial symbiosis is also used in China and south-east Asian regions as a rich fodder for pig, chicken, fish and ducks, supplied either fresh, cooked or fermented. The dried powder of *Azolla* comprises of 27 per cent of assimilable protein and the same has been demonstrated to support a significant increase in carotene content and egg production in poultry (Nierzwicki-Bauer 1990). It is important, however, to mention at this point that mass culture (vegetative propagation) of *Azolla* is still restricted generally to the coastal regions like Orissa and Tamil Nadu (India) and other South-East Asian regions with persisting relative humidity (> 60 per cent) all the year round. This is because of the high thermo-sensitivity of *Azolla* which poses a challenging problem to raise its mass culture in the North Indian rice growing belt during sowing and transplantation, when the summer temperature rises >40°C with a corresponding decline in relative humidity <60 per cent, being detrimental to the viability as well as propagation and N_2 fixation. In addition, *Azolla* requires large amounts of phosphate for growth and metabolic activity, and its spores are not enough and do not readily germinate into sporophytes. In order to find a solution to some of these problems related to the year round use of *Azolla* in wider paddy growing sectors as an effective bio-N fertilizer, genetic improvement of this fern-cyanobacterial symbiosis has been initiated with some success through meristem mutagenesis and culture.

Mutagenesis of *Azolla-Anabaena* Symbiotic N_2-fixing Complex

The *Azolla-Anabaena* symbiotic N_2-fixing complex has attracted a great deal of attention in order to understand the nature of whole life perfect endosymbiotic association of the cyanobiont, *Anabaena azollae*, with its eukaryotic partner, *i.e.*, the macrosymbiont, *Azolla*. Obviously, a free-living form of *Anabaena azollae* becomes a pre-requisite for an alround study of this kind. Although possibilities of its occurrence in a free-living condition exist, yet it has not been possible to reassociate it with the fern (see Peters and Meeks 1989, Vaishampayan *et al.*, 1998b). Efforts made on the isolation of *A. azollae* from the various *Azolla* species have also been well documented (Huneke 1933, Venkataraman 1962, Ashton and Walmsley 1976, Becking 1979, Newton and Herman 1979, Tel-Or *et al.*, 1983, Berliner and Fisher 1987, Meeks *et al.*, 1988), but none of these has been reconstituted with an endophyte-free *Azolla*. In this context, the report of Lin *et al.* (1988) is worth mentioning on an artificial reconstitution of *Azolla-Anabaena* symbiosis from heterologous *Azolla* megasporocarps. Actually, when the apical cap and indusium were excized from a fertile megaspore apparatus, the plants emerging from such megaspores were cyanobiont-free. However, if an apical cap from a different fertile megasporocarp was placed on the top of the excixed megasporocarp, reinfection of the cyanobacterium associated with the transplanted cap could be demonstrated and verified through SEM, monoclonal antibodies and nitrogen fixation assay (Lin *et al.*, 1988). However, using this technique, when free-living form of *Anabaena azollae* or any other filamentous, heterocystous and N_2-fixing cyanobacterium was allowed to infect an endophyte-free *Azolla* megaspore apparatus, the difference in growth rate eventually resulted in one partner outgrowing the other (Peters and Meeks 1989, Vaishampayan 1993a,b). Thus, reconstitution of symbiosis by replacing the cyanobiont chamber with a cultured cyanobacterial isolate has yet to be strengthened for a conclusive output.

As a stable marker for *Anabaena azollae*, the use of DNA endonuclease RFLP detected by southern hybridization, primarily with cloned nitrogenase (*nif*) structural genes from the free-living *Anabaena* sp. PCC7120, was introduced (Franche and Cohen-Bazire 1985). A detailed analysis of sequence divergence based on such hybridization studies with leaf cavity cyanobionts (microsymbionts) of four species from the sub-section Euazolla and five strains of *Azolla pinnata* (belonging to the sub-section Rhizosperma) indicated that *Anabaena azollae* associated with the different *Azolla* species belong to a common ancestor owing to a great internal similarity, yet these show slightly divergent evolutionary lines in relation to the Euazolla and Rhizosperma sub-sections (Franche and Cohen-Bazire 1987). Other investigations on *Anabaena azollae* isolated from *Azolla filiculoides* (Tel-Or *et al.*, 1983) and two independent isolates from *Azolla caroliniana* (Newton and Herman 1979, Meeks *et al.*, 1988) indicated that these isolates were identical neither to the associated *Anabaena azollae* in any of the *Azolla* species, nor to each other (Nierzwicki-Bauer and Haselkorn 1986). Meeks *et al.* (1988) have shown as to how the cyanobiont *Anabaena azollae* from *Azolla caroliniana* has a uniformly contiguous organization of *nif HDK* genes in all the cells. This is quite distinctive if one compares the *nif HDK* genetic pattern of the other *Anabaena azollae* isolates as well as the free-living heterocyst-forming cyanobacteria (Franche and Cohen-Bazire 1985, 1987, Saville *et al.*, 1987), since in all these cases, the *nif D* gene (coding for the a sub-unit of nitrogenase) in vegetative cells is nearly 11 kbp apart from *nif K* (Golden *et al.*, 1985, Meeks *et al.*, 1988). However, alike to the situation in free-living cyanobacteria, this gap is excized with the formation of transcripted *nif HDK* operon in the later stages of heterocyst maturation (Haselkorn 1995).

In an exponentially growing culture of *Azolla* whole plant under controlled condition, *Anabaena azollae* accounts for nearly 16 per cent of the total chlorophyll and protein of the association (Ray *et al.*, 1978) as well as the optimum contents of phycobiliproteins, phycocyanin, phycoerythrin and allophycocyanin (Tyagi *et al.*, 1980, Kaplan *et al.*, 1986). In *ex-planta* condition, *Anabaena azollae* exhibits a rate of photosynthesis close to that of the free-living cyanobacterial cultures along with an action spectrum in which the optimum quantum yield occurs in the region of phycobiliprotein absorption (Ray *et al.*, 1979, Kaplan and Peters 1988). The action spectra of *Azolla* whole plant and endophyte-free macrosymbiont with respect to photosynthesis are so close to green plants that cyanobiont's contribution is hardly detectable (Ray *et al.*, 1979), as the latter fixes no greater than 5 per cent of the total CO_2 fixed by the whole plant (Kaplan and Peters 1988). However, an alteration in photosynthetic assimilation of inorganic carbon by *Anabaena azollae* in symbiosis has been shown to occur by hitherto unknown mechanism(s) affecting the non-cyclic electron transport and/or activity of ribulose-1,5-biphosphate carboxylase. Intermediates of the reductive pentose phosphate cycle during CO_2 photoassimilation are manufactured by both micro–and macro-symbionts (Ray *et al.*, 1979). Sucrose is the primary photosynthetic end product in the whole plant as well as the endophyte-free macrosymbiont but not in the microsymbiont, although it is present as an important component of the soluble carbohydrate pool extracted with boiling water from freshly isolated *Anabaena azollae* (Peters *et al.*, 1985), thereby suggesting that the cyanobiont may be having a photoheterotrophic or mixotrophic metabolism in the mature leaf cavities (see Vaishampayan *et al.*, 1998b). Although the biochemical mechanism(s) of transport/uptake of photosynthate from *Azolla* to *Anabaena azollae* is not known, yet transfer of photosynthetically fixed carbon from former to the latter has been demonstrated (Kaplan and Peters 1988).

It is solely at the cost of photosynthetically generated energy and reductant that the nitrogen-fixing (nitrogenase) activity progresses inside the *Azolla-Anabaena* system. Even the aerobic nitrogenase activity in the dark is fully dependent on the endogenous photosynthetic reserve (Peters and Calvert 1983). The ultimate dependence of dinitrogen fixation upon photosynthesis has also been demonstrated by the action spectrum for nitrogenase-mediated acetylene reduction where the rate per incident quantum in both the association and *ex-planta Anabaena azollae* is as great in the region of phycobiliprotein absorption as it is in the region of chlorophyll absorption (Tyagi *et al.*, 1981). Thus, nitrogenase activity of the cyanobiont in the mature leaf cavities of macrosymbiont is presumably supported by photosystem-I and cyclic photophosphorylation mediated energy generated inside the N_2-fixing compartments of *Anabaena azollae, i.e.*, the heterocysts (up to 30 per cent of vegetative cells develop into heterocysts during N_2 fixation, as compared to only 5-6 per cent in the case of free-living cyanobacteria) which absorb light energy least effectively harvested by *Azolla* pigments.

Regarding assimilation of fixed nitrogen, it has been established through the use of N isotopes in pulse-chase experiments that the freshly separated *Anabaena azollae* releases around 40 per cent of fixed nitrogen in the form of ammonia which is translocated from the mature cavities to the stem apices of *Azolla* macrosymbiont to support the undifferentiated filaments devoid of nitrogenase activity (Kaplan *et al.*, 1986). Ray *et al.* (1978) have demonstrated the activities of three N-assimilating enzymes, *i.e.*, glutamine synthetase (GS), glutamate synthase (GOGAT) and glutamate dehydrogenase (GDH) in crude extracts of the *Azolla* whole plant as well as the endophyte-free microsymbiont. Apparently, therefore, *Azolla* has the capacity to assimilate N_2-derived NH_4^+ into glutamate either by GS-GOGAT pathway (Rhodes *et al.*, 1980) or by concurrent GS-GOGAT-GDH activities. However, now it is known, based on kinetics of the incorporation of radioisotopic N into glutamine and glutamate as also by the effects of GS/GOGAT inhibitor(s) on this process, that *Azolla* can assimilate well both dinitrogen-derived as well as exogenous ammonium via GS-GOGAT with little or no contribution from biosynthetic GDH (Meeks *et al.*, 1987, Peters and Meeks 1989). It is important to mention here that the catalytic activities of all these three N-assimilating enzymes have been detected in the crude extracts of *Anabaena azollae* (Stewart *et al.*, 1980, Uheda 1986). Through the use of radioisotopic N it has been established that GS-GOGAT activity is operative inside the microsymbiont, while GDH activity is localized in the leaf cavity trichomes of the macrosymbiont (Meeks *et al.*, 1985). In general, the specific catalytic activity of GS in *Anabaena azollae* preparations has been relatively low as compared to the diverse cultures of free-living *Nostoc* or *Anabaena*, which obviously is an index of lower amount of GS protein in the cyanobiont (Peters and Meeks 1989). However, despite all our knowledge about the microsymbiont, *Anabaena azollae*, the lack of a fool proof technique so far for its re-association with the macrosymbiont, *Azolla*, prompted to find ways of mutation in the macrosymbiont itself rather than the cyanobiont for genetic improvement with respect to thermo-insensitivity, lop P requirement and easy sporulation/ spore germination.

It is important to point out here that the effect of combined application of free-living and symbiotic (*Azolla-Anabaena*) cyanobacterial biofertilizers on rice variety 'CO-41' has been noted to result in a tremendous increase in grain yield of rice, significantly better than the individual application of either of the two forms of cyanobacterial biofertilizer, and the values in each case further enhanced by the addition of inorganic N fertilizers (see Kannaiyan 1993). *Azolla-Anabaena* system is, nevertheless, a better organic N supplier than the free-living cyanobacteria (Singh and Singh 1989). Yet, genetic improvement through mutagenesis, which is now a wide spread phenomenon with respect to the free-living cyanobacteria (see Vaishampayan *et al.*, 1998c), is presently in a state of infancy in relation to the symbiotic cyanobacterial biofertilizer, *i.e.*, the *Azolla-Anabaena* symbiotic N_2-fixing complex. The main problem in case of the latter is the selection of an appropriate starting material for clonal mutagenesis purpose, since spores of *Azolla*, enlodging the N_2-fixing cyanobacterial germ plasm seldom germinate under the controlled condition (Vaishampayan *et al.*, 1998b). Some success has, however, been achieved in this direction through the use of shoot/frond meristem mutagenesis in *Azolla-Anabaena* symbiotic N_2-fixing complex (Vaishampayan *et al.*, 2005a). Mutants isolated have had some definite objectives to tide over the non-congenial stress factors (except in tropical/coastal areas), limiting multiplication and N_2 fixation by this symbiosis, as being described below.

In fact, despite the well documented role of *Azolla-Anabaena* N_2-fixing complex as a biofertilizer, discussed above, the irony is that its up-to-the-mark cultivation and mass culture has still remained confined to the coastal regions like Orissa and Tamil Nadu in India (Manna and Singh 1989) as well as the other South-East Asian sectors owing to the requisite high relative humidity for its vegetative propagation, prevalent specifically at those places. Likewise, there are a variety of limitations with respect to the wider adoption of this fern-cyanobacterial N_2-fixing symbiotic complex as a biofertilizer, particularly in Brazil, India and Pakistan (IRRI 1987). Mutagenesis of this fern-cyanobacterial N_2-fixing complex is a relatively new area of work through which some success has been achieved to make possible the versatile use of this symbiotic cyanobacterial biofertilizer in areas not compatible for the growth of the wild material due to one or the other reason, being described below.

Growth of *Azolla* is dependent upon the presence of the requisite concentrations of key nutrients. The most important macronutrients are K, Ca, Mg and P (Watanabe 1982). The effects of these nutrient deficiencies on growth and nitrogen fixation rates of *Azolla pinnata* have been reported and accordingly the threshold concentration of P for growth is determined to be 0.5 mM (Yatazawa *et al.*, 1980). The reported minimum P requirement of *Azolla* is 0.4 per cent of its dry weight and 20 ppm available Olsen P in soil (Ali and Watanabe 1986). Sah *et al.* (1989) reported medium P concentration between 0 and 0.5 μ mol and about 1.0 g tissue P kg^{-1} dry weight for optimum growth of *Azolla*. Such P-rich soils are uncommon. Thus, P fertilization limits the growth of *Azolla*.

The P per cent and N per cent of *Azolla* increased with the increase of flooded water P until it reached about 0.4 mg l^{-1} P. At a flood water P content below 0.15 mg l^{-1} *Azolla* suffered severely from P deficiency (Watanabe *et al.*, 1989a,b). Rakotonaivo and Schramm (1988) observed the requirement of phosphate for *Azolla* growth and deduced that accumulated P in *Azolla pinnata* could be the source of available P. Phosphorus requirement of *A. pinnata* is though lower as compared to the other *Azolla* species (Kushari and Watanabe 1991). Although interest in *Azolla* research is expanding in different global sectors, use of this source of bio-N is severely limited by its high P requirement (Watanabe and Ramirez, 1984, Kushari and Watanabe 1991). Vaishampayan *et al.* (1992b) conducted laboratory experiments to culture *Azolla* populations with reduced P requirement in trays, following N-methyl-N′-nitro-N-nitrosoguanidine (MNNG) mutagenesis procedure on excised leaf meristems of the fern. The resulting isolate was not only remarkable for its normal rates of multiplication with a 50 per cent reduced P supplement under N_2-fixing condition, but also proved useful in view of its versatility to respond equally well to the various P sources used (Vaishampayan and Awasthi 1996).

Temperature is one of the most important environmental factors governing the adaptability of *Azolla*, its growth and N_2 fixation. The problems encountered with the use of *Azolla-Anabaena* N_2-fixing complex is primarily that of high temperature (above 40°C) during summer that coincides with rice cultivation in Northern India and other rice-growing countries. The relative humidity also goes down below 60 per cent that causes severe dessication problem to *Azolla* culture. Standing water is thus required throughout its growth cycle, and since it has to be propagated vegetatively, its inoculum has to be maintained in nurseries all the year round and multiplied for distribution before field cultivation. Thus, in summer the mass culture of *Azolla* and its use with rice as a biofertilizer becomes a difficult proposition.

Numerous studies have been conducted on the effect of temperature on the growth and/or nitrogen fixation rates of *Azolla*. The differences in temperature responses of *Azolla* species and their eco-physiological strains have been reported (Peters *et al.*, 1980, Li *et al.*, 1982, Tung and Watanabe 1983, Watanabe and Berja 1983, Zimmermann 1985a,b). The optimum temperature requirement for *Azolla* is 20-30°C that is below the average temperature in tropics (Lumpkin 1987). Singh (1977a,b, 1979, 1981) suggested that the most favourable temperature for growth and N_2 fixation of *Azolla pinnata* is between 20°-30°C. Hechler and Dawson (1995) found that the peak of N_2-fixing activity in *Azolla* species was at 25°C, which sharply decreased at higher and lower temperatures.

Peters *et al.* (1980) found that tolerance of an elevated temperature was highest in *Azolla mexicana*, followed by *A. pinnata*, *A. caroliniana* and *A. filiculoides*. In another report by Tung and Watanabe (1983), the relative heat tolerance of *Azolla* strains to a 37°/29°C thermo-period was *A. pinnata* > *A. microphylla* > *A. mexicana* > *A. caroliniana* > *A. filiculoides*. Watanabe *et al.* (1989a) reported that *A. microphylla* and *A. pinnata* from Paraguay were most tolerant to high temperature and had highest N content.

In Norther India, *Azolla pinnata* grows from July to December. In February, sporulation starts. It vanishes from the ponds in hot summer at a temperature above 38°C, the period between April and June (Gopal 1967, Vaishampayan *et al.*, 1998c,d). Kannaiyan and Somporn (1989) studied the effect of high temperature on growth, N_2 fixation and chlorophyll content of five species of *Azolla-Anabaena* complex and reported species dependent variation in chlorophyll content, N content and growth rate at higher temperatures.

The *Azolla-Anabaena* symbiotic diversity and relatedness of neo-tropical host taxa has been reported (Zimmermann *et al.*, 1991). Lin (1992) reported that *Anabaena azollae* plays some role in tolerance of *Azolla* to higher temperatures. However, Watanabe *et al.* (1989b) opined that the determining factor(s) for heat is possibly present in both host fern and the microsymbiont, *Anabaena*. Tung and Watanabe (1983) reported that exposure of *Azolla* to 37°C/29°C progressively decreased the heterocyst frequency (nitrogen-fixing sites) in *Anabaena* and hence N_2-fixing ability of the association, as compared to the situation at 28°C/20°C, suggesting thereby that high temperature directly affects the microsymbiont. Lai *et al.* (1988) stated that protein synthesis in the test organism eventually succumbed to higher temperatures.

In order to solve the above problems with *Azolla*, mutagenesis work has been initiated in this symbiotic complex not only with a view to genetically improving them to be reduced phosphorus-requiring (Vaishampayan *et al.*, 1992), but also for making them resistant to higher temperatures (Vaishampayan *et al.*, 1995b,c). Through laboratory tests these novel mutants have proved their enhanced efficiency of use as a better N_2-fixing biofertiolizer across many of the above mentioned stress barriers (Vaishampayan and Dey 1998).

Azolla pinnata mutant (Vaishampayan *et al.*, 1992c) grows well in the laboratory with the 50 per cent reduced P supplement (10 mg l^{-1}) while the wild type strain grows strictly with the normal (20 mg l^{-1}) P supplement. P levels are not so absolute for plants (Tisdale and Nelson, 1975) but the same appears to be very specific for *Azolla*, as the reduction in P concentration caused the typical P-deficiency symptom (elongation of rhizoids and browning of the material, followed by death of the organism) in the wild type material. However, no such P-deficiency symptom was noted in the mutant *Azolla* at the reduced P supplement (Vaishampayan and Banerjee 1995). The mutant, therefore, appeared to be a definitely an improved material with respect to its P requirement. Whether the mutant has a high P utilization efficiency of higher P uptake has not yet been worked out and thus it seems to be a fertile area of future biochemical work in *Azolla-Anabaena* symbiotic N_2-fixing complex.

The mutant strain of *Azolla* was found to grow well up to a temperature of 40°C, while the wild type strain showed a sharp decline in growth with the temperature rising above 25°C (Vaishampayan *et al.*, 1998b,c, 2000a). It appears, therefore, that alike to the thermophilic forms of bacteria and cyanobacteria, the mutant *Azolla* has possibly had a considerable degree of intracellular physiological and biochemical specializations as a mutational event, enabling it to tolerate higher temperatures. In fact thermostability has been reported to be brought about by a high stability of cellular proteins and enzymes (Koike and Katoh 1979) and membrane systems (Fork and Murata 1977, Yamaoka *et al.*, 1978) for the proper functioning of the vital life processes necessary for survival, particularly photosynthesis and nitrogen fixation (Watanabe *et al.*, 1989a,b). A large variety of such stable enzymes, *e.g.*, glyceraldehydes-3-phosphate dehydrogenase, alcohol dehydrogenase, phosphofructokinase, superoxide dismutase and a restriction endonuclease have been isolated from *Bacillus stearothermophillus* growing at a temperature rising up to 60-65°C, the functional and molecular properties of which are quite similar to those of the respective counterparts from mesophillic microbes, with the only difference that they are more resistant to heat denaturation in case of the former one (Fontana 1984). However, qualitative and quantitative estimations of the possible existence of such heat tolerant system still need to be worked in the mutant *Azolla* at the molecular and biochemical levels as an important area of future study.

The improved features of mutant *Azolla* enabled it to multiply in field nurseries with the 50 per cent reduced P input at the normal rate just prior to and during rice sowing and planting when summer temperature in the North Indian belt (Varanasi and adjoining areas) rises much above 40°C (with the water temperature touching 40+1°C). Rather, the rate of multiplication of this mutant strain of *Azolla* with 7 kg P ha^{-1} was much better than that of wild type *Azolla* cultured in field nurseries with the P input of 14 kg ha^{-1} (Dey 1999). The reason being the temperature-sensitivity of the latter which, unlike the low P-requiring temperature tolerant mutant, could not thrive the summer temperature of North Indian area just prior to and during the rice sowing/planting season. This was the reason why WT *Azolla* nurseries requires a higher labour input for daily twice cool watering of the *Azolla* plants to safeguard their viability and multiplication even at a lower rate, as the same was essential to run parallel experiments with the mutant *Azolla* for assessing their possible biofertilization activities in the two different rice genotypes, tested by Bhan (2000).

The comparative results obtained on biofertilization of the rice varieties, Saket-4 (a traditional variety) and HUR-36 (a high yielding variety), with wild type and thermo-tolerant mutant *Azolla* strains at graded N levels have been quite interesting, as these have shown supportive effects of biofertilization on grain yield and related physiological traits in both the rice varieties (Dey 1999, Bhan 2000). Table 1.5 demonstrates increase in grain yield through biofertilization with wild type and mutant *Azolla-Anabaena* bio-N fertilizer. Table 1.6 obviates the biofertilization effect (in terms of facilitating per cent increase in rice grain yield), and represents the N economy achievable as a result of wild type or mutant *Azolla* application (Vaishampayan 1998c,d, Vaishampayan *et al.*, 2005a,b).

Table 1.5: Per cent Increase in Grain Yield of Saket-4 (after Vaishampayan *et al.*, 1998c) and HUR-36 (after Vaishampayan *et al.*, 1998d) Rice Varieties Over Respective Controls on the Application of Wild Type or mutant *Azolla*

Rice Variety	*Inorganic N Level (kg ha^{-1})*	*Per cent Increase in Grain Yield with*	
		Wild Type Azolla	*Mutant Azolla*
Saket-4	0	51.30	73.57
	40	24.37	32.92
	60	27.41	38.22
	80	6.02	13.21
HUR-36	0	39.04	60.88
	40	11.78	19.83
	60	13.40	24.08
	80	2.56	5.64

Comparative response of two rice genotypes to Azolla:

Both the rice varieties responded to *Azolla* maximally at 60 kg N ha^{-1} level. Although, HUR-36 rice variety is a better yielder than Saket-4, yet the per cent increase in grain yield of Saket-4 rice variety through *Azolla* application was higher than that of HUR-36 by:

(i) 12.26 per cent (with WT *Azolla*), 12.69 per cent (with mutant *Azolla*) at 0 N level;

(ii) 12.59 per cent (with WT *Azolla*), 13.09 per cent (with mutant *Azolla*) at 40 N level;

(iii) 14.01 per cent (with WT *Azolla*), 14.14 per cent (with mutant *Azolla*) at 60 N level;

(iv) 3.46 per cent (with WT *Azolla*), 7.57 per cent (with mutant *Azolla*) at 80 N level.

Table 1.6: Saving of Inorganic N Fertilizer Due to the Application of Mutant *Azolla* to Obtain the Optimum Grain Yield in Saket-4 (after Vaishampayan *et al.*, 2005a) and HUR-36 (after Vaishampayan *et al.*, 2005b) Rice Varieties

Rice variety	*Treatment*	*Grain Yield ($q\ ha^{-1}$)*	*Saving of Fertilizer N ($kg\ ha^{-1}$)*	*Additional Gain Due to Azolla ($q\ ha^{-1}$)*
Saket-4	$N_{40}A_0$	16.49		
	$N_{40}A_1$	20.51		4.02
	$N_{40}A_2$	21.92		5.43
	$N_{60}A_0$	23.31		
	$N_{60}A_1$	29.70		6.39
	$N_{60}A_2$	32.22	20 $kg\ ha^{-1}$	8.91
	$N_{80}A_0$	32.02*		
	$N_{80}A_1$	32.89		1.87
	$N_{80}A_2$	35.12		4.10
HUR-36	$N_{40}A_0$	32.04		
	$N_{40}A_1$	35.85		3.78
	$N_{40}A_2$	38.43		6.36
	$N_{60}A_0$	39.77		
	$N_{60}A_1$	45.10		6.33
	$N_{60}A_2$	49.35	20 $kg\ ha^{-1}$	9.58
	$N_{80}A_0$	49.07*		
	$N_{80}A_1$	50.33		1.26
	$N_{80}A_2$	51.84		2.77

N_0 = 0 kg N ha^{-1} N_1 = 40 kg N ha^{-1} N_2 = 60 kg N ha^{-1} N_3 = 80 kg N ha^{-1}

A_0 = Without *Azolla* A_1 = Wild type *Azolla* A_2 = Mutant *Azolla*

The optimum grain yield obtainable with 80 kg inorganic fertilizer N ha^{-1} (without biofertilization); the same is obtainable with only 60 kg N ha^{-1} N when biofertilized with mutant *Azolla*, thus making possible a saving of 20 kg N ha^{-1} in both the rice varieties through mutant *Azolla* application.

A significant increase in rice grain yield as a result of biofertilization with *Azolla* may be attributed to the higher biomass build up of *Azolla* inoculated to the experimental rice field which, on mineralization, may have added sufficient organic matter and N to the soil to improve the C:N ratio of the soil. The diazotrophic (N_2-fixing) character of *Azolla-Anabaena* symbiosis is the other important reason through which *Azolla* adds to soil N fertility, thereby facilitating a better availability of assimilable N to rice crop for yield improvement. Moreover, *Azolla* canopy prevents light from penetrating to the flooded water; thus the growth of other phototrophs is inhibited and photodependent CO_2 uptake is depressed. *Azolla* may be expected to reduce N losses from inorganic sources by lowering ammonia volatilization (Singh 1977a,b, Hamdi *et al.*, 1980, Varghese 1990, Vlek *et al.*, 1995). Organic nitrogen from *Azolla*, due to its slow mineralization, might have supplied additional amount of available nitrogen which is reflected through higher grain yield. Supporting evidences towards increase in grain yield through *Azolla* application are already on record (Singh, 1980, Barthakur and Talukdar 1983, Mandal and Bharti 1983, Roy 1984, Watanabe 1985, Mahapatra and Sharma 1989, Kannaiyan

1993, Rathore *et al.*, 1995). Further, *Azolla* has been reported to have a lower K absorption threshold than rice in flood water (Liu 1987), which enables it to become a source of K for rice when incorporated, and there has been indications that K is generally a direct contributor to grain formation (Liu 1987). All these may collectively add to an enhanced grain yield in the two rice varieties tested through biofertilization with two strains of *Azolla*.

The use of mutant *Azolla* has given a much better mean performance over WT *Azolla* in both the rice genotypes which may be stated as statistically significant (Dey 1999, Bhan, 2000). This enhanced yield performance with mutant *Azolla* may be due to its reduced susceptibility to higher temperature (prevalent during rice cultivation in this area), resulting into greater organic matter supply/N-contribution. In N-unfertilized condition gains in per cent grain yield were, however, quite substantial through biofertilization with WT *Azolla* (51.30 per cent in Saket-4, and 39.04 per cent in HUR-36), but even higher through mutant *Azolla* (73.57 per cent in Saket-4, and 60.88 per cent in HUR-36). Such a substantial increase in grain yield with *Azolla* under N-unfertilized condition may be due to the fact that the N released by *Azolla* has been the only available external N source that may have directly contributed towards yield enhancement. Similar results have been reported earlier on grain yield enhancement in rice with *Azolla* by 36.6-38.0 per cent (Barthakur and Talukdar 1983) and 32.30 per cent (Singh and Mandal 1997). Under N supplemented condition the picture of per cent gain in grain yield with *Azolla* has been, however, different. At 40 N level, the gain in per cent grain yield was 24.37 per cent (Saket-4) and 11.78 per cent (HUR-36) with WT *Azolla*, and 32.92 per cent (Saket-4) and 19.83 per cent (HUR-36) with mutant *Azolla*. The maximum per cent gain in grain yield with *Azolla* under N-fertilized condition was noted at 60 N level (27.41 per cent and 13.40 per cent by wild type *Azolla*; and 38.22 per cent and 24.08 per cent by mutant *Azolla* in Saket-4 and HUR-36 rice varieties, respectively). Apparently, this has been the result of an appropriate combination of inorganic and organic N, contributing quite sufficiently towards grain yield increases in both the rice genotypes. Accordingly, at 80 N level quite lower gains in per cent yield could be obtained (6.02 per cent and 2.56 per cent by wild type *Azolla*; and 13.21 per cent and 5.64 per cent by the mutant *Azolla* in Saket-4 and HUR-36 rice varieties, respectively). Such a narrow range of increase in per cent grain yield at 80 N level may be due to (i) physiological yield stagnation at this N level in the two rice genotypes tested, and (b) a considerable repression of the N_2-fixing machinery of the *Azolla-Anabaena* symbiotic complex, as the enzyme responsible for N_2 fixation is sensitive to higher concentrations of assimilable N present in the exterior (Braun-Howland and Nierzwicki-Bauer 1990). Evidently, thus, a combination of 60 kg inorganic N ha^{-1} with mutant *Azolla* was found to be judicious enough to support the maximum possible gain in additional grain yield of both the rice varieties along with a saving of 20 kg N ha^{-1} hand in hand.

Azolla was also found to support an increase in total plant biomass at all N levels in the two rice varieties, and the corresponding values were even higher with mutant *Azolla*, suggesting thereby that biofertilization with *Azolla* has a supporting effect on an increase in straw yield as well as grain yield since crop biomass includes both (Dey 1999). This is quite possible as *Azolla* contributes a very high quantity of N together with plenty of organic matter to the soil and ultimately to the rice crop that may not only improve grain yield in rice but also contribute towards dry matter production needed for the nutrition of rice plants (Vlek *et al.*, 1995, Paramanik and Mahapatra 1997, Sarkar *et al.*, 1997). Evidently, the chlorophyll content and protein as well as other metabolic processes depend upon N supply which ultimately influences growth and yield. Reports in support of such valid interpretations are on record (Liu1987, Singh 1989). Nevertheless, higher activity of roots, greater cell division and rapid accumulation of protein up to panicle primordial stage may be some of the other possible reasons for greater biomass build up and efficient N absorption at early stages of plant growth. An increase in the number of effective tillers and panicle length in the case of Saket-4 rice variety (Bhan 2000) through biofertilization with WT and mutant *Azolla* at all N levels (mutant *Azolla* supporting better than the WT *Azolla* at most of the places) suggests an enhanced level of available N to rice through both inorganic and organic (*Azolla*) N. This may be correlated to some extent with the increase in per cent grain yield of both the rice varieties (Table 5). This finding disproves the earlier general belief of farmers that *Azolla* mat during the initial stages of crop growth may have hindred tiller formation. However, in support to these results, increase in the number of effective tillers through *Azolla* application has been on record (Mandal *et al.*, 1993).

Clearly thus, irrespective of the genotypic difference of the two rice varieties, *i.e.*, Saket-4 and HUR-36 both the varieties responded qualitatively alike to *Azolla* biofertilization (showing yield increases that were small, but significant, with WT *Azolla*, and greater with mutant *Azolla* application). At this juncture, a scientific question that naturally arises is that when grain yield increases in both the rice varieties on application of wild type *Azolla* (which is significantly superior over no *Azolla* at all N levels), what necessitates the use of mutant *Azolla*. Is it only due to the fact that per cent yield increases supported by mutant *Azolla* is significantly higher over the WT *Azolla* in both the rice genotypes? The answer to this is easily drawn when we have a comprehensive view of the overall results of this work, *e.g.* (i) to acquire the vegetative mass culture of the requisite population size of wild type *Azolla* (2 tonnes ha^{-1} for application to the rice field) in rice sowing (hot) season involves a double labour input (in daily cool watering of the nursery, etc.), as compared to the mutant *Azolla*, due to the former's high temperature-sensitivity, and (ii) even after the inoculation of identical population size of WT and mutant *Azolla* in rice field (in parallel sets of experiments), sustenance of the former may not be as smooth as that of the latter, as temperature persists to be higher which is injurious to growth and N_2 fixation by WT *Azolla*. Clearly thus, raising the requisite population size of WT *Azolla* for field use in summer (coinciding the rice sowing/planting season) is apparently uneconomic/less economic.

Consequently, the saving of N fertilizer and additional gain in rice grain yield by the use of WT *Azolla* on record may not be that attractive in view of the high labour input involved in raising its requisite vegetative population for this purpose. On the other hand, data on record on the saving of N fertilizer and additional gain in rice grain yield through biofertilization with mutant *Azolla* is doubly attractive in view of (i) low inputs involved to raise the requisite vegetative population of mutant *Azolla* for this purpose, and (ii) a significantly higher gain in grain yield along with a saving of 20 kg N ha^{-1} for the culture of both the rice genotypes with the mutant *Azolla* strain. Well concerted efforts in this direction will help not only to screen out the various high yielding and traditional rice varieties responsive to biofertilization with the natural as well as the genetically improved *Azolla-Anabaena* symbiotic N_2-fixing strains, but also to the wider use of this extremely potent symbiotic cyanobacterial biofertilizer in areas where its cultivation is limited owing to one or the other stress factors (particularly P-deficiency and temperature extreme).

Spores of *Azolla* do not readily germinate that could have been otherwise the easiest starting material for its mass propagation, *i.e.*, a hand full of spores (bearing the germ plasm of the N_2-fixing cyanobacterial endophyte, *Anabaena azollae*) thrown to the water-logged fields could have germinated and developed into the needful quantity of vegetative *Azolla* for inoculation as a bio-N fertilizer with rice without any problem of labour-intensive mass plant propagation and transportation (Braun-Howland and Nierzwicki-Bauer 1990, Nierzwicki-Bauer 1990, Vaishampayan 1994b). Fortunately, the *Azolla pinnata* mutants described above, isolated through alkylation mutations has helped in strain improvement not only with respect to increased temperature tolerance and decreased P-dependence, but also, remarkably in inducing a significant enhancement of sporulation and germination (Banerjee 1994, Dey, 1999), as compared to the corresponding activities in the wild type strain (see Table 1.7).

Table 1.7: Per cent Aporulation [sporocarps produced per 25 g fresh fronds (after Vaishampayan and Banerjee, 1995)] and Spore Germination [per 100 megasporocarps (after Vaishampayan and Awasthi, 1996)] in Wild Type and Mutant *Azolla pinnata* Whole Plant (bearing the N_2-fixing endosymbiotic cyanobacterium, *Anabaena azollae*) at Graded P Levels and Varying Temperatures (the values presented are the means of 30 random samples + standard errors)

Character	*Temperature (°C)*	*Wild Type*			*Mutant*		
		P-0	*P-10*	*P-20*	*P-0*	*P-10*	*P-20*
Sporulation	25.0	0	65±1.80	104±0.70	0	116±1.67	120±1.70
	30.0	0	38±0.74	77±1.70	0	111±0.70	116±2.13
	35.0	0	12±0.63	48±0.95	0	106±0.70	110±1.67
	37.5	0	0	0	0	98±1.12	105±0.40
	40.0	0	0	0	0	71±1.00	84±1.58
Spore Germination	25.0	0	10.0±0.81	30.0±0.70	0	38.0±1.38	39.6±0.67
	30.0	0	2.0±0.70	17±0.83	0	25.8±0.66	35.8±1.28
	35.0	0	0	2±0.66	0	19.6±0.74	31.6±0.74
	37.5	0	0	0	0	14.2±0.79	22.4±0.86
	40.0	0	0	0	0	5.2±0.37	8.0±0.32

P-0: Phosphate unsupplemented; P-10: 10 mg l^{-1} phosphate; P-20: 20 mg l^{-1} phosphate.

In nature, sporulation has been shown to be induced by the interacting effects of a variety of environmental and cultural factors, such as plant density (Kannaiyan 1978), nutrients (Lales and Murte 1986), light intensity (Becking 1979), temperature (Kannaiyan *et al.*, 1988), mat formation (Talley *et al.*, 1977) and strain specificity (Dovan Cat 1985). The positive effect of some growth hormones in inducing sporulation/spore germination has also been shown (Banerjee 1994, Kannaiyan 1994, Singh *et al.*, 1996). In case of the mutant, being described here, a proliferative increase in the number of sporocarps coupled with the higher incidence of germination was noted not only at the higher temperatures but also at the 50 per cent reduced P level (Table 1.7). This indicates that an alkylation mutation in *Azolla-Anabaena* symbiotic N_2-fixing complex favourably interacts with the factor responsible for either removing the repressor or producing inducer related to the emergence of spores in this symbiotic cyanobacterial biofertilizer, to be substantiated through future sophisticated molecular genetic studies. Moreover, a detailed biochemical analysis on the nature of repressor or inducer in this natural nitrogen resource would ascertain the exact site of alkylation and molecular basis of alkylation-induced mutation in relation to sporulation and spore germination as the life-cycle phase of choice in this fern-cyanobiont N_2-fixing mutualistic complex.

Earlier, due to the lack of such critical studies, laboratory experiments on the physiology of spore germination have been confined only to *Azolla mexicana* (Kannaiyan 1985), *A. microphylla* (Kannaiyan *et al.*, 1988) and *A. caroliniana* (Singh *et al.*, 1990). Zhang *et al.* (1990) examined the process of establishing symbiotic relationship of *Anabaena azollae* with its host during the megaspore germination and sporeling development. Most of the *Anabaena* spores adher to the hair cells arising out of the sporelings.

Germinating *Anabaena* spores are found at shoot region and the cavities of the sporeling (92 per cent of them being onto or near the hair cells which exhibit the ultrastructural characteristic of transfer cells), suggesting that *Anabaena* spores might get the chemical signal stimulating germination or necessary substance to support cell multiplication from the host. Some of the vegetative cells derived from the *Anabaena* spores were differentiated into N_2-fixing heterocysts within the cavity. Hybrids of *Azolla microphylla* (obtained through crossing the male and female sporocarps of different strains within the species) produced significantly higher biomass, chlorophyll content, along with cyanobiont's increased heterocyst frequency, nitrogenase activity, nutrient content and ammonia assimilating enzymes (Gopalaswamy and Kannaiyan 1998). However, *Azolla pinnata* was found tough to be worked out from this angle due to the complicated internal structure (glochidia and massullae) of its sporocarps (Braun-Howlland and Nierzwicki-Bauer 1990). But, of course, the increased incidence of sporulation and spore germination in case of the latter, achieved as a result of alkylation mutation (Banerjee 1994, Dey, 1999) magnifies the horizon and scope of fundamental and applied studies in *Azolla pinnata* based cyanobacterial biofertilizer, and the day appears to be quite close when sporocarps of this widely distributed prevalent species of *Azolla-Anabaena* symbiotic N_2-fixing complex may be used to culture sporophytes as an instant symbiotic cyanobacterial biofertilizer under field condition, alike to the normal practice with *Azolla filiculoides* confined to China (Shuying 1987, Quing-Yuan *et al.*, 1987).

Efforts Towards Raising Artificial Plant-Cyanobacterial Symbioses

We do have now a variety of natural and mutant isolates of N_2-fixing cyanobacteria (NH_4^+ liberating, pesticide-resistant, etc.) which would theoretically maximize the combined N available for rice (see Vaishampayan *et al.*, 1998d). Of particular importance are the strains with improved traits of agronomic importance, *e.g.*, pesticide-resistance with especially high rates of nitrogenase activity in laboratory studies (see Vaishampayan *et al.*, 1998d) and those, which release much of the fixed dinitrogen (NH_4^+) extracellularly (Spiller *et al.*, 1986). Although the mutants (or the rare natural isolates) may be resistant against many abiotic stresses and may release enough of ammonia leading to transfer of ^{15}N and hence inproved growth of rice in pot experiments (Kamuru *et al.*, 1998), their capacity to compete with the indigenous strains in soil is still questionable under the field condition. As a possible approach, Kannaiyan *et al.* (1997) has tried to give such special strains a competitive advantage under flood conditions by immobilizing them on polyurethane foam, with the idea that foam-immobilized cyanobacterium could excrete considerable amounts of ammonia into the flood water leading to an increase in rice grain and straw yields. But in these experiments, the strain (*Anabaena azollae*) used by him has been questionable as to whether it was an endosymbiont of the N_2-fixing water-fern *Azolla* (Whitton 2000). Thus, in spite of the fact that some of the genetically improved N_2-fixing cyanobacterial strains perform well under *in situ* condition, these do need further improvement to adapt to the various biotic and abiotic stress factors operative in the fields, particularly with respect to strain competition during changes in edaphic and climatic conditions (Häder *et al.*, 1995, Häder and Worrest 1997). Obviously, it is a difficult task to develop strains resistant to a host of known (as well as many unknown) biotic and abiotic stress factors which come in way of their thorough exploitation as an improved bio-N fertilizer in agricultural fields. Nevertheless, it is suggested under the situation to use such ecologically restricted strains as efficient N_2-fixing microsymbionts in artificial symbiosis with wet-field cereal crops, *i.e.*, to develop plant-cyanobacterial artificial symbiosis to shelter genetically improved cyanobacterial strains for constitutive N_2 fixation in close proximity to and for the maximum benefit of the host plant (Sinha *et al.*, 1998, Vaishampayan *et al.*, 1998d).

Construction of several artificial plant cyanobacterial symbioses is on record (Gusev and Korzhenevskaya 1990, Sinha *et al.*, 1998), including those of (a) *Nicotiana tabacum* with *Anacystis nidulans, Anabaena variabilis* and *Nostoc* sp.; (b) *Panex ginseng* with *Chlorogloeopsis fritschii, Anabaena variabilis, Calothrix elenkinii* and *Nostoc muscorum*; (c) *Solanum laciniatum* with *Anabaena variabilis* and *Chlorogloeopsis fritschii*; (d) *Papaver somniferum* with *Anabaena variabilis*; (e) *Dioscorea deltoidea* with *Anacystis nidulans*; and (f) *Daucus carota* with *Anabaena flos-aquae, Plectonema boryanum* and *Gloeothece* sp. Attempts have been made in the successful cases to establish interactions for nitrogen metabolites of cyanobionts (cyanobacterial microsymbionts) with the complete exclusion of combined nitrogen from the medium for facilitating gain of fixed nitrogen by plant cells in the presence of artificially introduced cyanobiont (Gorelova *et al.*, 1985).

Still, due to an inadequacy of introduction techniques, many attempts to introduce cyanobacteria into isolated plant protoplasts, including cereals, *e.g.*, rice, the wet field crop, failed to produce viable systems (Vitousek *et al.*, 1997, Ladha and Pareek 2000). In fact, while attempting to construct an artificial plant-cyanobacterial symbiosis at the intracellular level, it has to be ascertained first as to whether it is possible to introduce a N_2-fixing cyanobacterial cell into a plant cell and what is the fate of both the partners after fusion. At the same time it is important to assess whether the cyanobacterial cell introduced into a plant protoplast will supply fixed N to the plant cell, and if so, whether it is possible to regenerate a stable N_2-fixing plant from cell. The latter is to be considered even in the case of intercellular interactions where cyanobacterial cells mixed with plant callus can occupy intercellular space inside the plant tissues. In case of intercellular interactions the plant seldom contains a microsymbiont in its cells, and thus its association with an N_2 fixer will take place only in individual zones or organs of the plants (Vasil *et al.*, 1977), like the natural symbiotic systems where the cyanobacterial biomass constitutes only 1-2 per cent of the total biomass (Stewart *et al.*, 1983). However, there are both ease and limitations in developing an artificial plant-cyanobacterial symbiosis.

The benefits of introducing the intact cells of N_2-fixing cyanobacteria into plant cells and tissues are immense, however, in the case of intracellular symbiosis. The pronounced most advantage is that the oxyphobic N_2-fixing enzyme, *i.e.*, nitrogenase, will remain protected from the irreversibly damaging effect of O_2, which is generated by the plant cells. Moreover, cyanobacterial photosynthetic energy will be available for performing the function of N_2 fixation as well as photoassimilation of fixed nitrogen with subsequent transfer into organic compounds of the cells. In view of a variety of specific properties that they possess, cyanobacteria are, indeed, treated to be a suitable partner of plant cells in artificial associations (Sinha *et al.*, 1998). Margulis (1981) discussed the possible role played by the various cyanophytes in the establishment of eukaruotic cells through forming symbiosis with non-photosynthetic organisms in the evolutionary process. Evidently, with the exception of green algae, cyanobacteria are the most prevalent phototrophs to enter into symbiotic relationship with other organisms in nature, establishing exo–or endo-symbiotic associations, a truth which presents valid grounds for their introduction into both isolated protoplasts (Reisser 1984) and cultured cells (Fogg *et al.*, 1973). In addition to acting as an N_2-fixing component, the cyanobacteria perform a number of metabolic functions in symbiosis with plants which involve excretion of carbohydrates, polysaccharides, peptides, amino acids, organic acids, glycerol, 'gibbereline-like' growth hormones and vitamins (Wolk 1973, Shah and Vaidya 1977, Gibson and Smith 1982) to help plant cell culture. Oxygen is generated during cyanobacterial photosynthesis, which is useful for the respiration of the host plant in the symbiosis (Pearl 1982). Further, cyanobacterial growth rates are comparable to those of plant cells in batch cultures (Street 1977, Suleimanova and Mineyeva 1981) which is, indeed, important considering the necessary synchrony in growth of the two partners during the establishment of symbiosis (Sinha *et al.*, 1998).

Despite the points in favour of establishing an artificial plant-cyanobacterial N_2-fixing symbiosis, described above, there are certain still unresolved limitations, which come in way of developing all sorts of desired associations. The first serious limitation in this context is the fact that only certain cyanobacterial species, belonging to the genera *Nostoc* and *Anabaena*, occur in natural associations with only a limited number of higher plants, displaying a high degree of specificity (Stewart *et al.*, 1983, Gusev and Korzhenevskaya 1990). In fact toxins produced by most of the cyanobacteria may affect the higher plant partner (Gibson and Smith 1982), and thus selection of host plant capable of avoiding the effects of such toxins is important (Sinha *et al.*, 1998). The optimum pH values are acidic (5.0-5.5) in plant cells, and neutral to alkaline (7-10) in cyanobacteria. The optimum temperature for the growth of plant cells is 30-40°C, which is higher than that of cyanobacteria (24-27°C). The concentration of mineral salts in the medium for growing plant cells is higher than that used for culturing cyanobacteria (Vaishampayan, 1995a, 1996). Further, cyanobacteria do not require an organic source in the medium while plant cells require sucrose and other organic compounds (Stanier *et al.*, 1971, Yeomann and Macleod 1977). In the cases where these limitations could not be a barrier, isolated plant protoplasts have been used to introduce various cyanobacterial species, including *Anacystis nidulans* (Davey and Power 1975), *Gloeocapsa* sp. (Burgoon and Bottino 1976, 1977, Hughes *et al.*, 1978), *Anabaena variabilis* (Venkataraman 1981) and *Nostoc muscorum* (Hughes *et al.*, 1978), of which the latter three are N_2 fixers. Of these, *Anabaena variabilis* is although a phototroh, yet it is capable of chemoheterotrophic growth in the dark (Wolk and Shaffer 1976). Possibly this ensures the growth of this cyanobiont in the dark, such as during germination of seeds (Meeks *et al.*, 1978). For forming an artificial association in this case, *Anabaena variabilis* filaments are fragmented and cells are treated with lysozyme to obtain spheroplasts (Agafodorova *et al.*, 1982, Semenova *et al.*, 1982) for their introduction in the pre-isolated plant protoplasts. Rather, following this technique, even an auxotrophic mutant of this N_2-fixing cyanobacterium has been successfully introduced into plant protoplasts to establish a possible dependence of the cyanobiont on the growth and metabolism of the host plant (see Sinha *et al.*, 1998). This clearly indicates that it is possible to combine not only the N_2-fixing natural cyanobacterial isolates, but also their rare mutant strains (resistant to various stress factors and capable of constitutive N_2 fixation) which otherwise fail to compete with the native species on inoculation to the fields, in artificial symbiosis with a variety of fruit, vegetable, cereal and oilseed plants. It would be of significant benefit at the applied level to maintain viability and effective use of the improved cyanobacterial fixed N suppliers through establishing such functional symbiosis.

Concluding Remarks

Of the entire lot of cyanobacterial symbioses with animals, non-photosynthetic protists, bacteria and various classes of plants, the *Azolla-Anabaena* symbiotic N_2-fixing complex is currently the most attractive from the agricultural point of view, as nitrogen fixed by this complex is extremely efficient in rice paddy fields, contributing the agronomically significant levels of nitrogen, thereby offering nearly self sufficiency in rice agriculture and generating plenty of organic matter leading to an overall improvement of the physico-chemical properties of the field soils. But, of course, its high thermo-sensitivity, excessive P requirement and lack of spore-mediated culture (burdening with the problems of its mass vegetative propagation and transportation to fields) come in way of its round the year use as a green manure/bio-N fertilizer to wider sectors. Initiation of success in genetic improvement of this symbiotic N_2-fixing complex through meristem mutagenesis has nevertheless been quite encouraging but lot of work still remains to be done as to the acclimatization and prevention of disease infestation of the mutant *Azolla* in fields for its maximum utilization by rice soil. In fact the interactive effects of the various biotic and abiotic stress factors on *Azolla* multiplicity, spore culture and reestablishment of symbiosis with improved microsymbiont (cyanobiont) need to be investigated stepwise at the fundamental level before harvesting the applied potentials of the mutant strains. This may not only enhance the wider acceptability of *Azolla-Anabaena* symbiosis as a seasoned biofertilizer, but may also give insight to the creation of fruitful artificial symbiosis of some of the extremely efficient free-living N_2-fixing cyanobacteria and their genetically improved mutants with the non-leguminous plants.

Indeed, *Azolla-Anabaena* symbiotic N_2-fixing complex is an experimental verification of the endosymbiotic origin of eukaryotic cell, which suggests that photosynthetic eukaryotes have evolved through the formation of associations between photosynthetic prokaryotes (protoplasts/sphaeroplasts or cyanobacterial cells) and heterotrophic microorganisms (Gusev and Korzhenevskaya 1990). If this is understood to be true, then to make associations artificially between an O_2-evolving photosynthetic prokaryote and a photosynthetic eukaryote may be rather an easy task. It is quite encouraging in this regard that after 2-3 decades of intensive basic work, sophisticated biotechniques have been standardized to form associative somaclones of *Nicotiana tabacum, Panex ginseng, Medicago sativa, Solanum laciniatum, Papaver somniferum, Discorea deltoidea* and *Daucus carota* with the cyanophytes, *viz. Anacystis nidulans, Chlorogloeopsis fritschii, Anabaena variabilis, Calothrix elenkinii, Nostoc muscorum, Plectonema boryanum* or *Gloeocapsa* sp. as microsymbionts (see Sinha *et al.*, 1998, Vaishampayan *et al.*, 2001). Localization of cyanobacteria and relation of partners in artificial associations have well characterized similarities with the natural symbioses, like (a) the stimulating effects of plant cells and tissues on the formation of heterocysts, (b) supply of fixed N by the cyanobacterial microsymbiont (cyanobiont) to support the growth of plant tissues, and (c) transfer of carbon compounds from plant tissues to cyanobacteria as an added source of photosynthetic energy to perform more efficiently the physiological and biochemical functions of nitrogen fixation for a higher output of NH_4^+. However, there still exist considerable differences between artificial and natural associations of cyanobacteria. In the case of natural associations, cyanobiont growth is inhibited or reduced in a plant host, whereas in an artificial association it is stimulated by metabolites from the eukaryotic partner. Heterocyst frequency of the cyanobiont is not as high as that in the microsymbiont of natural symbioses. In the absence of a specialized space for the cyanobiont inside the host plant body (as is well defined in the case of natural symbiosis) the cyanobionts of artificial symbioses create their own habitats in the population of plant cell/tissue cultures. Mutant strains of N_2-fixing cyanobacteria with chains of heterocysts (see Prasad and Vaishampayan 1987) may thus prove to be a better microsymbiotic partner in creating an artificial symbiosis. If agronomically important (herbicide resistant and derepressed) cyanobacterial mutants are used as constitutive N_2-fixing microsymbionts in such artificial associations, these are hoped not to face the problems of competition and exposure to the various non-congenial biotic and abiotic stess factors in the fields. It is, therefore, the need of the time to (i) systematically expand our genetic, physiological, biochemical and molecular research at the fundamental level in order to stepwise remove the various limitations of the plant-cyanobacterial artificial symbiosis while maintaining the benefits of the association, (ii) transfer this phycotechnology to the cereal, fruit, vegetable and oilseed plants, to be followed by laboratory to land transfer technology through a long-term integrated bionutrien management research under the available 'R' and 'D' plans.

Acknowledgements

A major portion of research reported in this review has been worked out at this Centre under the financial support from UGC (in the form of *Career Award* Project to the first author) and ICAR (in the form of an All-India Coordinated Research Project on Biological Nitrogen Fixation to the Department of Genetics and Plant Breeding, Institute of Agricultural Sciences, Banaras Hindu University.

References

Agafodorova, M.N., Gorelova, O.A., Baulina, O.I., Semionova, L.R., Korzhenevskaya, T.G., Butenko, R.G. and Gusev, M.V., 1982. The study of the initial stages of the cyanobacterial cells and spheroplasts' penetration into the isolated protoplasts of tobacco, induced by polyethylene glycol. *Izv. Akad. Nauk. USSR Ser. Biol.*, 4: 510–518.

Ali, S. and Watanabe, I., 1986. Response of *Azolla* to P, K, Zn in different wetland rice soils in relation to chemistry of flood water. *Soil Sci. Pl. Nut.*, 32: 239–254.

Anand, D., 1990. *Handbook of Blue-green Algae of Rice Fields of South India*. Bishen Singh-Mahendra Pal Singh, 23-A, Connaught Place, Dehra Dun.

Anand, N., 1992. Physiological responses on nitrogen-fixing blue-green algae (Cyanobacteria) to commercial biofertilizers. *Proc. Nat. Symp. Cyanobact. Nitrogen Fixation*, New Delhi, pp. 83–91.

Anand, N. and Karuppuswamy, A., 1987. Response of nitrogen-fixing and non-nitrogen fixing blue-green algae (Cyanobacteria) in the presence of certain common fertilizers. *Phykos.*, 26: 22–26.

Anand, N. and Parmeswaran, P., 1992. Growth and nitrogen fixation by six cultures of blue-green algae in the presence of NPK fertilizers. In: *Proc 31st Meet Assoc. Microbiologists of India on Biological Nitrogen Fixation and Biogas Technol.*, Tamil Nadu Agricultural University, Coimbatore, India, pp. 107–122.

Antarikanonda, P., 1982a. Effect of salinity on growth, nitrogen fixation and sodium uptake of rapidly growing N_2-fixing blue-green alga *Anabaena* sp. TA_1. *Microbios.*, 34: 177–184.

Antarikanonda, P., 1982b. Influence of pretreating the seeds with extract of *Anabaena siamensis* on germination and growth of some rice varieties from Thailand. *FAO International Comm.*, 31: 37–39.

Antarikanonda, P., 1984. Production of extracellular free amino acids by cyanobacterium *Anabaena siamensis*. *Current Microbiol.*, 11: 191–196.

Astier, C., Vernotte, C., Der-Vartenium M. and Joset-Espardellier, F., 1980. Isolation and characterization of two DCMU-resistant mutants of the blue-green alga *Aphanocapsa* 6714. *Pl. Cell Physiol.*, 20: 1501–1510.

Barthakur, H.B. and Talukdar, H., 1983. Use of *Azolla* and commercial nitrogen fertilizer in Jorhat, India. *Internat. Rice Res. Newslett.*, 8: 20–21.

Becking, J.H., 1979. Environmental requirements of *Azolla* for use in tropical rice production. In: *Nitrogen and Rice*. Internat Rice Res Insit, Manila, Philippines, pp. 245–274.

Berliner, M.D. and Fisher, R.W., 1987. Surface lectine binding to *Anabaena variabilis* and to cultured and freshly isolated *Anabaena azollae*. *Curr. Microbiol.*, 16: 150–152.

Bhan, U., 2000. Biofertilization of two rice genotypes with N_2-fixing *Azolla-Anabaena* symbiotic complex and its thermo-tolerant mutant strain. *M.Sc.(Ag.) Thesis* in Genetics and Plant Breeding, Bananas Hindu University, India pp. 1–100.

Boussiba, S., 1997. Ammonia assimilation and its biotechnological aspects in cyanobacteria. In: *Cyanobacterial Nitrogen Metabolism and Environ. Biotechnol.*, (Ed.) A.K Rai. Springer-Verlag/Narosa Publishing House, New Delhi, India, pp. 35–72.

Boussiba, S., Dilling, W. and Gibson, J., 1984. Methylammonium transport in *Anacystis nidulans* R-2. *J. Bacteriol.*, 160: 204–210.

Braun-Howland, E.B. and Nierzwicki-Bauer, S.A., 1990. *Azolla-Anabaena* symbiosis: Biochemistry, physiology, ultrastructure and molecular biology. In: *Hand Book of Symbiotic Cyanobacteria,* (Ed) A.N. Rai. CRC Press, Florida, USA, pp. 119–136.

Brouers, M., 1986. Hydrogen production by immobilized *Scenedesmus* cells and ammonia production by immobilized *Anabaena* filaments. In: *Process Engineer Aspects Immobilized Cell Systems,* (Eds.) C Webb, G.M. Black and B. Atkins. The Institute of Chemical Engineers, Rugby, England.

Burgoon, A.C. and Bottino, P.J., 1977. Uptake of nitrogen-fixing blue-green alga *Gloeocapsa* by plant protoplasts. In: *Basic Life Sciences,* (Eds) A. Hollaender, R.H. Burris, P.R. Day, R.W.F. Hardy, D.R. Helinsky, M.R. Lomborg, L. Owens, R.C. Valentine. Plenum Press, New York, USA, pp. 213–238.

Canini, A., Caiola, M.G. and Mascini, M., 1990. Ammonium content, nitrogenase activity and heterocyst frequency within the leaf cavities of *Azolla filiculoides* Lam. *FEMS Microbiol. Lett.,* 71: 205–210.

Castorph, H. and Kleiner, D., 1984. Some properties of a *Klebsiella pneummoniae* ammonium transport negative mutant. *Arch. Microbiol.,* 139: 245–247.

Dalton, D.A. and Chatfield, J.M., 1985. A nitrogen-fixing cyanophyte-hepatic association: *Nostoc* and *Porella. Am. J. Bot.,* 72: 781–784.

Das, B., 1977. Effect of herbicides and pesticides on the fresh water blue-green algae. *Ph.D. Thesis,* Utkal University, Bhuwaneshwar, India.

Da Silva, E.J., Henrickson, E., Haselkorn, E., 1975. Effect of pesticides on blue-green algae and nitrogen fixation. *Arch. Environ. Contam. Toxicol.,* 3: 193–204.

Davey, M.R. and Power, J.B., 1975. Polyethylene glycol-induced uptake of microorganisms into higher plant's protoplasts: An ultrastructural study. *Pl. Sci. Lett.,* 5: 269–288.

Dey, T., 1999. Induction and characterization of *Azolla-Anabaena* symbiotic N_2–fixing complex and their assessment in rice (*Oryza sativa* L.). *Ph.D. Thesis* in Genetics and Plant Breeding, Banaras Hindu University, Varanasi, India pp. 1–123.

Dodge, A.D., 1975. Some mechanisms of herbicide action. *Sci. Prog.,* 62: 447–466.

Dovan, Cat, 1985. *Studies on Sexual Reproduction and Ability to Hybridize between Species of Azolla.* Internat. Rice Res. Inst. Publ., Manila, Philippines.

Fernandez-Valentine E., Quesada, A., Prosperi, C., Nieva, M., Leganes, F. and Ucha, A., 1997. Short- and long-term effects of ammonium on photodependent nitrogen fixation in wetland rice fields of Spain. *Biol. Fertility of Soils,* 24: 353–357.

Fogg, G.E., Stewart, W.D.P., Fay, P. and Walsby, A.E., 1973. *The Blue-Green Algae.* Academic Press, London and New York, pp. 1–559.

Fontana, A., 1984. Thermophilic enzymes and their potential use in biotechnology. Proc. *Third Europ. Cong. Biotechnol.,* 1: 221–232.

Fork, D.C. and Murata, N., 1977. Studies on the effect of transition of the physical phase of membrane lipids on electron transport in the extreme thermophile *Synechococcus lividus. Carnegie Inst. Year Book,* 76: 222–226.

Franche, C. and Cohen-Bazire, G., 1985. The structural *nif* genes of four symbiotic *Anabaena azollae* show a highly conserved physical arrangement. *Pl. Sci.,* 39: 125–131.

Franche, C. and Cohen-Bazire, G., 1987. Evolutionary divergence in the *nif HDK* gene region among nine symbiotic *Anabaena azollae* and some free-living heterocystous cyanobacteria. *Symbiosis,* 3: 159–178.

Gibson, C.E. and Smith, R.V., 1982. Freshwater plankton. In: *Botanical Monographs,* (Eds.) N.G. Carr and B.A. Whitton. Blackwell Scientific Publishers, Oxford, UK, 19: 464–487.

Golden, J.W., Robinson, S.J.and Haselkorn, R., 1985. Rearrangement of nitrogen fixation genes during heterocyst differentiation in the cyanobacterium *Anabaena. Nature (Lond.),* 314: 419–423.

Gopal, B., 1967. Contribution of *Azolla pinnata* R. Br. to the productivity of temporary ponds at Varanasi. *Tropical Ecol.,* 8: 126–139.

Gopalaswamy, G. and Kannaiyan, S., 1998. Biomass production, nitrogen fixation and biochemical characterization of *Azolla* hybrids. *Indian J. Microbiol.,* 38: 81–84.

Gorelova, O.A., Rerabek, J., Korzhenevskaya, T.G., Butenko, R.G. and Gusev, M.V., 1985. Influence of the cyanobacterium *Chlorogloeopsis fritschi* on growth and biosynthetic activity of *Solanum laciniatum* in mixed culture. *Fiziol Rast.,* 32: 1158–1172.

Goyal, S.K., 1989. Stress compatability in Cyanobacteria. *Phykos,* 28: 267–273.

Gusev, M.V. and Korzhenevskaya, T.G., 1990. Artificial associations. In: *Handbook of Symbiotic Cyanobacteria,* (Ed.) A.N. Rai, CRC Press, Boca Raton, Florida, USA, pp. 173–230.

Häder, D-P. and Worrest, R.C., 1997. Consequences of the effects of increased solar ultraviolet radiation on aquatic systems. In: *The Effects of Ozone Depletion on Aquatic Ecosystems,* (Ed.) D-P. Häder. R.G. Landes Company, Academic Press, USA, pp. 11–30.

Häder, D-P., Worrest, R.C., Kumar, H.D. and Smith, R.C., 1995. Effects of increased solar ultraviolet radiation on aquatic ecosystem. *Am. Bio. J.,* 24: 174–180.

Hall, D.O., Affolter, D.A., Broouers, M., Shi, D-J., Yang, L-W. and Rao, K.K., 1985. Photobiological production of fuels and chemicals by immobilized algae. *Annual Proc. Phytochem. Soc., Europe,* 26: 161–185.

Hamdi, Y.A., Allaa, El-Din, M.N., Shalan, S.N. and Hassan, M.E., 1980. *Report on Methodology of Azolla Preservation and Transport.* Ministry of Agriculture, Giza (Egypt) FAO, Rome.

Hamdi, Y.A., El-Nawawy and Tewfi, A.S., 1970. Effect of herbicides on growth and nitrogen fixation of alga *Tolypothrix tenuis. Acta Microbiol. Pollenol.,* 2: 53–56.

Haselkorn, R., 1992. Developmentally regulated generearrangements in prokaryotes. *Ann. Rev. Genet.,* 26: 111–128.

Haselkorn, R., 1995. Molecular genetics of nitrogen fixation in photosynthetic prokaryotes. In: *Nitrogen Fixation: Fundamentals and Applications,* (Eds.) I.A. Tikhonovich, N.A. Provorov, V.I. Romanov and W.E. Newton. Kluwer Academic Publishers, Dordrecht-Boston-London, pp. 29–36.

Hechler, W.D. and Dawson, J.O., 1995. Factors affecting nitrogen fixation in *Azolla caroliniana. Transactions of Illinois State Acad. Sci.,* 88: 97–107.

Henriksson, E., Henriksson, L.E., Norman, J.O. and Nyman, P.O., 1987. Biological nitrogen fixation (acetylene reduction) exhibited by blue-green algae (cyanobacteria) in association with mosses gathered on Surtsey, Iceland. *Arct. Alp. Res.,* 19: 432–436.

Hien, N.T., Kerby, N.W., Machray, G.C., Rowell, P. and Stewart, W.D.P., 1988. Expression of glutamine synthetase in mutant strains of the cyanobacterium *Anabaena variabilis* which liberate ammonia. *FEMS Microbiol. Lett.*, 56: 337–342.

Hughes, B.G., White, F.G., Vernon, L.P. and Smith, M.A., 1978. Uptake of cyanelles and blue-green algae into barley and tobacco protoplasts. *Pl. Physiol. Suppl.*, 61: 54–67.

Huneke, A., 1933. Beitrage zur Kenntinis des Symbiose zwischen *Azolla* and *Anabaena*. *Beitr. Biol. Pflanzen*, 20: 315–341.

Ibrahim, A., 1972. Effect of certain herbicides on growth of nitrogen-fixing algae and rice plants. *Symbiotic Biol. (Hungary)*, 11: 445–448.

IRRI, 1987. *Azolla Utilization*. International Rice Research Institute, Manila, Philippines, pp. 296–299.

Jäger, K.M., Johansson, C., Kunz, U. and Lehman, H., 1997. Sub-cellular element analysis of a cyanobacterium (*Nostoc* sp.) in symbiosis with *Gunnera maniculata* by ESI and EELS. *Bot. Acta.*, 110: 151–157.

Jayakumar, A., Hwang, S.J., Fabiny, J.M., Chinault, A.C. and Barnes, E.M., 1989. Isolation of an ammonium or methylammonium ion transport mutant of *Escherichi coli* and complementation of the cloned gene. *J. Bact.*, 171: 996–1001.

Kamuru, F., Albrecht, S.L., Allen, L.H. and Shanmugam, K.T., 1998. Growth and accumulation of ^{15}N in rice inoculated with the parent and a nitrogenase-derepressed mutant strain of *Anabaena variabilis*. *Appl. Soil Ecol.*, 5: 189–195.

Kannaiyan, S., 1978. Nitrogen fixation by *Azolla*: A brief review. *Farm Sci.*, 5: 19–26.

Kannaiyan, S., 1985. *Algal Bio-fertilizers for Lowland Rice*. Tamil Nadu Agricultural University, Coimbatore, India pp 12–24

Kannaiyan, S., 1993. Nitrogen fixation by *Azolla* and its contribution of nitrogen to the rice crop. In: *Biological NitrogenFixation*, (Eds.) S.N. Tata, A.M. Wadhwani and M.S. Mehdi. Indian Council of Agricultural Research, New Delhi, India pp. 77–111.

Kannaiyan, S., 1994. Sporulation and spore germination in symbiotic nitrogen-fixing water fern *Azolla*. In: *Biology and Application of Nitrogen-fixing Organisms: Problems and Prospects*, (Eds.) A.B. Prasad and A. Vaishampayan. Scientific Publishers, Jodhpur, (India), pp. 71–94.

Kannaiyan, S., Aruna, S.J., Kumari, M.P. and Hall, D.O., 1997. Immobilized cyanobacteria as a biofertilizer for rice crops. *J. Appl. Phycol.*, 9: 167–174.

Kannaiyan, S., Goyal, S.K., Goyal, S.S. and Rains, D.W., 1988. Some factors influencing the growth and sporulation in *Azolla*. *Phykos*, 27: 197–205.

Kannaiyan, S. and Somporn, C., 1989. Effect of high temperature on growth, N_2 fixation and chlorophyll content of five species of *Azolla-Anabaena* symbiosis. *Biol. Fertil. Soils*, 7: 168–172.

Kaplan, D., Calvert, H.E. and Peters, G.A., 1986. Nitrogenase activity and phycobiliproteins of the endophyte as a function of leaf age and cell type. *Pl. Physiol.*, 80: 884–890.

Kaplan, D. and Peters, G.A., 1988. The *Azolla-Anabaena* relationship. XIV. Chemical composition of the association and soluble carbohydrates of the association, endophyte-free *Azolla*, and the freshly isolated endophyte. *Symbiosis*, 24: 35–50.

Kerby, N.W., Rowell, P. and Stewart, W.D.P., 1985. Ethylenediamine uptake metabolism in the cyanobacterium *Anabaena variabilis*. *Arch. Microbiol.*, 141: 244–248.

Kerby, N.W., Rowell, P. and Stewart, W.D.P., 1986. The uptake and metabolism of methylamine by N_2-fixing cyanobacteria. *Arch. Microbiol.*, 43: 353–358.

Kerby, N.W., Rowell, P. and Stewart, W.D.P., 1987. Cyanobacterial ammonium transport, ammonium assimilation and nitrogenase regulation. *New Zealand Marine Freshwater Res.*, 21: 447–456.

Koike, H. and Katoh, S., 1979. Heat stability of cytochromes and ferredoxin isolated from a thermophilic blue-green alga. *Pl. Cell Physiol.*, 20: 1157–1161.

Kushari, D.P. and Watanabe, I., 1991. Differential response of *Azolla* to phosphorus deficiency. I. Scanning methods in quantity controlled condition. *Soil Sci. Pl. Nut.*, 37: 271–282.

Ladha, J.K. and Pareek, R.P., 2000. Monitoring nitrogen equilibrium for sustainable crop production in rice-based cropping systems. In: *Proc Internat. Cong. Managing Natural Resources*, (Ed.) G.B. Singh. Indian Agricultural Research Institute, New Delhi (India), 1: 71–72.

Lai, Y.K., Li, C.W., Hu, C.H. and Lee, M.L., 1988. Quantitative and qualitative analysis of protein synthesis during heat shock in marine diatom *Nitzschia alba* (Bacillariophyceae). *J. Phycol.*, 24: 509–514.

Lales, J.S. and Murte, R.S., 1986. Nitrogen accumulation and biomass production of *Azolla microphylla* (kaulfuss). *Philippines Agricultutrist*, 64: 535–542.

Lattore, C., Lee, J.H., Spiller, H. and Shanmugam, K.T., 1986. Ammonium ion-excreting cyanobacterial mutants as a source of nitrogen for growth of rice: A feasibility study. *Biotechnol. Lett.*, 8: 507–512.

Li, Z.X., Zu, S.X., Mao, M.F. and Lumpkin, T., 1982. Study on the utilization of 8 *Azolla* species in agriculture. *Zhunggua Nongye Kexue*, 1: 19–27.

Lin, C., 1992. Effects of *Anabaena azollae* on the tolerance of *Azolla* to high temperature. *J. Fujian Acad. Sci.*, 7: 30–35.

Lin, C., Watanabe, I., Tan, L.F. and Liu, C.C., 1988. Reestablishment of symbiosis to *Anabaena*-free *Azolla*. *Zhunguo Kexue (Sceintia Sinica)*, 30: 1–5.

Lindblad, P., Atkins, C.A. and Pate, J.S., 1991. N_2 fixation by freshly isolated *Nostoc* from coralloid roots of the cycad *Macrozamia riedlei* (Fisch ex Gaud). *Garden Pl. Physiol.*, 95: 753–759.

Liu, C.C., 1987. Re-evaluation of *Azolla* utilization in agricultural production. In: *Azolla Utilization*. Internat Rice Res. Inst., Manila, Philippines, pp. 67–76.

Lumpkin, T.A., 1987. Environmental requirements for successful *Azolla* growth. In: *Azolla Utilization*. Internat Rice Res. Inst., Manila, Philippines, pp. 89–97.

Lundkvist, I., 1970. Effect of two herbicides on nitrogen fixation by blue-green algae. *Svensk Bot Tidsk.*, 64: 460–461.

Mahapatra, B.S. and Sharma, G.L., 1989. Integrated management of *Sesbania, Azolla* and urea nitrogen in lowland rice under rice wheat cropping system. *J. Agric. Sci. (Camb)*, 113: 203–206.

Mandal, B.K. and Bharti, A.K., 1983. *Azolla* as an organic manure for rice in west Bengal. *Ind. J. Agric. Sci.*, 53: 472–475.

Mandal, R., Tahmida Begum, Z.N., Khan, Z.U.M. and Hossain, M.Z., 1993. N_2-fixing blue-green algae in rice fields and their relationship with soil fertility. *Bangladesh J. Bot.*, 22: 73–79.

Manna, A.B. and Singh, P.K., 1989. Rice yields as influenced by *Azolla* N_2 fixation and urea-N fertilization. *Pl. Soil,* 14: 63–68.

Margulis, L., 1981. *Symbiosis in Cell Evolution.* W.H. Freeman Publications, San Francisco, USA

Martinez, M.R., 1984. *Algae: Biofertilizer for Rice.* Philippines Council Agric. Res. and Resource Develop. (PCARRD), 12: 9–12.

Meeks, J.C., Joseph, C.M. and Haselkorn, R., 1988. Organization of *nif* genes in cyanobacteria in symbiotic association with *Azolla* and *Anthoceros. Arch. Microbiol.,* 150: 61–71.

Meeks, J.C., Malmberg, L. and Wolk, C.P., 1978. Uptake of auxotrophic cells of a heterocyst-forming cyanobacterium by tobacco protoplasts. *Planta,* 139: 55–63.

Meeks, J.C., Steinberg, N.A., Enderlin, C.S., Joseph, C.M. and Peters, G., 1987. *Azolla-Anabaena* relationship. XIII. Fixation of (^{13}N) dinitrogen. *Pl. Physiol.,* 84: 883–886.

Meeks, J.C., Syeinberg, N.A., Joseph, C.M., Enderlin, C.S., Jorgensen, P.A. and Peters, G.A., 1985. Assimilation of exogenous and dinitrogen derived $^{13}NH_4^+$ by *Anabaena azollae* separated from *Azolla caroliniana* Wild. *Arch. Microbiol.,* 142: 229–233.

Mishra, A.K., Vaishampayan, A. and Singh, V.P., 1991a. Isolation and physiological characterization of a herbicide-resistant mutant of a nitrogen-fixing cyanobacterium. In: *Physiological Strategies for Crop Improvement,* (Eds.) D.N. Tyagi, B. Bose, A. Hemantaranjan and T.M. Devi. Banaras Hindu University, Varanasi, pp. 63–68.

Mishra, A.K., Vaishampayan, A. and Singh, V.P., 1991b. A herbicide-resistant cyanobacterial mutant derepressed for nitrogen fixation under ammonia-mediated condition. In: *Proceedings of Golden Jubilee Symposium of the Indian Society of Genetics and Plant Breeding* (Genetic Research and Education: Current Trends and the Next Fifty Years), Indian Agricultural Research Institute, New Delhi, pp. 98–100.

Musgrave, S.C., Kerby, N.W., Codd, G.A. and Stewart, W.D.P., 1982. Sustained ammonia production by immobilized filaments of the nitrogen-fixing cyanobacterium *Anabaena* 27893. *Biotechnol. Lett.,* 8: 507–512.

Newton. J.W. and Cavins, J.F., 1985. Liberation of ammonia during nitrogen fixation by a facultative heterotrophic cyanobacterium. *Biochem. Biophys. Acta.,* 809: 44–50.

Newton, J.W. and Herman, I., 1979. Isolation of cyanobacteria from the aquatic fern, *Azolla. Arch. Microbiol.,* 120: 161–165.

Nierzwicki-Bauer, S.A., 1990. *Azolla-Anabaena* symbiosis: Use in agriculture. In: *Hand Book of Symbiotic Cyanobacteria,* (Ed.) A.N. Rai. CRC Press, Florida, USA, pp. 119–136.

Nierzwicki-Bauer, S.A. and Haselkorn, R., 1986. Differences in mRNA levels in *Anabaena* living freely or in symbiotic association with *Azolla. EMBO J.,* 5: 29–35.

Paramanik, S.C. and Mahapatra, B.S., 1997. Effect of integrated use of inorganic and organic N sources on mineralization, uptake and grain yield of rice. *Oryza,* 34: 181–184.

Pearl, H.W., 1982. Interactions with bacteria. In: *Botanical Monographs,* (Eds.) N.G. Carr and B.A. Whitton. Blackwell Scientific Publishers, Oxford, UK, 19: 441–452.

Peters, G.A., 1991. *Azolla* and other plant plant-cyanobacterial symbioses: aspects of form and function. *Pl. Soil,* 137: 25–36.

Peters, G.A. and Calvert, H.E., 1983. The *Azolla-Anabaena* symbiosis. In: *Algal Symbiosis,* (Ed.) L.J. Goff. Cambridge University Press, New York, USA, pp. 109–145.

Peters, G.A., Kaplan, D., Meeks, J.C., Buzby, K.N., Marsh, B. and Corbin, J.L., 1985. Aspects of nitrogen and carbon interchange in the *Azolla-Anabaena* symbiosis. In: *Nitrogen Fixation and CO_2 Metabolism* (Eds.) P.W. Ludden, J.E. Burris. Elsevier Science Publishers, New York, USA, pp. 213–222.

Peters, G.A. and Meeks, J.C., 1989. The *Azolla-Anabaena* symbiosis: Basic biology. *Ann. Rev. of Pl. Physiol. Pl. Molec. Biol.,* 40: 193–210.

Peters, G.A., Toia, R.E. Jr., Evans, W.R., Crist, D.K., Mayne, B.C. and Poole, R.E., 1980. Characterization and comparisons of five N_2–fixing *Azolla-Anabaena* associations. I. Optimization of growth conditions for biomass increase and N content in a controlled environment. *Pl. Cell Environ.,* 3: 261–269.

Prasad, A.B., Samanta, R., Vishwakarma, M.L. and Vaishampayan, A., 1986. Biological effects of a mercury fungicide on a nitrogen-fixing blue-green alga *Nostoc muscorum*: Isolation and preliminary characterization of a Hg-resistant mutant. *New Phytol.,* 102: 45–50.

Prasad, A.B. and Vaishampayan, A., 1984. Genetic recombination in *Nostoc muscorum* strains.In: *Perspectives in Cytology and Genetics,* (Eds.) G. Manna and U. Sinha. Typographers Rashtravani Press, New Delhi, pp. 179–184.

Prasad, A.B. and Vaishampayan, A., 1987. Genetic dissections in cyanobacteria. In: *Mutagenesis: Basic and Applied,* (Ed.) A.B. Prasad. Print House, Lucknow, pp. 253–284.

Prasad, A.B. and Vaishampayan, A., 1994a. *Biology and Application of Nitrogen-fixing Organisms: Problems and Prospects* (Edited Book), Scientific Publishers, Jodhpur, pp. 1–332.

Prasad, A.B. and Vaishampayan, A., 1994b. Prospects of using pesticide-resistant cyanobacterial mutants as a viable bio-N fertilizer and a source material for genetic engineering studies. In: *Biology and Application of Nitrogen-fixing Organisms: Problems and Prospects,* (Eds.) A.B. Prasad and A. Vaishampayan. Scientific Publishers, Jodhpur, pp. 47–70.

Prasad, A.B., Vaishampayan, A., Vishwakarma, M.L. and Shrivastava, D.K., 1991. Development of fungicide-resistance, coupled with N_2-fixing ability during NH_4^+ mediated growth, as a mutational event, in a cyanobacterium *Nostoc muscorum. Biomed. Lett.,* 46: 245–251.

Quing-Yuan X., Yan-Ru, S., Guang-Li, Y. and Ke-Lin, P., 1987. Germination of *Azolla filiculoides* Lam.: Sporocarps and factors affecting their growth. In: *Azolla Utilization.* Internat Rice Res Inst, Manila, Philippines, pp. 33–38.

Rai, A.N., 1990. Cyanobacteria in symbiosis. In: *Handbook of Symbiotic Cyanobacteria,* (Ed.) A.N. Rai. CRC Press, Boca Raton, Florida (USA), pp. 1–7.

Rai, A.N., Rowell, P. and Stewart, W.D.P., 1983. Interactions between cyanobacteria and fungus during $^{15}N_2$-incorporation and metabolism in the lichen *Peltigera canina. Arch. Microbiol.,* 134: 136–142.

Rakotonaivo, G. and Schramm, M., 1988. Influence of P, K, micronutrients and dolomite on *Azolla* growth. *IRRI Newslett.,* 13: 23–24.

Ramos, J.L., Guerrero, M.G. and Losada, M., 1984. Sustained photoproduction of ammonia from dinitrogen and water by the nitrogen-fixing cyanobacterium *Anabaena* sp. strain 33047. *Appl. Environ. Biotechnol.,* 48: 114–118.

Ramos, J.L., Guerrero, M.G. and Losada, M., 1987. Factors affecting the photoproduction of ammonia from dinitrogen and water by the cyanobacterium *Anabaena* sp. strain ATCC 33047. *Biotechnol. Bioengineer.*, 29: 566–571.

Rathore, A.L., Chipe, S.J. and Pal, A.R., 1995. Direct and residual effects of bio–organic and inorganic fertilizers in rice-wheat cropping system. Ind. J. Agron., 40: 14–19.

Ray, T.B., Mayne, B.C., Toia, Jr. R.E and Peters, G.A., 1979. *Azolla–Anabaena* relationship. VIII. Photosynthetic characterization of the association and the individual partners. *Pl. Physiol.*, 64: 791–795.

Ray, T.B., Peters, G.A., Toia, Jr. R.E. and Mayne, B.C., 1978. *Azolla-Anabaena* relationship. VI. Distribution of ammonia assimilating enzymes, proteins and chlorophyll between host and symbiont. *Pl. Physiol.*, 62: 463–467.

Reisser, W., 1984. Endosymbiotic cyanobacteria and cyanellae. In: *Encyclopedia of Plant Physiology,* (Eds.) H. Linskens, J. Heslop-Harrison. Springer-Verlag, Berlin, Germany, New Series, Vol. 17, pp. 91–133.

Rhodes, D., Sims, A.P. and Folkes, B.F., 1980. Pathway of ammonium assimilation in illuminated *Lemna minor. Phytochem.*, 19: 357–365.

Rippka, R., Deruelles, J., Waterbury, J.B., Herdman, M. and Stanier, R.Y., 1979. Generic assignments, strain histories and properties of pure cultures of cyanobacteria. *J. Gen. Microbiol.*, 111: 1–11.

Rodgers, G.A., 1982. Effect of ammonia or nitrate on nitrogen fixation by a terrestrial blue-green alga. *Pl. Soil*, 64: 263–266.

Roger, P.A. and Kulasooriya, S.A., 1980. *Blue-Green Algae and Rice.* International Rice Research Institute, Los Banos, Laguna, Philippines, pp. 1–112.

Roy, B., 1984. Manuring of rice with *Azolla. Oryza*, 21: 238–241.

Sah, R.N., Goyal, S.S. and Rains, D.W., 1989. Effect of light on NO_3-transport by *Azolla pinnata. J. Exp. Bot. (Oxford)*, 214: 543–549.

Saleh, M.A., Kea, T.L. and Osman, G., 1989. Plasmid-mediated resistance to paraquat in soil isolate of *Pseudomonas* species. *Malayas Appl. Biol. J.*, 18: 1–8.

Sarkar, R.K., Saha, A. and Furutani, T., 1997. Photosynthesis and productivity of rice at two levels of nitrogen. *Oryza*, 34: 28–33.

Saville, B., Straus, N. and Coleman, J., 1987. Contiguous organization of nitrogenase genes in a heterocystous cyanobacterium. *Pl. Physiol.*, 85: 26–29.

Semenova, L.R., Mineyeva, L.A. and Gusev, M.V., 1982. The effect of osmotic stabilizing agents on the formation of spheroplasts in cyanobacteria and their photosynthetic activity. *Mikrobiologiya*, 52: 259–271.

Shah, A.K. and Vaidya, B.S., 1977. Detection of vitamin B-12 and pantothenic acid in cell exudates of blue-green algae. *Biol. Plant*, 19: 426–431.

Shuying, L., 1987. Methods of using *Azolla filiculoides* sporocarps to culture sporophytes in the field. In: *Azolla Utilization*. Internat. Rice Res. Instit., Manila, Philippines, pp. 27–32.

Singh, A.L. and Singh, P.L., 1989. *Nitrogen Fixation in Indian Rice Fields (Azolla and Blue-green Algae).* Agro-Botanical Publishers, Bikaner, India, pp. 1–236.

Singh, D.T., Nirmala, K., Modi, D., Katiyar, S. and Singh, H.N., 1987. Genetic transfer of herbicide-resistance gene(s) from *Gloeocapsa* spp. to *Nostoc muscorum*. *Molec. Gen. Genet.*, 208: 436–438.

Singh, H.N. and Singh, H. Nath, 1978. An azide-resistant mutant of the blue-green alga *Nostoc muscorum* producing heterocysts and nitrogenase in the presence of fixed inorganic nitrogen source. *Arch. Microbiol.* 119: 197–201.

Singh, H.N., Singh, H.R. and Vaishampayan, A., 1979. Toxic and mutagenic action of the herbicide alachlor (*lasso*) on various strains of a N_2-fixing blue-green alga *Nostoc muscorum* and characterization of herbicide-induced mutants resistant to methylamine and L-methionine-DL-sulfoximine. *Environ. Exp. Bot.*, 19: 5–12.

Singh, H.N. and Vaishampayan, A., 1978. Biological effects of rice-field herbicide *machete* on various strains of the N_2-fixing blue-green alga *Nostoc muscorum*. Environ. Exp. Bot., 18: 87–94.

Singh, H.N., Vaishampayan, A. and Singh, R.K., 1978a. Evidence for the Involvement of a genetic determinant controlling functional specificity of group VI-B elements in N_2 and NO_3-metabolism in a blue-green alga *Nostoc muscorum*. *Biochem. Biophys. Res. Commun.*, 81: 67–74.

Singh, H.N., Vaishampayan, A. and Sonie, K.C., 1978b. Mutation from Mo-dependent growth to W-dependent growth and further evidence for a genetic determinant common to nitrogenase and nitrate reductase in a blue-green alga *Nostoc muscorum*. Mut. Res., 50: 427–432.

Singh, P.K., 1975. Fertilizer tolerance of blue-green algae and their effect on heterocyst differentiation. *Phykos*, 14: 81–88.

Singh, P.K., 1977a. Multiplication and utilization of fern *Azolla* containing nitrogen-fixing algal symbiont as green manure in rice cultivation. *Riso*, 26: 125–136.

Singh, P.K., 1977b. *Azolla* plant as fertilizer and feed. *Ind Farm.*, 27: 19–21.

Singh, P.K., 1979. Use of *Azolla* in rice production in India. In: *Nitrogen and Rice*. Internat. Rice Res. Inst., Los Banos, Laguna, Philippines, pp. 407–418.

Singh, P.K., 1980. Introduction of green *Azolla* biofertilizer in India. *Curr. Sci.*, 49: 155–156.

Singh, P.K., 1981. Use of *Azolla* and blue-green algae in rice cultivation in India. In: *Associative N_2 Fixation*, (Eds.) P.B. Vose and A.P. Puschel. CRC Press, Boca Raton, Florida, USA, pp. 183–195.

Singh, P.K., 1989. Use of *Azolla* in Asian Agriculture. *Appl. Agric. Res.*, 4: 149–161.

Singh, P.K., Singh, D.P. and Singh, R.P., 1996. Germination of water fern *Azolla caroliniana* sporocarps at varied light, amino acid, sugars ans ABA. *Proc. Ind. Nat. Sci. Acad.*, 62: 352–358.

Singh, R.N., 1961. *Role of Blue-green Algae in Nitrogen Economy of Indian Agriculture*. Indian Council of Agricultural Research, New Delhi, India, pp. 1–175.

Singh, S., Prasad, R., Singh, B.V., Goyal, S.K. and Sharma, S.N., 1990. Effect of green manuring, blue-green algae and neem-cake coated urea on wetland rice. *Biol. Fertil. Soils*, 9: 235–238.

Singh, Y.V. and Mandal, B.K., 1997. Nutrition of rice, *Oryza sativa*, through *Azolla* organic materials and urea. *Ind. J. Agron.*, 42: 623–633.

Sinha, R.P. and Häder, D.-P., 1996a. Photobiology and ecophysiology of rice field cyanobacteria. *Photochem Photobiol.*, 64: 887–896.

Sinha, R.P., Häder, D-P., 1996b. Response of a rice field cyanobacterium *Anabaena* sp. to physiological stressors. *Environ. Exp. Bot.*, 36: 147–155.

Sinha, R.P. and Häder, D-P., 1997. Impacts of UV-B irradiation on rice field cyanobacteria. In: *The Effects of Ozone Depletion on Aquatic Ecosystems,* (Ed.) D-P. Häder. R.G. Landes Company, Academic Press, USA, pp. 189–197.

Sinha, R.P., Klisch, M., Vaishampayan, A. and Häder, D-P., 1999. Biochemical and spectroscopic chatracterization of the cyanobacterium *Lyngbya* sp. inhabiting mango (*Mangifera indica*) trees: Presence of an ultraviolet-absorbing pigment, scytonemin. *Acta Protozool.,* 38: 291–298.

Sinha, R.P. and Vaishampayan, A., 1997. Photobiological ammonium [NH_4^+] excretion by immobilized N_2-fixing rice-field cyanobacteria. In: *Proc. Nat. Symp. Microbes for Health, Wealth and Environment,* Jawaharlal Nehru University, New Delhi pp. 27–45.

Sinha, R.P., Vaishampayan, A. and Häder, D-P., 1998. Plant-cyanobacterial symbiotic somaclones as a potential bionitrogen fertilizer for paddy agriculture: Biotechnological approaches. *Microbiol. Res.,* 153: 297–307.

Smith, D.C. and Douglas, A.E., 1987. *The Biology of Symbiosis.* Edward Arnold, London, Uk

Söderbäck, E., Lindblad, P. and Bergman, B., 1990. Developmental patterns related to nitrogen fixation in the *Nostoc-Gunnera magellanica* Lam. Symbiosis. *Planta,* 182: 355–362.

Spiller, H., Lattore, H.C., Hassan, M.E. and Shanmugam, K.T., 1986. Isolation and characterization of nitrogenase derepressed mutants of the cyanobacterium *Anabaena variabilis. J. Bacteriol.* 165: 412–419.

Stanier, R.Y., Kunisawa, R., Mandel, M., and Cohen-Bazire, G., 1971. Purification and properties of unicellular blue-green algae (order Chrococcales). *Bacteriol. Rev.,* 35: 171–183.

Stewart, W.D.P. and Rowell, P., 1975. Effects of L-methionine-DL-sulfoxcimine on the assimilation of newly fixed NH_3, acetylene reduction and heterocyst production in *Anabaena cylindrica. Biochem. Biophys. Res. Commun.,* 65: 846–856.

Stewart, W.D.P., Rowell, P. and Rai, A.N., 1980. Symbiotic nitrogen-fixing cyanobacteria. In: *Nitrogen Fixation,* (Eds.) W.D.P. Stewart and J.R. Gallon. Academic Press, New York, USA, pp. 239–277.

Stewart, W.D.P., Rowell, P. and Rai, A.N., 1983. Cyanobacteria-eukaryotic plant symbioses. *Annals Microbiol. (InstitutePasteur),* 134: 205–214.

Street, H.E., 1977. Cell (suspension) culture-techniques. In: *Botanical Monographs,* (Ed.) H.E. Street. Blackwell Scientific Publishers, Ofxord, UK, 11: 61–89.

Suleimanova, S.S. and Mineyeva, L.A., 1981. Effects of high light intensity on growth and pigment content of cyanobacteria under conditions of photoauto- and photoheterotrophic cultivation. *Vestn Mosk Uni. Ser. Biol.,* 1: 42–53.

Subramanian, G. and Shanmugasundaram, S., 1986a. Influence of herbicide 2,4-D on nitrogen fixation and ammonia excretion by the cyanobacteria *Anabaena. Proc. Ind. Acad. Sci., B*52: 308–312.

Subramanian, G. and Shanmugasundaram, S., 1986b. Uninduced ammonia released by nitrogen-fixing cyanobacteria *Anabaena. FEMS Microbiol. Lett.,* 37: 151–154.

Suseela, M.R. and Goyal, S.K., 1995. Effect of ammonium nitrogen on growth and nitrogen fixation by Cyanobacteria. *Phykos,* 34: 123–130.

Talley, S.N., Talley, B.J. and Rains, D.W., 1977. Nitrogen fixation by *Azolla* in rice fields. In: *Genetic Engineering for Nitrogen Fixation* (Ed.) A. Hollaender. Plenum Press, New York, USA, pp. 259–281.

Tang, L.F., Watanabe, I. and Liu, C.C., 1990. Limited multiplication of symbiotic cyanobacteria of *Azolla* spp. On artificial media. *Appl. Environ. Microbiol.*, 56: 3623–3626.

Tel-Or, E., Sandovsk, T., Kobiler, D., Arad, C. and Weinberg, R., 1983. The unique symbiotic properties of *Anabaena* in the water fern *Azolla*. In: *Photosynthetic Prokaryotes: Cell Differentiation and Function*, (Eds.) G.C. Papageorgiou, L.P. Packer. Elsevier Science Publishers, New York, USA, pp. 303–314.

Thomas, S.P., Zaritsky, A. and Boussiba, S., 1990. Ammonium excretion by an L-methionine-DL-sulfoximine-resistant mutant of the rice-field cyanobacterium *Anabaena siamensis*. *Appl. Environ. Microbiol.*, 56: 3499–3504.

Tisdale, S.L. and Nelson, W.L., 1975. *Soil Fertility and Fertilizers*. Collier Macmillan International Publishing Co., New York, USA, pp. 189–242.

Tiwari, D.N., Pandey, A.K. and Mishra, A.K., 1981. Action of 2,4-D and rifampicin on growth and heterocyst formation in *Nostoc linckia*. *J. Biol. Sci.*, 3: 38–39.

Tung, H.F. and Watanabe, I., 1983. Differential response of *Azolla-Anabaena* association to high temperature and minus phosphorus treatments. *New Phytol.*, 93: 423–431.

Tyagi, V.V.S., Mayne, B.C. and Peters, G.A., 1980. Purification and initial characterization of phycobiliproteins from the endophytic cyanobacterium of *Azolla*. *Arch. Microbiol.*, 128: 41–44.

Tyagi, V.V.S., Ray, T.B., Mayne, B.C. and Peters, G.A., 1981. The *Azolla-anabaena azollae* relationship. XI. Phycobiliproteins in the action spectrum for nitrogenase-catalyzed acetylene reduction. *Pl. Physiol.*, 68: 1479–1484.

Uheda, 1986. Isolation of empty packets from *Anabaena*-free *Azolla*. *Pl. Cell. Physiol.*, 27: 1187–1190.

Vaishampayan, A., 1981. Nutritional screening of amino acids in an auxotroph mutant of the cyanobacterium *Nostoc muscorum*. *Microbios. Lett.*, 18: 111–116.

Vaishampayan, A., 1982a. Amino acid nutrition in the blue-green alga *Nostoc muscorum*. *New Phytol.*, 90: 545–549.

Vaishampayan, A., 1982b. Amino acids metabolism in the cyanobacterium *Nostoc muscorum*. In: *Proc XII Internat Cong Biochem, Perth (Western Australia)*, August 15–21, p. 252.

Vaishampayan, A., 1982c. A methylamine-resistant mutant of the blue-green alga, *Nostoc muscorum*: possible involvement of glutamine synthetase in methylamine metabolism. *New Phytol.*, 91: 607–613.

Vaishampayan, A., 1982d. Cu-Fe interactions in the nitrogen-fixing cyanobacterium *Nostoc muscorum*. *Microbios. Lett.*, 21: 17–23.

Vaishampayan, A., 1982e. Acetate reversal of the inhibition of photosynthetic assimilation of inorganic carbon by DCMU in the cyanobacterium *Nostoc muscorum*. In: *Proc. 1st Internat. Phycol. Cong., St.John's (Newfoundland, Canada)*, August 8–14, p. 52.

Vaishampayan, A., 1982f. Toxic and mutagenic action of the pesticide sevin on the blue-green alga *Nostoc muscorum*. In: *Proc. 23rd Ann. Meet. Europ. Soc. Toxicol. (Tel–Aviv, Israel)*, p. 105.

Vaishampayan, A., 1982g. Mutagenicity of the herbicide paraquat in the cyanobacterium *Nostoc muscorum*. In: *Proc. XIII Internat. Cong. Microbiol., Boston (Massachusetts, USA)*, August 8–13, p. 177.

Vaishampayan, A., 1983a. Mo-V interactions during N_2 and NO_3-metabolism in a N_2-fixing blue-green alga *Nostoc muscorum*. *Experintia*, 39: 358–360.

Vaishampayan, A., 1983b. Vanadium as a trace element in the blue-green alga, *Nostoc muscorum*: influence on nitrogenase and nitrate reductase. *New Phytol.*, 95: 55–60.

Vaishampayan, A., 1983c. Genetics of methylamine-resistance leading to carbon auxotrophy in the cyanobacterium *Nostoc muscorum*. In: *XVth Internat Cong Genet, New Delhi (India)*, December 12–21, p. 812(1462).

Vaishampayan, A., 1983d. UV-induced wide-range mutations in a cyanobacterium *Nostoc muscorum*. In: *Proc VIIth Internat. Conf. Radiation Res., Rijswijk*, July 3–8, p. 4–41.

Vaishampayan, A., 1983e. Heterocyst differentiation and speciation in the N_2-fixing cyanobacterium *Nostoc muscorum* as influenced by a bipyridylium ion. In: *Proc XXIXth Internat Cong of Physiol, Sydney*, 6: 47–136.

Vaishampayan, A., 1983f. Pesticide-resistance in a N_2-fixing cyanobacterium. In: *Proc Internat Cong Environ Hazards of Agrochem in Developing Countries, University of Alexandria*, 3: 7–102.

Vaishampayan, A., 1983g. Metabolic utilization of a C_1 compound in a nitrogen-fixing cyanobacterium *Nostoc* sp. In: *Proc VIth Internat Symp Environ Biogeochem, New Mexico*, 2: 3–15

Vaishampayan, A., 1983h. Isolation of a carbon-requiring strain of a nitrogen-fixing cyanobacterium. In: *Proc IIIrd Internat Symp Microbial Ecol, Michigan State University*, August 7–12, V: 5–81.

Vaishampayan, A., 1983i. An ammonia liberating strain of the N_2–fixing cyanobacterium *Nostoc muscorum*. *Transac Meet Biol Process and Soil Fertility, University of Reading*, July 4–8, 130.

Vaishampayan, A., 1984a. Screening of amino acids for carbon and nitrogen enrichments in different strains of a cyanobacterium *Nostoc muscorum*. *Biochem. Physiol. Pflanzen.*, 179: 411–417.

Vaishampayan, A., 1984b. Copper-iron interactions in a nitrogen-fixing cyanobacterium *Nostoc muscorum*. *J. Pl. Nutr.*, 7: 567–573.

Vaishampayan, A., 1984c. Genetic manipulation and recombination leading to pesticide-resistance in a nitrogen-fixing cyanobacterium. In: *Proc. IIIrd Europ. Cong. Biotechnol.*, 3: 353–358.

Vaishampayan, A., 1984d. Collection of a carbon-requiring strain of a nitrogen-fixing cyanobacterium. In: *Proc. V[th] Internat. Cong. Culture Collections*, Thailand Centre for Culture Collections, Bangkok, pp. 63–72.

Vaishampayan, A., 1984e. Biological effects of a herbicide in a nitrogen-fixing cyanobacterium (blue-green alga): an attempt for introducing herbicide-resistance. *New Phytol.*, 96: 7–11.

Vaishampayan, A., 1984f. Biological effects of the rice-field herbicide monuron on a nitrogen-fixing cyanobacterium *Nostoc muscorum*. *Microbios. Lett.*, 28: 105–111.

Vaishampayan, A., 1984g. Studies on diuron uptake in a blue-green alga *Nostoc muscorum*. *J. Exp. Bot., (Oxford)*, 35: 897–904.

Vaishampayan, A., 1984h. Powerful mutagenicity of a bipyridylium herbicide in a nitrogen-fixing blue-green alga *Nostoc muscorum*. *Mut. Res.*, 138: 39–46.

Vaishampayan, A., 1984i. Strong mutagenicity of a bipyridylium herbicide in a nitrogen-fixing blue-green alga. *Experientia*, 40: 1016–1019.

Vaishampayan, A., 1984j. Mutagenicity of a bipyridylium herbicide in a N_2-fixing cyanobacterium *Nostoc muscorum*. *Environ. Internat.*, 10: 18–27.

Vaishampayan, A., 1985a. Growth of *Nostoc muscorum* mutants with diuron (DCMU) and L-methionine-DL-sulfoximine. *Experientia*, 41: 137–139.

Vaishampayan, A., 1985b. Potent mutagenicity of a bipyridylium herbicide in a nitrogen-fixing blue-green alga, *Nostoc muscorum. Biochem. Physiol. Pflanzen.*, 180: 327–331.

Vaishampayan, A., 1985c. Mutagenicity of bipyridylium salts in a nitrogen-fixing cyanobacterium. *Microbios.*, 43: 53–65.

Vaishampayan, A., 1985d. Mutagenic activity of alachlor, butachlor and carbaryl to a nitrogen-fixing cyanobacterium *Nostoc muscorum. J. Agric. Sci. (Cambridge)*, 104: 571–576.

Vaishampayan, A., 1988a. Evidence for an inter-strain conjugational transfer of *nif* (nitrogen-fixing) genes in a cyanobacterium *Nostoc muscorum*. In: *Proc XVI Internat Cong Genetics, Toronto*, p. 216–218.

Vaishampayan, A., 1988b. Collection and reconstruction of a non-heterocystous filamentous cyanobacterial culture from a sugar factory effluent, fixing nitrogen with a remarkably higher rate under a simple microaerophilic condition. In: *Proc VIth Internat Cong Culture Collections: Microbial Gene Pool*, University of Maryland, pp. 15–19.

Vaishampayan, A., 1989. A non-sporulating nitrogen-fixing cyanobacterium *Nostoc muscorum*: An efficient tool for a short-term mutagenicity screening of unknown chemicals. In: *Proc. XIVth Nat. Symp. Environ. Mutagen. Soc., India*, (Ed.) A.B. Prasad. L.N.M. University, Darbhanga, pp. 79–102.

Vaishampayan, A., 1993a. Two-step mutation for reduced phosphorus requirement in *Azolla-Anabaena* symbiotic nitrogen–fixing complex. In: *Proc XVIIth Internat Cong Genetics*, (Birmingham), pp. 4–7.

Vaishampayan, A., 1993b. Mutation breeding for phosphate independence in the nitrogen-fixing water-fern *Azolla* sp. In: *Proc XVth Internat Bot Cong*, Yokohama, pp. 522–526.

Vaishampayan, A., 1994a. Microsymbiotic cyanobiont of the low phosphate-requiring *Azolla-Anabaena* symbiotic nitrogen-fixing complex and its use in agriculture. In: *Proc IV International Cyanobacterial Workshop*, (Eds.) A.R. Grossman and M.R. Shaffer. Asilomer (California), pp. 11–12.

Vaishampayan, A., 1994b. Recent advances in the molecular biology of *Azolla-Anabaena* symbiotic nitrogen-fixing complex and its use in agriculture. In: *Microbes and Environment*, (Eds.) A. Prasad and R.S. Bilgrami. Narendra Publishing House, New Delhi, pp. 121–143.

Vaishampayan, A., 1994c. Molecular genetics of nitrogen fixation in cyanobacteria. In: *Biology and Application of Nitrogen-fixing Organisms: Problems and Prospects*, (Eds.) A.B. Prasad, A Vaishampayan. Scientific Publishers, Jodhpur pp. 1–18.

Vaishampayan, A., 1995a. Siderophore-mediated iron uptake in nitrogen-fixing cyanobacteria. In: *Advancements in Iron Nutrition Research*, (Ed.) A. Hemantaranjan. Scientific Publishers, Jodhpur, pp. 233–254.

Vaishampayan, A., 1995b. Somaclonal mutagenesis in *Azolla-Anabaena* symbiotic N_2-fixing complex for agronomically important traits. In: *Proc 10th Internat Cong Nitrogen Fixation*, All-Russia Research Institute of Agricultural Microbiology, Saint Petersburgh, pp. 7–9.

Vaishampayan, A., 1995c. *Azolla* micropropagation with functional generative filaments of the N_2-fixing endosymbiont *Anabaena azollae*. In: *Proc. Internat. Workshop Associat. Interactions Nitrogen-fixing Bacteria*, Institute of Biochemistry and Physiology of Plants and Microorganisms, Russian Academy of Sci., pp. 17–26.

Vaishampayan, A., 1995d. Molecular biology of herbicide-resistant cyanobacterial mutants in relation to photosynthesis, dinitrogen fixation and N assimilation. In: *Proc. Internat. Cyanobactl. Molecular Biol. Workshop,* (Ed.) Donald P Bryant, Asilomer (California), pp. 18–24.

Vaishampayan, A., 1996. Mineral requirements of the free-living and cymbiotic cyanobacteria. In: *Advancements in Micronutrient Research,* (Ed.) A. Hemantaranjan. Scientific Publishers, Jodhpur, pp. 103–126.

Vaishampayan, A., 1998. Physiological responses of genetically improved nitrogen-fixing cyanobacteria to agro-chemicalization in relation to paddy culture: Prospects as a source material for engineering herbicide sensitivity and resistance in plants. In: *Advances in Plant Physiology,* (Ed.) A. Hemantaranjan. Scientific Publishers, Jodhpur, pp. 191–220.

Vaishampayan, A., 2002a. Application of genetically improved nitrogen-fixing cyanobacteria to agro-chemicalization in relation to paddy culture: Prospects as a source material for engineering herbicide sensitivity and resistance in plants. In: *Advances in Plant Physiology,* (Ed.) A. Hemantaranjan. Scientific Publishers, Jodhpur, pp. 191–220.

Vaishampayan, A., 2002b. Recent phycotechnological advances concerning bio-N fertilization of rice. In: *Proc. Nat. Workshop Recent Developments in Biofertilizers for Rice Based Cropping System,* (Eds.) S. Kannaiyan, K. Govindarajan. Tamil Nadu Agricultural University, India, 1: 29–30.

Vaishampayan, A., 2005a. A comprehensive data-sheet on the phycotechnological advances towards the mass production of genetically engineered photobio-N fertilizers for organic rice farming. In: *Conservation and Utilization of Greenchip Wealth of Jharkhand,* (Ed.) Rashmi. Centre for Bioinformatics, Society for Nature Education and Health, Ranchi (In Press).

Vaishampayan, A., 2005b. Phycotechnological production of genetically improved associative bio-N fertilizers for rice. In: *Proc. Nat. Cong. Molecular Symbiosis and Biotechnol,* (Eds.) Ajit Varma, Springer-Verlag, Berlin (In Press).

Vaishampayan, A. and Awasthi, A.K., 1996. Micropropagation of *Azolla-Anabaena* symbiotic nitrogen-fixing mutant somaclones for an economic rice culture. In: *Proc. First Internat. Symp. Microbial Exploit.,* Centre for Advanced Studies in Botany, Banaras Hindu University, Varanasi, pp. 9–10.

Vaishampayan, A., Awasthi, A.K. and Dey, T., 2003. An ecological overview of *Nostoc*: The experimentally suitable most genus of nitrogen-fixing blue-green algae. In: *Phycology: Nature and Nurture,* (Eds.) B.N. Verma, A.N. Kargupta and R.N. Yadava. Kalyani Publ., Ludhiana, 9: 163–196.

Vaishampayan, A. and Banerjee, K.K., 1995. Genetic approaches to accomplish reduced phosphate-dependence of *Azolla-Anabaena* symbiotic nitrogen-fixing complex in wet agriculture. In: *Algal Ecology: An Overview,* (Eds.) A.N. Kargupta and E.N. Siddique. International Book Publishers and Distributors, Dehradun, pp. 161–193.

Vaishampayan, A. and Dey, T., 1998. A protocol for the genetic improvement of *Azolla* as an efficient dinitrogen fixer with rice at a higher thermal regime. *Transactions Egyptian Microbiol Soc: IIIrd Annual Meeting,* University of Alexandria, pp. 3–4.

Vaishampayan, A., Dey, T. and Awasthi, A.K., 1996. Genetic improvement of nitrogen-fixing cyanobacteria in response to modern rice agriculture. In: *Cytology, Genetics and Molecular Biology of Algae,* (Eds.) B.R. Chaudhary and S.B. Agrawal. SPB Academic Publishing bv, Amsterdam, pp. 395–421.

Vaishampayan, A., Dey, T. and Awasthi, A.K., 1998a. Advances in molecular biology, agronomics and somaclonal mutagenesis of *Azolla-Anabaena* symbiotic N_2-fixing complex. In: *Microbes: For Health, Wealth and Sustainable Environment*, (Ed.) Ajit Varma), Malhotra Publishing House, New Delhi, pp. 81–114.

Vaishampayan, A., Dey, T., Singh, R.P. and Prasad, B.K., 1998b. Genetically improved *Azolla* biofertilizer increases the rice production at low N fertilizer. In: *Proc. Nat. Seminar Management Strategies for North-Eastern Hilly Ecosystem*, ICAR Research Complex for N.E.H. Region, Umiam, pp. 46–48.

Vaishampayan, A., Dey, T., Singh, R.P., Rao, A.L. and Gupta, A.K., 2005a. Enhanced rice yields with genetically manipulated thermo-tolerant *Azolla-Anabaena* symbiotic N_2-fixing complex as an efficient bio-N fertilizer in the hot wet-fields of Varanasi. In: *Proc. Nat. Seminar and Southern Regional Conference on Microbial Inoculants*, (Ed.) M.D. Sundaram. Annamalai University (In Press).

Vaishampayan, A., Dey, T., Sinha, R.P. and Häder, D.P., 1998c. Successful rice cultivation with genetically manipulated thermotolerant *Azolla* as a bio-N fertilizer. *Acta Hydrobiol.*, 40: 207–213.

Vaishampayan, A., Dey, T., Sinha, R.P. and Singh, R.P., 1998d. Successful rice varietal trials with genetically manipulated thermotolerant *Azolla* as a bio-N fertilizer. In: *Proc Nat Symp Rice Res for 21st Century: Challenges, Priorities and Strategies*, Central Rice Research Institute, Cuttack, pp. 80–90.

Vaishampayan, A., Gupta, A.K., Singh, R.P. and Rao, A.L., 2005b. Increase in rice production at low "N" by the use of genetically improved *Azolla-Anabaena* symbiotic N_2-fixing biofertiliser. In: *Past Present and Future Perspectives in Biofertiliser Technology*, (Eds.) A.K. Yadav, S. Raychaudhuri and N.C. Talukdar. Regional Biofertiliser Development Centre, Ministry of Agriculture (Imphal) Govt. of India (In Press).

Vaishampayan, A. and Hemantaranjan, A., 1984. Physiological significance of vanadium uptake during N_2 and NO_3-metabolism in various strains of a N_2-fixing cyanobacterium *Nostoc muscorum*. *Pl. Cell Physiol.*, 25: 845–850.

Vaishampayan, A. and Mishra, A.K., 1989. Mutation for herbicide-resistance, coupled with metabolic utilization of the herbicide as a carbon and nitrogen source in a nitrogen-fixing cyanobacterium. In: *49th National Symposium on Recent Trends in Genetics and Plant Breeding Research in India*, (Indian Society of Genetics and Plant Breeding), Banaras Hindu University, Varanasi, November, 15–, pp. 4–6.

Vaishampayan, A. and Mishra, A.K., 1990. Cyanobacteria: A gene bank for the pesticide–resistance markers. In: *Proc IIIrd World Cong Environ Hazards Pesticides*, Cairo, pp. 1–51.

Vaishampayan, A., Mishra, A.K. and Singh, V.P., 1992a. Uptake of monuron in a cyanobacterial mutant. *Biomed. Lett.*, 47: 39–46.

Vaishampayan, A. and Mishra, P.K., 2003. A cyanobacterial mutant resistant against a uracil herbicide. In: *Proc Sixth Internat Workshop PGPR*, (Eds.) M.S. Reddy, M. Anandaraj, S.J. Eapen, Y.R. Sarma and J.W. Kloepper. Auburn University, Alabama and Indian Spice Research Institute, Kozhikode, 84: 248–256.

Vaishampayan, A. and Prasad, A.B., 1980. Biological effects of the fungicide blitox on *Nostoc muscorum* and *Nostoc linckia*. In: *Proc. Golden J. Session Nat. Acad. Sci.*, Allahabad, pp. 166–168.

Vaishampayan, A, and Prasad, A.B., 1981a. A pesticide-resistant mutant of the N_2-fixing blue-green alga *Nostoc muscorum*. *Experientia*, 37: 1285–1286.

Vaishampayan, A. and Prasad, A.B., 1981b. Mutagenic effects of two pesticides in the blue-green alga *Nostoc muscorum*. In: *Proc. Third Internat. Cong. Environ. Mutagens*, Tokya 2: 38–118.

Vaishampayan, A. and Prasad, A., 1982a. Blitox-resistant mutants of the N_2-fixing blue-green algae *Nostoc linckia* and *Nostoc muscorum*. *Environ. Exp. Bot.*, 22: 427–435.

Vaishampayan, A. and Prasad, A.B., 1982b. Mutation for resistance to two metal pesticides in a N_2-fixing blue-green alga. In: *Proc. Nat. Seminar Environ. Agents on Genetical Systems*, University of Calcutta, pp. 4–11.

Vaishampayan, A. and Prasad, A.B., 1982c. Glucose reversal of pesticide-inhibited CO_2 photoassimilation in a cyanobacterium. In: *Proc. All India Seminar New Approaches Pl. Sci. Res.*, University of Kashmir, pp. 16–39.

Vaishampayan, A. and Prasad, A.B., 1984. Inter-strain transfer of a pesticide-resistance marker during N_2 and NO_3-metabolism in various strains of a N_2-fixing cyanobacterium *Nostoc muscorum*. *Mol. Gen. Genet.*, 193: 195–197.

Vaishampayan, A., Reddy, Y.R., Singh, B.D. and Singh, R.M., 1992b. Reduced phosphorus requirement of a mutant *Azolla-Anabaena* symbiotic N_2-fixing complex. *J. Exp. Bot.*, 43: 851–856.

Vaishampayan, A. and Sahay, A., 1984. Changes in cell morphology as a result of UV-irradiation in a cyanobacterium *Nostoc muscorum*. In: *Proc. Internat. Cong. Cell Biol.*, Tokyo pp. 430–438.

Vaishampayan, A. and Singh, H.N., 1981a. Mutation in the blue-green alga *Nostoc muscorum*. I. Methylamine-resistant mutants. *Biochem. Physio. Pflanzen*, 176: 625–630.

Vaishampayan, A. and Singh, H.N., 1981b. Mutation in the blue-green alga *Nostoc muscorum*. II. L-methionine-DL-sulfoximine resistant mutants. *Biochem. Physio. Pflanzen*, 176: 631–637.

Vaishampayan, A., Singh, H.R. and Singh, H.N., 1978. Biological effects of rice-field herbicide *stam f-34* on various strains of the N_2-fixing blue-green alga *Nostoc muscorum*. *Biochem. Physiol. Pflanzen*, 173: 410–419.

Vaishampayan, A., Singh, R.P., Dey, T. and Prasad, B.K., 2000a. Use of improved phototrophic and heterotrophic N_2 fixers in rice, wheat and barley cultivation. In: *Proc Internat Cong Natural Resource Management*, (Ed.) G.B. Singh. Indian Agricultural Research Institute, Indian Council of Agricultural Research, New Delhi, 2: 662–664.

Vaishampayan, A., Sinha, R.P., Gupta, A.K. and Häder, D-P., 2000b. A cyanobacterial mutant resistant against a bleaching herbicide. *J. Basic Microbiol.*, 4: 1–10.

Vaishampayan, A., Sinha, R.P., Gupta, A.K. and Häder, D-P., 2000c. A cyanobacterial recombination study, involving an efficient N_2-fixing non-heterocystous partner. *Microbiol. Res.*, 155: 1–5.

Vaishampayan, A., Sinha, R.P., Gupta, A.K. and Häder, D-P., 2004. A N_2-fixing rice-field cyanobacterial mutant resistant against uracil herbicide. *Ecohydrol. Hydrobiol.*, 3: 417–423.

Vaishampayan, A., Sinha, R.P., Häder, D-P., 1998e. Use of genetically improved nitrogen-fixing cyanobacteria in rice paddy-fields: Prospects as a source material for engineering herbicide sensitivity and resistance in plants. *Botanica Acta*, 111: 176–190.

Vaishampayan, A., Sinha, R.P, Häder, D-P., Dey, T., Gupta, A.K., Bhan, U. and Rao, A.L., 2001. Cyanobacterial biofertilizers in rice agriculture. *Bot. Rev. (New York)*, 67: 453–516.

Varghese, A., 1990. Effect of organic manure or natural occurrence of *Azolla pinnata* and its effect on rice yield. *IRRI Newslett.*, 15: 16.

Vasil, I.K., Vasil, V. and Hubbell, D., 1977. Engineered plant cell or fungal association with bacteria that fix nitrogen. In: *Basic Life Sciences,* (Eds.) A. Hollaender, R.H. Burris, P.R. Day, R.W.F. Hardy, D. Helinsky, M.R. Lomborg, L. Owens and R.C. Valentine. Plenum Press, New York, USA, pp. 197–221.

Venkataraman, G.S., 1962. Studies on nitrogen fixation by blue-green algae. III. Nitrogen fixation by *Anabaena azollae. Ind. J. Agricl. Sci.,* 32: 22–24.

Venkataraman, G.S., 1975. The role of blue-green algae in tropical rice cultivation. In: *Nitrogen Fixation by Free-living Microorganisms,* (Ed.) W.D.P. Stewart. Cambridge University Press, U.K., pp. 207–218.

Venkataraman, G.S., 1981. Blue-green algae: A possible remedy to nitrogen scarcity. *Curr. Sci.,* 50: 253–256.

Venkataraman, G.S., 1993. Blue-green algae (cyanobacteria). In: *Biological Nitrogen Fixation,* (Eds.) S.N. Tata, A.M. Wadhwani and M.S. Mehdi. Indian Council of Agricultural Research, New Delhi, India, pp. 45–76.

Verma, S.K., Singh, A.K., Katiyar, S. and Singh, H.N., 1990. Genetic transformation of glutamine auxotrophy to prototrophy in the cyanobacterium *Nostoc muscorum. Arch. Microbiol.,* 154: 414–416.

Vitousek, P.K., Aber, J.D., Howarth, R.W., Likens, G.E., Matson, P.A., Schindler, D.W., Schlesinger, W.H. and Timan, D.G., 1997. Human alteration of the global nitrogen cycle sources and consequences. *Ecolo. Appl.,* 7: 737–750.

Vlek, P.L.G., Diakite, M.Y. and Mueller, H., 1995. The role of *Azolla* in curbing ammonia volatilization from flooded rice systems. *Fertilizer Res.,* 42: 165–174.

Watanabe, I., 1982. *Azolla-Anabaena* symbiosis, its physiology and use in tropical agriculture. In: *Microbiology of Tropical Soil and Plant Productivity,* (Eds.) Y.R. Dommergues and H.G. Dien. Martinus Nijhoff, Dr. W. Junk Publisher, The Hague, Boston, London, pp. 169–185.

Watanabe, I., 1984. Use of symbiotic and free-living blue-green algae in rice culture. *Outlood Agricl.,* 13: 166–172.

Watanabe, I., 1985. Limiting factors in increasing N_2 fixation in rice fields. In: *Biological Nitrogen Fixation in Africa: Proc 1st Cong. Afr. Asso. Biol. Nitrogen Fixation,* pp. 436–464.

Watanabe, I. and Berja, N.S., 1983. Growth of four species of *Azolla* as affected by temperature. *Aquatic Bot.,* 15: 175–185.

Watanabe, I., Lin, C., Ramirez, C. and Liu, C.C., 1989a. Physiology and agronomy of *Azolla-Anabaena* symbiosis. In: *Nitrogen Fixation with Non-Legumes,* (Ed.) F.A. Skinner. Kluwer Academic Press, The Netherlands, pp. 57–62.

Watanabe, I. and Ramirez, C.M., 1984. Relationship between soil phosphorus availability and *Azolla* growth. *Soil Sci. Pl. Nut.,* 30: 595–598.

Watanabe, I., Ventura, G., Mascarina, G. and Eskew, D.L., 1989b. Fate of *Azolla* sp. and urea nitrogen applied to wetland rice (*Oryza sativa*). *Biol. Fertil. Soils,* 8: 102–110.

Whitton, B.A., 2000. Soils and rice-fields. In: *The Ecology of Cyanobacteria,* (Eds.) B.A. Whitton and M. Potts. Kluwer Academic Publishers, The Netherlands, pp. 233–255.

Wolk, P., 1973. Physiology and cytological chemistry of blue-green algae. *Bacteriol. Rev.*, 37: 32–101.

Wolk. C.P. and Schaffer, P., 1976. Heterotrophic micro- and macro-cultures of a nitrogen-fixing cyanobacterium. *Arch. Microbiol.*, 110: 145–149.

Yamaoka, T., Satoh, K. and Katoh, S., 1978. Photosynthetic activities in a thermophillic blue-green alga. *Pl. Cell Physiol.*, 19: 934–954.

Yatazawa, M., Tomomatsu, N., Hasoda, N. and Nunome, K., 1980. Nitrogen fixation in *Azolla-Anabaena* symbiosis as affected by mineral nutrient status. *Soil Sci.Pl. Nut.*, 26: 415–416.

Yeomann, M.M. and Macleod, J., 1977. Tissue (callus) culture techniques. In: *Botanical Monographs*, (Ed.) H.E. Street. Blackwell Scientific Publishers, Oxford, UK, 11: 31–57.

Zhang, W.W., Lin, Y.H., Lu, P.J., Liu, Z.Z. and Huang, J.H., 1990. Electron microscopic observation of symbiotic relationship between *Azolla* and *Anabaena* during megaspore germination and sporeling development of *Azolla. Acta. Sin.*, 37: 514–520.

Zimmermann, W.J., 1985a. Biomass and pigment production in three isolates of *Azolla.* I. Response to water stress. *Annals Bot.*, 56: 689–700.

Zimmermann, W.J., 1985b0 Biomass and pigment production in three isolates of *Azolla.* I. Response to light and temperature stress. *Annals Bot.*, 56: 701–710.

Zimmermann, W.J., Watanabe, I. and Lumpkin, T.A., 1991. The *Anabaena-Azolla* symbiosis diversity and relatedness of neotropical host taxa. *Pl. Soil*, 137: 167–170.

© R.K. Gupta & V.D. Pandey (*Editors*)
Advances in Applied Phycology, 2007 42–55
Daya Publishing House, New Delhi

Chapter 2
Bioprospecting of Marine Algae

J. Rath and S.P. Adhikary*

P.G. Department of Biotechnology, Utkal University, Bhubaneswar–751004, Orissa

ABSTRACT

Marine organisms represent a valuable source of new compounds. The biodiversity of the marine habitat and the associated chemical diversity constitute a unlimited resource of bioactive substances in the field of development of bioactive products. Bioprospecting of marine algae as a source of novel bioactive compounds is presented along with issues related to conservation of biodiversity, bioprospecting and sustainable utilization of bioresources

Keywords: Marine algae, Biodiversity conservation, Bioprospecting

Introduction

The investigation of chemicals produced by plants and animals (*i.e.* natural products chemistry) has resulted in the discovery of several organic chemicals, many of which have found applications as pigments, fragrances, insecticides, pharmaceuticals or biomedical tools. Early studies which focused on terrestrial plants and microorganisms proved extremely fruitful yielding many useful organic compounds. By the early 1960s researchers began to view the ocean as a new and untouched source of potentially useful compounds–perhaps not surprising considering that more than 95 per cent of the Earth's biosphere is ocean. The sea hosts the richest diversity of life-forms and offers a largely untapped source of molecules with almost unknown physiological and ecological functions. Since then, marine natural product chemists have determined the chemical structure of over 6000 new compounds (Harvey, 2000). Research targeting organisms such as macro–as well as micro-algae, sponges, soft corals, ascidians, bryozoans, and molluscs (McCarthy and Pomponi, 2004; Donia and Hamann, 2003) has

* Corresponding Author: Tel: +91-674-2354054; Fax: +91-674-2587389; Email: adhikary2k@hotmail.com.

demonstrated that marine organisms produce unique secondary metabolites unlike those found in terrestrial organisms (Haefner, 2003).

The Marine Environment as a Source of Bioprospecting

Bioprospecting is the examination of biological resources (*e.g.* plants, animals, microorganisms) for features that may be of value for commercial development. These features may include chemical compounds, genes and their products or, in some cases, the physical properties of the material in question. Its main distinguishing feature from other biotechnology research is the concept of "prospecting"–the search for biological material for as-yet undiscovered applications.

The marine environment is characterized by physical and chemical properties that are markedly different from those of the terrestrial environment. Oceans cover approximately 71 per cent of the earth's surface (362 million square Km) and contain 95 per cent of the habitat space on the planet. The marine environment comprises approximately half of the total global biodiversity. Although the diversity of life on land is great, the marine environment is the centre of global diversity. For bioprospecting research, biodiversity translates to genetic uniqueness and diversity, which in turn relates to new biosynthetic pathways and chemical diversity. The vast genetic variety available in marine plants and microorganisms therefore offers a wealth of different and distinctive chemical compounds with biological activities. The oceans encompass habitats ranging from highly productive coastal regions to lightless, high-pressure and low temperature deep-sea environments (Vries and Hall, 1994). With ecosystems spanning cold polar waters through temperate latitudes to the tropics, marine biodiversity is far more extensive than popularly perceived. As a result of these physiological differences, marine organisms have evolved unique enzymatic systems producing many novel chemical structures unknown from terrestrial life forms. Seawater usually contain 30-35 grams of dissolved salts per litre, while freshwater has a total salt content of about 1 gram per litre. Marine natural products are often modified by the composition of the halide-rich seawater with chlorine, bromine and iodine in the ratio 19000, 65 and 0.0005 mg/l (Bohlin, 1989). In addition marine algal extracts have shown to enhance the growth of specific bacteria found on their surfaces. This may be an example of a complex interaction, whereby the host in some way attracts specific bacteria to its surface, which in turn protects the host against harmful microorganisms by producing antimicrobial compounds. Several such surface-associated bacteria have shown to produce antibiotics.

Bioprospecting Potential of Marine Algae

India has a long coastline of about 7,500 kilometers, a sizable exclusive economic zone (2.5 million km^2) and a large shelf area (0.13 million km^2). Exclusive Economic Zone (EEZ) is about two third area of the mainland. This biogeographically important area harbor a variety of natural resources, including flora and fauna widely used as fuel, food and feed of mankind. Marine algae are one of the constituents of these resources, globally used for phycocolloids and recently for the extraction of bioactive substances (Crasta *et al.*, 1997; Padmakumar and Ayyakkannu, 1997 and Sastry and Rao, 1994). Seaweeds or marine macro-algae are an important source of industrially important thickening agents and gels such as, agar, carrageenan and alginates and there has been rapid demand of these products in recent years. They are also a source of fine chemicals such as natural pigments, mannitol, iodine and therapeutically active products. Seaweed-based fertilizers, which contain a range of growth promoting hormones are also gaining importance as Seaweed liquid fertilizer (SLF) and their efficacy has been proven for a variety of crops. Several metabolites with antimicrobial and cytotoxic activities have been isolated and characterized from marine algae.

Bioprospecting of Marine Micro-algae

Marine micro-algae comprise the largest group of living organisms in the ocean, constituting an estimated 10000 species. Cyanobacteria (blue green algae) in general and marine algal forms in particular are one of the richest source of known and novel bioactive compounds including toxins with wide pharmaceutical applications is unquestionable. Among the five divisions of microalgae, studies on biomedical natural products have been concentrated only on two divisions, *i.e.* Cyanophyta (Cyanobacteria) and Pyrrophyta (dinoflagellates). Although several metabolites have been isolated from cyanophytes (Moore *et al.*, 1988; Beltron and Nielan, 2000), most of them were isolated from fresh water species, which can be cultured easily in comparison to marine organisms. Lyngbyatoxin-A and debromoaplysiatoxin are two highly inflammatory but structurally different metabolites isolated from toxic strains of *Lyngbya mausculata* (Cardillina *et al.*, 1979) and anatoxin-A from *Anabaena cincinalis* (Beltron and Nielan, 2000). Some of the marine cyanobacteria appear to be potential sources for large-scale production of vitamins of commercial interest such as vitamins of the B complex group and vitamin-E (Plavsic *et al.*, 2004). The carotenoids and phycobiliprotein pigments of cyanobacteria have commercial value as natural food coloring agents, as feed additives, as enhancer of the color of egg yolk, to improve the health and fertility of cattle as drugs and in cosmetic industries. Anti-HIV activity has also been observed with the compounds extracted from *Lyngbya lagerhaimanii* and *Phormidium tenue*. More than 50 per cent of the 100 cyanobacteria isolates from marine sources possess potentially exploitable bioactive substances. The substances tested were either the ones that kill cancer cells by inducing apoptotic death, or those that affected cell signaling through activation of the members of the protein kinase-C family (signaling enzymes) (Fujiki and Sugimura, 1989; Wender *et al.*, 1986). Cultured *Fusarium chlamydosporum* isolated from the Japanese marine red alga *Carpopeltis affinis* is the source of fusaperazines A and B, two new sulphur containing dioxopiperazine derivatives and two known compounds which had been originally isolated from the fermentation by the fungus *Tolypocladium* spp. (Lin *et al.*, 2002). Chalcomycin-B exhibited activity against a variety of microorganisms and microalgae. Four new epi-polysulphanyl-dioxopiperazines were isolated from a culture of the fungus *Leptosphaeria* spp. originating from the Japanese brown alga *Sargassum tortile* (Yamada *et al.*, 2002). Absolute stereochemistries were determined by chemical analyses and transformations. Each compound possess significant cytotoxic activity against the P388 cell-line while one of the leptosins also exhibited appreciable cytotoxicity against a disease-oriented panel of 39 human cancer cell-lines, and specifically inhibited two protein kinases and topoisomerase-II. The anti-inflammatory and anti-proliferative properties of scytonemin, an extracellular sheath pigment originally isolated from the cyanobacterium *Scytonema* spp. have been reported (Proteau *et al.*, 1993; Stevenson *et al.*, 2002 a, b). An immuno-suppressive linear peptide microcolin-A which at nano molar concentrations suppresses the two way murine mixed lymphocyte reaction has been isolated from *Lyngbya majusculata* (Koehn *et al.*, 1992). A unique thiozoline-containing compound, curacin–A has been purified from the organic extract of a Curacao collection of *L. majusculata* (Gerwick *et al.*, 1994). This compound has been found to be an exceptionally potent anti-proliferative agent as it inhibits the polymerization of tubulin which shows some selectivity for colon, renal and breast cancer-derived cell lines (Carte, 1996).

The dinoflagellates are well known sources of marine toxins, including the brevetoxins (the toxic principle of the Florida red tide), saxitoxin (the paralytic shell-fish poison) and the ciguatoxins (responsible for ciguatera sea food poisoning) (Murakami *et al.*, 1988). In addition, maitotoxin exhibits multiple activities including hormone stimulation, neurotransmitter release, activation of phosphoinositide degradation and potentiation of protein kinase, all of which appear to be linked to elevation of intracellular Cap+ concentration. Thus, it has become a valuable probe for the study of Ca2+ regulation. A series of noval antibiotic agents have been isolated from dianoflagellates, antifungal agents from *Gambierdiscus toxicus* (Nagai *et al.*, 1992) and brevitoxin from *Ptychodiscus brevis*. As they depolarize the excitable membranes and their binding sites on sodium channel the mechanism seems to be different from that of other activators (Shimizu, 1993; Carte, 1996). Okadaic acid, a polyether fatty acid produced by *Prorocentrum* spp., has been a key molecule in studying signal transduction pathways in eukaryotic cells since it is a selective protein phosphatase inhibitor (Cohen *et al.*, 1990). Goniodomin-A, an antifungal polyether macrolide from the dinoflagellate *Goniodoma pseudogoniaulax* (Murakami *et al.*, 1988) has been shown to inhibit angiogenesis by the inhibition of endothelial cell migration and basic fibroblast growth factor (BFGF)-induced tube formation.

Bioprospecting of Marine Macro-algae (Seaweed)

Seaweeds are abundant in the intertidal zones and in clear tropical waters. These algae have received comparatively less bioassay attention. Presently the seaweed industry largely consists of two kelps, three *Gelidium* species one *Gracilaria/Gracilariopsis* species (Anderson *et al.*, 2003). In addition, there are a number of seaweeds with economic potential (Critchley *et al.*, 1998). It will be of great significance if these species could be the major role players in drug development. Alternatively, findings from academic laboratories could result in new cultivation initiatives. Nonetheless, the red alga *Sphaerococcus coronopifolius* has shown to have antibacterial activity (Donia and Hamann, 2003); the green alga *Ulva lactuca* posses an anti-inflammatory compound; and an anti-tumor compound was isolated from *Portieria hornemannii* (Failkner, 2002). *Ulva fasciata* producing a novel sphingosine derivative has been found to have antiviral activity *in vivo* (Garg *et al.*, 1992). A cytotoxic metabolite, stypoldione, which inhibit microtubule polymerization and prevents mitotic spindle formation has been isolated from tropical brown alga *Stypodium zonale* (Gerwick and Fenical, 1981; Jacobs *et al.*, 1985). *P. hornemannii* was found to be a novel source of cytotoxic penta-halogenated-monoterpene, Halomon, which exhibited one of the most extreme examples of differential cytotoxicity in the screening conducted by the National Cancer Institute (NCI), USA. Halomon has been selected for preclinical drug development since this compound show toxicity to brain, renal and colon tumor cell-lines, and the preliminary *in vivo* evaluations have been encouraging (Carte, 1996). An iodinated novel nucleoside has been isolated from *Hypnea valitiae*, which is a potent and specific inhibitor of adenosine kinase. It can be used in the studies of adenosine receptors in a variety of systems, and in studies on nucleotide metabolism and regulation (Ireland *et al.*, 1999).

The green alga *Codium iyengarii* from the Karachi coast of the Arabian Sea has been found as the source of a steroid, iyengadione and two new steroidal glycosides, iyengarosides A and B. Iyengaroside-A displayed moderate activity against a range of bacteria (Ali *et al.*, 2002). *Sargassum carpophyllum* from the South China Sea is the source of two new bioactive sterols. These sterols induced morphological abnormality in the plant pathogenic fungus *Pyricularia oryzae*; also exhibited cytotoxic activity against several cultured cancer cell lines (Tang *et al.*, 2002). *Sargassum polycystum* collected in the North China Sea yielded a new sterol, Stigmast (Xu *et al.*, 2002). The fact that there are many algae that can convert simple poly-unsaturated fatty acids such as arachidonic acids into complex eicosanoids and related oxylipins had been an exiting development (Gerwick and Bernart, 1993). Derivatives of arachidonic acids are important in maintaining homeostasis in mammalian systems. Aberrant production of metabolites of this class occurs in diseases such as psoriasis, asthma, arteriosclerosis, heart disease, ulcers and cancer (Carte, 1996).

Antibacterial Compounds from Marine Algae

Since ancient times early societies used natural medicines generally as crude extracts from plants to treat infection, inflammation, pain and a variety of other maladies. In China, marine algae have been used in traditional and folk medicine for more than two thousand years; among others the use of red algae *Gloiopeltis* sp. for treatment of tuberculosis. The presence of antibacterial materials in marine algae has been known since the late 1800's. The use of brown algal extracts as antiseptic was documented in 1937 (Emerson and Taft, 1945). Results of a number of screening studies aiming to establish the antibacterial activity in marine algae showed variations in activity which may be explained by the fact that marine algae display seasonal variations in the production of antimicrobial compound; and each alga has its own periodicity with maximum antibiotic activity (Chesters and Stott, 1956; Rao and Parekh, 1981; Padmakumar and Ayyakkannu, 1997). The antibiotic production is generally highest during the algae's active growth period. Other factors that may influence the antibiotic content are the reproductive state of algae (Moreau *et al.*, 1984; Hornsey and Hide, 1985), geographical location (Vidyavathi and Sridhar, 1991) and part of the algal thallus (Conover and Sieburth, 1964; Hornsey and Hide, 1976). Antibacterial activity has been detected in practically all marine divisions of the kingdom Plantae, including Phaeophyta, Rhodophyta, Chlorophyta, Pyrrophyphyta (dinoflagellates), Prasinophyta (green flagellates), Bacillariophyta (diatoms), Haptophyta and Cryptophyphyta (Ragan, 1984). The first partly identified antimicrobial compound isolated from algae termed chlorellin. Acrylic acid was the first antibiotic that was unambiguously identified (Lohsiri *et al.*, 1995). Since then a variety of antibacterial compounds have been isolated from marine algae. These includes phlorotannins (Nagayama *et al.*, 2002), fatty acids (Rosell and Srivastava, 1987), polysaccharides (Criado and Ferreiros, 1984), phenols (Ohta and Takagi, 1977), phthalates (Sastry and Rao, 1995), cyclic polysulphides (Wratten and Faulkner, 1976), terpenoids (Paul and Fenical, 1984), sterols (Awad,2000), steroidal glycosides (Ali *et al.*, 2002) and various aliphatic compounds (Glombitza, 1979). In addition, studies by Haas (950) and Berland *et al.* (1972) indicated presence of AMPs in marine algae. Some of the compounds isolated have shown potent antibacterial activity against multi-resistant bacteria (Nagayama *et al.*, 2002). It was not until after World War II that bioprospecting for antibacterial compounds in marine organisms began. Since then a large number of marine organisms from a broad range of algal classes have been screened for antimicrobial activity. However, by far the majority of marine algae are yet to be screened and the potential for discovering a useful antibiotic is sufficient to warrant further research.

Antioxidant Activity of Marine Algae

Reactive oxygen species (ROS) such as superoxide anion ($O^{\cdot -}_2$), hydroxyl radical (HO-) and hydrogen peroxide (H_2O_2) are physiological metabolites formed during aerobic life as a result of the metabolism of oxygen. DNA, cell membranes, proteins and other cellular constituents are target site of the degradation processes and consequently induce different kinds of serious human diseases including atherosclerosis, rheumatoid arthritis, muscular dystrophy, cataracts, some neurological disorders and some types of cancers as well as aging. Moreover, ROS are predominant cause of qualitative decay of foods, which lead to rancidity, toxicity and destruction of biomolecules important in physiologic metabolism. However, with safety concerns identified for these synthetic antioxidant (Kitts, 1996; Wichi *et al.*, 1998), considerable interest has arisen in finding alternative sources of antioxidants for use in food systems and increased in researches for finding natural antioxidants. The use of natural antioxidants has the advantage that the consumer is considered to be safe because of no chemical contamination readily accepts them and no safety tests are required by the legislation if the food component is Generally Recognized As Safe (GRAS) (Pokorny, 1991). Hence, the antioxidants are essentially needed in body system and to extend the storage time of food.

Almost all photosynthesizing plants including marine algae are exposed to a combination of light and high oxygen concentrations, which lead to the formation of free radicals and other strong oxidizing agents, but they seldom suffer any serious photodynamic damage during metabolism. This fact implies that their cells have some protective antioxidative mechanisms and compounds (Matsukawa *et al.*, 1997). Seaweeds are considered to be a rich source of antioxidants (Cahyana *et al.*, 1992). Recently, the potential antioxidant compounds were identified as some pigments (*e.g.* fucoxanthin, astaxanthin, carotenoid) and polyphenols (*e.g.* phenolic acid, flavonoid, tannins). Those compounds are widely distributed in marine algae and are known to exhibit higher antioxidative activities. The activities have been reported through various methods of reactive oxygen species scavenging activity and the inhibition of lipid peroxidation (Yan *et al.*, 1999; Athukorala *et al.*, 2003a,b; Heo *et al.*, 2003 a,b; Siriwardhana *et al.*, 2003, 2004).Over the past several decades, seaweeds or their extracts have been studied as novel sources which have been shown to produce a variety of compounds and some of them have been reported to possess biological activity of potential medicinal value (Konig *et al.*, 1994; Tutour *et al.*, 1998). Recently, much attention has been paid on the anti-tumor activity, anticholesterolemic activity and antioxidant activity of seaweed constituents. Consequently, antioxidant activity is intensively focused due to the currently growing demand from the pharmaceutical industry where there is interest in anti-aging and anticarcinogenic natural bioactive compounds.

Huang and Wang (2004) screened and evaluated antioxidant activities of 16 species of seaweeds collected along the Qingdao coastline using β-carotene-linoleate assay system and observed all the selected seaweeds exhibited various degrees of antioxidative efficacy. Tutour *et al.* (1998) investigated the antioxidant and pro-oxidative activities of *Laminaria, Himanthalia, Fucus* and *Ascophyllum* species to scavenge peroxyl radicals by kinetic studies in a model system and concluded that seaweed extracts exhibited antioxidant activities by extending the induction period, but they did not suppress the rate of oxygen uptake as did vitamin E. Heo *et al.* (2003) screened seven species of brown seaweeds for their antioxidant activity and found that enzymatic extracts of seaweeds possess a potent antioxidant activity. Matsukawa *et al.* (1997) examined extracts of over 100 species of microalgae both from nature and from laboratory cultures and screened their antioxidant activity and found that lipoxygenase and tyrosinase activities were inhibited by the extract of several microalgae and suggest for their use as potential antioxidants. Anggadiredja *et al.* (1997) analyzed fresh and dry specimens of *Sargassum* and *Laurencia* species and observed that the dry specimens did not show any antioxidant activity whereas the fresh materials did show activity. They also observed that in case of *Sargassum* methanolic extract was more active and in case of *Laurencia* n-hexane extract was more active than the diethylether and methanol extracts. Ruberto *et al.* (2001) examined antioxidant activity of the lipid extracts of eight marine algae belonging to the genus *Cystoseira* in a micellar model system and ascribed the antioxidant activity in the extract of tetraprenyltoluquinols and expressed the results as relative antioxidant efficiency (RAE) and deduced a composition-activity relationship. Heo *et al.* (2005) evaluated potential antioxidative activities of enzymatic extracts from seven brown seaweeds using four different reactive oxygen species scavenging assay and observed that the enzymatic extracts exhibited more prominent effects in hydrogen peroxide scavenging activity compared to other scavenging activity and concluded that enzymatic extracts of the brown seaweeds might be valuable antioxidative sources. Zhang *et al.* (2003) studied antioxidant activities of sulfated polysaccharide fractions from *Porphyra haitanesis* and showed that all the three polysaccharide fractions exhibited antioxidant activities and they had strong scavenging effect on superoxide radical and much weaker effect on hydroxyl free radical. Siriwardhana *et al.* (2004) examined the enzymatic hydrolysis for effective extraction of antioxidative compounds from *Hizikia fusiformis* using five carbohydrases. Zhao *et al.* (2004) studied the antioxidant and hepatoprotective activities of low molecular weight sulfated polysaccharide from *Laminaria japonica*.

Bioprospecting and its Relationship to the Conservation and Utilization of Bioresources

There is a huge interest world-wide in the bioprospecting and the issue arises on the conservation of biodiversity, its measurement and the effective and non-destructive utilization of this resource. There are many international policies that address various aspects of bioprospecting activities and conservation of biodiversity, these are principally the United Nations Convention on the Law of the Sea (UNCLOS), the Convention on Biological Diversity (CBD), the World Intellectual Property Organization (WIPO), International Treaty on Plant Genetic Resources for Food and Agriculture, The Microbial Diversity 21, BioNET-INTERNATIONAL, SPECIES 2000, Systematics Agenda 2000 International, and International Organization for Systematic and Evolutionary Biology (IOSEB) initiatives.

The natural loss due to species extinctions accompanying habitat loss is important to society for a number of reasons including the maintenance of ecosystem function and for ethical reasons. But species extinction is specifically important to pharmaceutical companies in their search for novel natural product. Amongst the species being lost will be some which have the ability to produce important, but as yet undiscovered, chemical molecules. Having said this, there are a variety of biodiversity, conservation and sampling issues associated with examining different types of biota for the production of novel natural product. The quantities required are not an issue with microbial samples where isolates may be stored under appropriate conditions for future use. However, there is a paucity of information available concerning the taxonomic relationships between microorganisms and information concerning their geographic distributions, as noted earlier. But, for other types of samples, there are clear collection and re-collection issues to be considered. With the increased difficulty in discovering new medicines there has been a change of emphasis within the natural products discovery units of pharmaceutical companies from searching for novel natural product that could be delivered to market and to search for novel lead compounds. These lead compounds would eventually find their way to market after chemical modification and/or a complete chemical synthesis being devised. This "philosophical" change to searching for new natural products has opened up opportunities for screening a variety of previously unscreened organisms for natural product production. It may be unrealistic to collect only samples where large amounts of material are available as a hedge against requiring further material in the near future. This would severely curtail the numbers and types of samples collected. However, in the event of a successful discovery, it is imperative that raw materials be available in sufficient quantities for follow-up work and/or compound production at a reasonable scale. A balance must be reached between small initial collections requiring further follow-up re-collections with their inherent problems, and larger initial collections that may restrict the diversity of samples able to be collected. Re-collection issues certainly do arise with marine algae and other marine organisms. For example, marine algae species are difficult to arise in artificial conditions. The synthetic media composition and the environment to grow marine algae are very difficult and not too many species lend themselves to tissue culture on an economic basis. This means that the natural source of algal materials must be judiciously conserved and sustainable managed to ensure continuing supply. Also, it may not be possible in every instance to use an associated organism to produce compounds of interest in the way that the use of the endophytic fungus *Taxomyces andreanae* from the Pacific view is being investigated for the production of taxol by fermentation. It should also be borne in mind that it has taken many years of work to get a natural product to come to the market.

It has been said that drug discovery based on sourcing and direct extraction of vegetative algal material is limited by two factors: restricted access to phytochemistry due to genetic and/or metabolic regulation of secondary metabolism reflecting specific growing conditions and, the uncertainty of re-accessing interesting chemistry in subsequently sourced samples of the same algae. Thus pharmaceutical industry's involvement in conserving biodiversity in situ is essential due to variations in secondary metabolite production between different samples of the same species from different habitats. It is not clear, however, whether these differences mirror the diversity of genotypes within a sampling site, or if the diversity is a result of habitat variability or epigenetic variability. Studies undertaken with marine algae to examine this have highlighted the complex relationship that exist between genotype, environment and phenotype as expressed in secondary metabolic products. It is thought, for example, that differences in macro–and microclimatic factors influencing the algal populations, differences in substrate patchiness, environmental disturbance within a habitat might allow more diverse populations of marine algae to evolve in some habitats. Production of these compounds was directly correlated with predation levels and with the compounds undergoing enzymatic conversion from inactive precursors following grazing. Such examples highlight the benefit of examining more than one population of a species and of having ecological, in addition to systematic knowledge of the organisms in order to maximize the chances of discovering new chemicals. A greater understanding of the variation in metabolic activity of populations of species in natural ecological settings may have value in optimizing methods of searching and screening for useful natural products. Re-collections might necessitate survey of the distribution and abundance of organisms, as well as determination of the variation of drug content in different organisms and the fluctuation of content under different biotic and abiotic conditions. However there is a perception amongst conservationists and the public that 'large quantities' of material are being collected from the oceans for screening for novel natural products. Whilst small amounts of material may be used for the initial stage of the drug discovery process, there is a clear desire within the pharmaceutical industry to conserve the world's biota so that more species can be examined for novel chemical molecules and, that compounds of interest are produced via routes that do not involve the destructive and costly harvesting of samples.

Benefits of Marine Algal Bioprospecting

Bioprospecting activities can create a wide range of potential benefits, when implemented within the proper framework. It is adding value through the chain of development. It means of securing benefit is to add as much value as possible throughout the chain of testing and development, rather than focusing on an end point of commercialization. This creates direct financial benefits, for example, through the sale of valuable knowledge. It also creates flow-on effects in employment and capacity-building in local biotechnology industries. Value-adding through the chain of development can also include benefit-sharing regimes with the local community. For example, bioprospecting activities undertaken the training and employment of local people to collect biological samples and there are opportunities for benefit-sharing arrangements with local communities. Thus there can be a connection between indigenous knowledge and the biological material in question. Bioprospecting also has the potential to lead to industries around the production of valuable substances. This could involve the production of the raw biological material. New industries could also involve production of a finished product, for example natural skin care products and this downstream production related to new scientific discoveries. It is also possible through bioprospecting to secure royalty streams for researchers through the retention of intellectual property in a downstream commercial product. Such benefits, however, rely on eventual commercialization. A small proportion of bioactive substances make it to the stage of commercial production. In addition, the complexity of inputs into commercialization means intellectual property around a substance discovered may be a proportionately small part of the end product. Involvement in bioprospecting projects enhances the skills and knowledge of researchers.

This also improves the downstream opportunities to add value through the development chain of a substance. Research can therefore contribute to the economic development of a country. Bioprospecting increases the understanding of our natural systems, regardless of whether the substance in question eventually leads to financial returns. This knowledge has intrinsic value in its own right. It may also contribute to conservation of biodiversity and environmental management practices.

Conclusion: Marine Bioprospecting–An Expanding Field

Recent reviews document the escalating numbers of marine natural products discovered annually (Faulkner, 2002; Mayer and Hamann, 2002; Blunt *et al.*, 2003). The increasing number of patents in the area of marine natural products provides an indication that marine bioprospecting is enjoying continuing growth with an expanding interest from pharmaceutical industry. According to the Marine Lit literature database a relatively small number of marine plants, animals and microbes have already yielded more than 15,000 novel chemical compounds. Today, marine natural products are emerging as excellent resources for novel antibacterial (Smith and Chisholm, 2001), antiviral (Tziveleka *et al.*, 2003), antitumor (Rinehart, 2000; Mayer *et al.*, 2003), antifungal (Fusetani, 1988), antimalarial (EI Sayed *et al.*, 1996) and other anti-infective lead compounds. Bioactive compounds from marine organisms have also proven to be interesting candidates for the development of novel agrochemicals (EI Sayed *et al.*, 1997; Peng *et al.*, 2003), antifouling compounds (Armstrong *et al.*, 2000) and also for other industrial applications. Today marine bioprospecting has an annual market value of $3-4 million. However the cost and technical difficulties of identifying and collecting marine samples, especially those occurring beyond the limits of conventional diving, the sometimes lengthy and highly skilled laboratory processes, the occasional need to develop new screening techniques and the difficulty associated in securing a sustainable source need to over come. Japan spends almost a billion dollars a year; 80 per cent of which comes from the private sector. Other countries spend far less but the search for new compounds marine bioprospecting is a big business, and is very likely to increase further. Sequencing of the total genome of marine organisms is a rapidly advancing yet new area of bioprospecting research. Genetic screening will become a lot more important in the near future and would lead to the development of gene probes for antibiotics and other molecular targets.

Acknowledgements

We thanks Council of Scientific and Industrial Research, New Delhi for a research project on bioactive substance from marine algae of Orissa and Research Associateship to one of us (JR).

References

Ali MS, Saleem M, Yamdagni R and Ali MA. (2002) Steroid and antibacterial steroidal glycosides from marine green alga *Codium iyengarii* Borgesen. *Nat Prod Lett*. 16: 407-413.

Anderson RJ, Bolton JJ, Molloy FJ and Rothmann KWG (2003) Commercial seaweeds in southern Africa. *Proceedings of the 17th International Seaweed Symposium*. *In* Chapman RO, Anderson RJ, Vreeland V J, Davison IR (ed), Oxford University Press: Oxford, 1–512.

Anggadiredja J, Andyani R, Hayati and Muawanah (1997) Antioxidant activity of *Sargassum polycystum* (Phaeophyta) and *Laurencia obtuse* (Rhodophyta) from Seribu Islands. *J. Appl. Phycol.* 9: 477-479

Armstrong E, Boyd KG, Pisacane A, Peppiatt CJ and Burgess JG (2000) Marine microbial natural products in antifouling coating. *Biofouling* 16: 215-224.

Athukorala Y, Lee KW, Song CB, Ahn CB, Shin TS, Cha YJ, Shahidi F and Jeon YJ (2003a) Potential antioxidant activity of marine red.alga *Grateloupia filicina* extracts. *J. Food Lipids* 10: 251–265.

Athukorala Y, Lee KW, Shahidi F, Heu MS, Kim HT, Lee JS and Jeon YJ, (2003b) Antioxidant efficacy of extracts of an edible red alga (*Grateloupia filicina*) in linoleic acid and fish oil. *J. Food Lipids* 10: 313–327.

Awad NW (2000) Biologically active steroid from the green algae *Ulva lactuca Phytotherapy Research* 14: 641-643.

Beltron E C and Nielan B A (2000) Geographical segregation of Neurotoxin-producing Cyanobacterium *Anabaena circinalis. Appl. and Environ. Microbiol.* 66: 4468-4474.

Berland BR, Bonin DJ, Cornu AL, Maestrin Sy and Marino JP (1972). Antibacterial substances of marine alga *Stichochrysis immobilis* (Chrysophyta). *J. Phycology* 8: 383-392.

Blunt JW, Copp BR, Munro MHG, Northcote PT and Prinsep MR (2003) Marine Natural products. *Nat Prod Rep.* 21:1-49.

Bohlin L (1989) Pharmacologically active compounds from marine organisms. *Acta Pharmaceutica Nordica* 1: 175-182.

Cahyana AH, Shuto Y and Kinoshita Y. (1992). Pyropheophytin a as an antioxidative substance from the marine alga, Arame (*Eisenia bicyclis*). *Biosci. Biotechnol. Biochem.* 56: 1533–1535.

Cardillina JH, Marner FJ and Moore RE (1979) Seaweed dermatitis: structure of lyngbyatoxin A. *Science.*204: 193-195.

Carte B K (1996) Biomedical potential of marine natural products. *Bioscience.* 46: 271-286.

Chesters CGC and Stott JA (1956) The production of antibiotic substancesby seaweeds. *In:* Braarud T and Sorensen NA (ed) *Second International Seaweed Symposium* Pergamon Press, London pp 49-53.

Cohen P, Holmes C and Tsukitani Y (1990) Okadaic acid: a new probe for the study of cellular regulation. *Trends Biochem. Sci.*15: 98-102.

Conover JT and Sieburth JM (1964) Effect of Sargassum distribution on its epibiota and antibacterial activity. *Bot. Mar.* 6: 147-157.

Criado MT and Ferreiros CM (1984) Toxicity of an algal mucopolysaccharide for *Escherichia coli* and *Neisseria meningitides* strains *Espanola de Fisiologia* 40: 227-230.

Crasta PJ, Raviraja NS and Sridhar KR (1997) Antimicrobial activity of some marine algae of southwest coast of India. *Indian J. Mar. Sci.* 26: 201-205.

Critchley AT, Gillespie RD and Rotman KWG. *In.* Critchley M and Ohno AT (ed) *Seaweed Resources of the World,* Japan International Cooperation Agency: Japan, 1998; pp 413-425.

Donia M and Hamann MT (2003) Marine natural products and their potential applications as antiinfective agents. *The Lancet.* 3: 338-348.

Emerson GA and Taft CH (1945) Pharmacologically active agents from the sea. *Texas Reports on Biology and Medicine* 3: 302-338.

Faulkner DJ (2002) Marine natural products. *Nat Prod Rep.*19: 1-48.

Fujiki H and Sugimura T (1987) New classes of tumor promoters: telocin, aplysiatoxin and palytoxin, *Adv Cancer Res.* 59: 223-264.

Fusetani N (1988) Antifungal substances from marine invertebrates. *Annals of the New York Academy of Sciences* 544: 115-127.

Garg HS, Sharma T, Bhakuni DS, Pramanik BN and Bose AK (1992) An antiviral sphingosine derivative from green alga *Ulva fasciata*. *Tetrahedron lett*. 33: 1641-1644.

Gerwick WH, Proteau PJ, Nagh DG, Hamel E, Blobhin A and Slate DL (1994) Structure of cruacin A, a novel antimitotic, antiproliferative and brine shrimp toxic natural product from the marine cyanobacterium *Lyngbya majusula*. *J. Org. Chem.* 59:1243-1245.

Gerwick WH and Bernart MW (1993) In. Attaway D, Zaborsky O (ed.) *Marine Biotechnology*; Plenum Press: New York, pp101-152.

Gerwick WH and Fenical W (1981) Ichthyotoxic and cytotoxic metabolites of the tropical brown alga *Stypopodium zonale*. *J. Org. Chem*.46: 21-27.

Glombitza KW and Schmidt A (1999). Trihydroxyphlorethols from the brown algae *Carpophyllum angustifolium*. *Phytochemistry* 51: 1095-1100.

Haas P (1950) On certain peptides occurring in marine algae. *Biochemical Journal* 46: 503-505.

Haefner B (2003) Drugs from the Deep. *Drug Discov. Today*. 8: 536-544.

Harvey A (2000) Strategies for discovering drugs from previously unexplored natural products. *Drug Discovery Today*. 5: 294-300.

Heo SJ, Park EJ, Lee KW and Jeon YJ (2005) Antioxidant activities of enzymatic extracts from brown seaweeds. *Bioresource Technology* (article in press).

Heo SJ, Jeon YJ, Lee J, Kim HT and Lee KW, (2003a). Antioxidant effect of enzymatic hydrolyzate from a Kelp, *Ecklonia cava*. *Algae* 18: 341–347.

Heo SJ, Lee KW, Song CB and Jeon YJ, (2003b) Antioxidant activity of enzymatic extracts from brown seaweeds. *Algae* 18: 71–81.

Hornsey LS and Hide D (1976) The production of antimicrobial compounds by British marine algae. II. Seasonal variation in production of antibiotics. *British Phycology Journal* 11: 63-67.

Hornsey LS and Hide D (1985) The production of antimicrobial compounds by British marine algae IV. Variation of antimicrobial activity with algal generation. *British Phycology Journal* 20: 21-25.

Huang HL, and Wang BG (2004) Antioxidant Capacity and Lipophilic Content of Seaweeds Collected from the Qingdao Coastline *J. Agric. Food Chem.* 52: 4993-4997

Ireland C, Copp B, Foster M, McDonald L, Radisky and D Swersey J. (1993) *In*: Attaway D and Zeborsky O (ed) *Marine Biology*; Plenum Press: New York, pp1-43.

Jacobs RS, Culver P, Langdon R, O'Brien T and White S (1985) Some pharmacological observations on marine natural products. *Tetrahedron*.41:981-984.

Kitts DD (1996) Toxicity and safety of fats and oils. *In*: Hui YH (ed.) Baileys Industrial Oil and Fat Products Wiley-Interscience, New York, pp. 215–280.

Koehn FE, Longley RE and Reed JK (1992) Microcolin A and B, new immunosuppressive peptides from the blue green alga *Lyngbya majuscula*. *J Nat Prod*. 55: 613-619.

Konig GM, Wright AD, Sticher O, Anghofer CK and Pezutto JM (1994) Biological activities of selected marine natural products. *Planta Med*. 60: 532–537.

Lin Y, Li H, Jiang G, Zhou S, Vrijmoed L L P and Jones E B G (2002) A novel g-lactone, eutypoid-A and other metabolites from marine fungus *Eutypa* sp. (#424) from the South China Sea. *Indian J Chem, Sect B: Org Chem Incl Med Chem*. 41B: 1542-1547.

Lohsiri W, Fusetani N, Matsunaga S, Sangkasila R, Monvises A and Kasiroek W (1995). Antimicrobial activities of marine sponges from the Gulf of Thiland and Andaman Sea. *In* Snidvongs A, Utoomprukporn W and Hungspreugs M (ed.) *Proceedings of the NRCT-JSPS Joint Seminar on Marine Science, December 2-3, 1993, Songkhla, Thiland* Bangkok, Thiland. pp. 238-246.

Matsukawa R, Dubinsky Z, Kishimoto E, Masak K, Masuda Y, Takeuchi T, Chihara M, Yamamoto Y, Niki E and Karube I (1997) A comparison of screening methods for antioxidant activity in seaweeds. *J. Appl. Phycol.* 9: 29–35.

Mayer AMS and Hamann MT (2002) Marine pharmacology: compounds with antibacterial, anticoagulant, antifungal, anthelmintic, anti-inflammatory, antiplatelet, antiprotozoal and antiviral activities affecting the cardiovascular, endocrine, immune and nervous systems, and other miscellaneous mechanisms of action. *Comparative Biochem. Physiol.* 132: 315-339.

Mayer AM and Gustafson KR (2003) Marine pharmacology: antitumor and cytotoxic compounds. *International J. Cancer* 105: 291-299.

McCarthy PJ and Pomponi SA (2004) A search for new Pharmaceutical Drugs from marine organisms. *Marine Biomed. Res.*22: 1-2.

Moore RE, Patterson ML and Carmichael WW (1988) In: Fautin DG, (ed.) *Biomedical Importance of MarineOrganisms*; California Academy of Sciences: San Francisco. Pp.143-150.

Moreau J, Pesando D and Caram B (1984) Antifungal and antibacterial screening of Dictyotales from the french mediterranean coast. *Hydrobiologia* 116: 521-524.

Murakami M, Makabe K, Yamaguchi S, Konosu S and Walchi R (1988) A novel polyether macrolide from the dinoflagellate *Goniodoma pseudogoniaulax*. *Tetrahedron Lett.*29: 1149-1152.

Nagai H, Murata M, Torigoe K, Satake M and Yasumoto T (1992) Gambieric acids, new potent antifungal substance with unprecedented polyether structures from a marine dinoflagellate *Gambierdiscus toxicus*. *J Org. Chem. Commun.* 57:,5448-5453.

Nagayama K, Iwamura Y, Shibata T, Hirayama I and Nakamura T (2002) Bactericidal activity of phlorotannins from the brown algae *Ecklonia kuroma*. *J. Anti. Microbial Chemotherapy* 50: 889-893.

Ohta K and Takagi M. (1977) Antimicrobial compounds of marine red algae *Marginisporum aberrans*. *Phytochemistry* 16: 1085-1086.

Padmakumar K and Ayyakkannu K (1997) Seasonal variation of antibacterial and antifungal activities of the extracts of marine algae from southern coasts of India. *Bot. Mar* 40: 507-515.

Paul VJ and Fenical W (1984) Bioactive terpenoids from Caribbean marine algae of the gerera *Penicillus* and *Udotea* (Chlorophyta). *Tetrahedron* 40: 2913-2918.

Peng J, Shen X, EI Sayed KA, Dunbar DC, Perry TL, Wilkins SP, Hamann MT, Bobzin S, Huesing J, Camp R, Prinsen M, Krupa D and Wideman M A (2003). Marine natural products as prototype agrochemical agents. *J. Agric. Food Chem.* 51: 2246-2252.

Plavsic M, Terzic S, Ahel M and Berg CMG (2004) Folic acid in coastal waters of the Adriatic Sea *Mar. Freshwater. Res.* 53:1245-1252.

Pokorny J (1991) Natural antioxidants for food use. *Trends Food Sci. Technol.* 2: 223–227.

Proteau PJ, Gerwick WH, Garcia-Pichel F and Castenholtz R (1993) The structure of scytonemin, an ultraviolet sunscreen pigment from the sheaths of cyanobacteria. *Experientia* 49: 825-829.

Ragan MA (1984) Bioactivities in marine genera of Atlantic Canada: The unexplored potential. *Proceedings Nova Scotian Inst. science* 34: 83-132.

Sreenivasa Rao P and Parekh K S (1981) Antibacterial activity of Indian seaweed extracts. *Bot. Mar.* 24: 577-582.

Rinehart KL (2000) Antitumor compounds from tunicates. *Medicinal Research Reviews* 20: 1-27

Rosell KG and Srivastava LM (1987) Fatty acids as antimicrobial substances in brown algae. *Hydrobiologia* 151: 471-475.

Ruberto G, Baratta MT, Biondi DM and Amico V (2001) Antioxidant activity of extracts of the marine algal genus *Cystoseira* in a micellar model system *J. Applied Phycology* 13: 403–407.

Sastry VMVS and Rao GRK (1994) Antibacterial substances from marine algae: Successive extraction using benzene, chloroform and methanol Bot. Mar. 37: 357-360.

EI Sayed KA, Dunbar DC, Goins DK, Cordova CR, Perry TL, Wesson KJ, Sanders SC, Janus SA, and Hamann MT (1996) The marine environment: A resource for prototype antimalarial agents. *J. Natural Toxins* 5: 261-285.

EI Sayed KA, Dunbar DC, Perry TL, Wesson SP and Hamann MT (1997) Marine natural products as prototype insecticidal agents. *J. Agricultural Food Chemistry* 45: 2735-2789.

Shimizu Y (1993) *In*: Attaway D, Zeborsky O (ed.) *Marine Biotechnology.*; Plenum Press: New York, pp 391-410.

Siriwardhana N, Lee KW, Kim SH, Ha JW and Jeon YJ (2003) Antioxidant activity of *Hizikia fusiformis* on reactive oxygen species scavenging and lipid peroxidation inhibition. *Food Sci. Tech. Int.* 9: 339–346.

Siriwardhana N, Lee KW, Kim SH, Ha JW, Park GT and Jeon YJ (2004) Lipid peroxidation inhibitory effects of *Hizikia fusiformis* methanolic extract on fish oil and linoleic acid. *Food Sci. Tech. Int.* 10: 65–72.

Smith VJ and Chisholm JRS (2001) Antimicrobial proteins in crustaceans. *Adv. Experimental Med. Biology* 484: 95-112.

Stevenson CS, Capper EA, Roshak A K, Marquez B, Grace K, Gerwick WH, Jacobs RS and Marshall LA (2002) Scytomenin-a marine natural product inhibitor of kinases key in hyperproliferative inflammatory diseases. *Inflammation Res.* 51: 112-118.

Stevenson CS, Capper EA, Roshak AK, Marquez B, Eichman C, Jackson JR Mattern M, Gerwick WH, Jacobs RS and Marshall LA (2002) The Identification and Characterization of the Marine Natural Product Scytonemin as a Novel antiproliferative pharmacophore. *J Pharmacol Exp Ther.* 303: 858-866.

Tang HF, Yi YH, Yao XS, Xu QZ, Zhang SY and Lin HW (2002) Bioactive steroids from the brown alga *Sargassum carpophyllum. J Asian Nat Prod Res.*, 4: 95–105.

Tutour BL, Benslimane F, Gouleau MP, Gouygou JP, Saadan B and Quemeneur F (1998) Antioxidant and pro-oxidant activities of the brown algae, *Laminaria digitata, Himanthalia elongata, Fucus vesiculosus, Fucus serratus* and *Ascophyllum nosum. J. Appl. Phycol.* 10, 121–129.

Tziveleka LA, Vagias C and Roussis V (2003) Natural products with anti-HIV activity from marine organisms. *Curr. Topics Medicinal Chemistry* 3: 1512-1535.

Vidyavathi N and Sridhar KR (1991) Seasonal and geographical variations in the antimicrobial activity of seaweeds from the Mangalore coast of India. *Bot. Mar.* 34: 279-284.

Vries De and Hall M R (1994) Marine biodiversity as a source of chemical diversity. *Drug Development Research* 33: 161-173.

Wender PA, Koehler KE, Sharkey NA, Dell'Aquilla HL and Blumberg PM (1986) Analysis of phorbol ester pharmocophor on protein kinase C as a guide to the retional design of new classes of anlogas. *Proc. Natl. Acad. Sci.*, USA. 83: 4214-4218.

Wichi CM, Kendall CWC, Stamp D, Medline A, Archer MC and Bruce WR (1998) Thermally oxidized dietary fat on colon carcinogenesis in rodents. *Nutr. Cancer.* 30:,69–73.

Wratten SJ and Faulkner DJ (1976) Cyclic polysulfides from the red algae *Chondria californica. J. Organic Chemistry* 41: 2465-2467.

Xu SH, Ding LS, Wang MK, Peng SL and Liao X (2002) Studies on the Chemical Constituents of the Algae *Sargassum polycystum* and *Youji Huaxue. Chinese J. Org. Chem.* 22: 138-140.

Yamada T, Iwamoto C, Yamagaki N, Yamanouchi T, Minoura K, Yamori T, Uehara Y, Andoh T, Umemura K and Numata A (2002) Leptosins M-N1, cytotoxic metabolites from a *Leptosphaeria* species separated from a marine alga. Structure determination and biological activities. *Tetrahedron* 58: 479-487.

Yan XJ, Chuda Y, Suzuki M and Nagata T (1999) Fucoxanthin as the major antioxidant in *Hizikia fusiformis*, a common edible seaweed. *Biosci. Biotechnol. Biochem.* 63: 605–607.

Quanbin Z, Pengzhan Y, Zhien L, Hong Z, Z and Pengcheng L (2003) Antioxidant activities of sulfated polysaccharide fractions from *Porphyra haitanesis. J. Applied Phycology* 15: 305–310, 2003.

Xue Z, Chang H X, Zhao JL, Yue-Piao C, Hong YL and Hong T Q (2004) Antioxidant and hepatoprotective activities of low molecular weight sulfated polysaccharide from *Laminaria japonica. J. Applied Phycology* 16: 111–115, 2004.

© R.K. Gupta & V.D. Pandey (*Editors*)
Advances in Applied Phycology, 2007
Daya Publishing House, New Delhi

56–74

Chapter 3

Fueling the Future: By Microalgae as a Source of Renewable Energy

C. Dayananda, R. Sarada* and G.A. Ravishankar

Plant Cell Biotechnology Department, Central Food Technological Research Institute, Mysore – 570 020, India

ABSTRACT

Solar power, wind energy, water currents, geothermal energy, tidal energy, fuel wood and fossil fuels like coal, lignite, petroleum, nuclear and natural gas have so far played a vital role for the industrial development and human civilization. Fossil fuel reserves are depleting rapidly across the world, intensifying the energy problems on existing reserves day-by-day due to increased demand. Global warming resulting from carbon dioxide has been discussed as an important environmental issue. Photosynthetic organisms are able to fix carbon dioxide to organic substances by solar energy; therefore, one way to solve this problem could be photosynthetic fixation of carbon dioxide using microalgae. However, it is meaningless without utilizing an algal biomass before degradation. Utilization of a photosynthetic biomass for fuel and chemicals is very important. A great deal of attention has been focused on the autotrophic microalga, *Botryococcus braunii, Dunaliella tertiolecta, Anabaena, Spirulina, Oscillatoria, Chlorogloea, Nostoc, Mestigocladus, Chlamydomonas reinharditii, Phormidium* and some diatoms for their application in biofuel production. The present chapter deals with the potentials and prospects of microalgae as a source of energy.

Keywords: Biomass, Biofuel, Fatty acids, Hydrogen, Hydrocarbon, Microalgae.

Introduction

Energy is an important currency of human's activity. In the early days, the energy demands were met primarily by muscular effort, wind and water currents, direct solar warming, fuel wood and other

* Corresponding Author: Scientist, Tel: +91-821–2516501; Fax: +91-821–2517 233; E-mail: pcbt@cftri.res.in.

sources. Fossil fuels like coal, lignite, petroleum, nuclear and natural gas have so far played a vital role for the industrial development.

Energy sources can be classified in to renewable and non-renewable sources. Non-renewable resources (fossil fuels and nuclear energy sources) will be depleted irreversibly with mining. The fossil fuels are non-renewable and are depleting day by day, so it is necessary to identify and develop alternative and renewable energy sources to meet the world's energy demand. In view of this there is an increasing quest throughout the world for the exploration and exploitation of renewable energy resources.

Biotechnology-based production of fuels continues to attract much attention. Bioethanol, fat and vegetable oil, firewood, biogas, biodiesel (Graboski and McCormick, 1998) and biohydrogen (Nandi and Sengupta, 1998) are examples of biofuels. Except for biohydrogen, commercial or pilot experimental use of the other biofuels is already established or emerging. Although bioconversion of lignocellulosic biomass to sugars for fermentation to ethanol has been extensively studied but it remains intractable. More successful and widely used is the bioconversion of starch to sugars for producing bioethanol. Similarly, fuel ethanol produced from residues of cane and beet sugar processing has been in use for several decades. Anaerobic digestion of organic waste to methane is another widely used technology. Modern biotechnology has already greatly impacted the traditional production of bioethanol. For example, the higher yielding genetically modified corn reduces cost of the main feedstock; the starch in gene engineered corn is more amenable to enzymatic bioconversion to sugars, than natural corn starch; microbial enzymes have been engineered for enhanced stability and ability to rapidly convert starch to fermentable sugars; microorganisms have been engineered to withstand higher levels of toxic ethanol and achieve rapid fermentation. These and other future improvements will make bioethanol more economic than it is today. Similar advances are being targeted for enhancing anaerobic digestion technologies. Blending of gasoline with bioethanol directly reduces consumption of fossil fuels and environmental pollution (*e.g.* volatile organic compounds, nitrous oxides, benzene and particulates) associated with combustion of unblended gasoline. Similarly, biodiesel is significantly less polluting than petrodiesel. Conversion of biomass to energy is highly attractive. Although in energy terms annual land production of biomass is about five times the global energy consumption, only 1 per cent of commercial energy originates from biomass at present. Organic waste from landfill sites and farms can be converted to combustible biogas (approximately 55 per cent methane and 45 per cent carbon dioxide) through anaerobic digestion. Liquid hydrocarbon fuels can be produced from plant, animal and microbial oils (Gavrilescu and Chisti, 2005).

Algae are one of the important groups of organisms in the kingdom planate. In the absence of microbial technology no progress could be made in the early centuries. Man's use of algae has long history and the use of marine algae as food dates back to 600–800 B.C. In twelth century in France algae were used as manure, later it was adopted in U K. Microalgal research and development was initiated five decades ago in US. Algae have found their place in modern day science and technology, due to their various applications in biotechnology (Figure 3.1), environment and rural development. *Botryococcus braunii* is one such alga known for the production of high levels of liquid hydrocarbons and lipids along with this other alga which are having their application in biofuel production are also dealt in the present chapter.

Microalgae as a Source of Bioenergy

Microalgae are the most primitive form of plants. While the mechanism of photosynthesis in microalgae is similar to that of higher plants, they are generally more efficient converters of solar

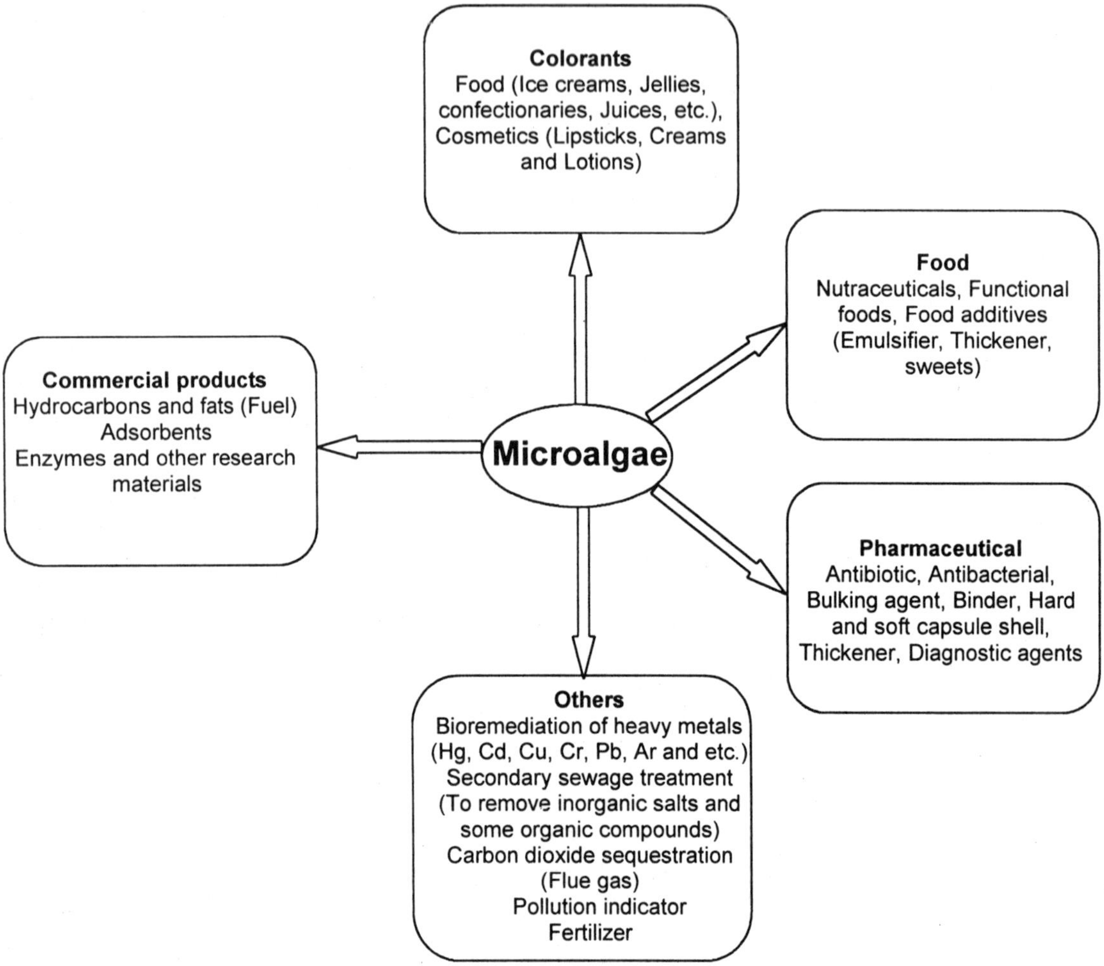

Figure 3.1: Microalgae Applications in Various Fields

energy because of their simple cellular structure. In addition, because the cells grow in aqueous suspension, they have more efficient access to water, CO_2, and other nutrients. For these reasons, microalgae are capable of producing 30 times the amount oil per unit area of land, compared to terrestrial oilseed crops.

It has been extensively studied by many researchers on the microalgae which can be used as a source of biofuels like methane gas, alcohol or hydrocarbons or vegetable oils (Benemann *et al.*, 1986; Cohen, 1986). In early 1970's, the US department of energy has spent 10 million US dollars towards developing low cost technology using microalgae for fuel production, through its solar energy research institute aquatic species programe (Neenan *et al.*,1986). The production of various kinds of biofuel was reported in different organisms. Cultures of *Anabaena, Spirulina, Oscilatoria, Chlorogloea, Nostoc, Mestigocladus* and *Chlamydomonas reinharditii* were reported for the production of hydrogen (Hallen beck *et al.*, 1978; Mitsui, 1980; Kumazawa and Mitsui, 1982; Karube *et al.*, 1986; Lindblad *et al.*, 1998; Hall, 1988; Robinson *et al.*, 1986). In 1977 Keenan reported the production of methane gas by *Anabaena flos–aquae*. Generation of electricity by immobilized cyanobacterial species *Mestigocladus* and *Phormidium* onto SnO_2 optically transparent electrodes (Robinson *et al.*, 1986). The alga *Botryococcus braunii* has been shown to yield 70 per cent of its extract is hydrocarbon liquid, which closely resembles crude oil.

In the late 1970s, the emphasis switched from methane production to algal oils as the fuel product. This was based on the known ability of some microalgae species to accumulate large amounts of algal lipids, in particular under conditions of nutrient (mainly N and Si) limitations. As before, 40-ha earthwork ponds were used; however, this time with paddle wheel mixing. Productivities were now projected of 67.5 mt/ha/yr for an algal biomass containing 40 per cent lipids (oils) by weight. This corresponded to about 90mt/ha/yr for conventional algal biomass; yielding almost 160 barrels of crude oil/ha/yr. Harvesting was again assumed to be by bioflocculation, followed by a centrifugation process to concentrate the biomass to a paste-like consistency. A solvent extraction process as used for soybean oil extraction was assumed, at three-time higher unit cost to account for the high moisture in the paste. (However, as was pointed out, direct solvent extraction was unlikely to be feasible for such high moisture biomass.). As before, the residual biomass was to be anaerobically digested in covered ponds to produce methane gas, with the nutrients (and C) from the digester (and digester gas) recycled to the ponds (Benemann *et al.*, 1982a). Four major research needs to achieve the objectives of high productivity in large-scale outdoor systems:

1. Photosynthetic efficiency for light energy and high lipid production.
2. Fundamentals of species selection and control in open pond systems.
3. Mechanisms (and control) of algal bioflocculation.
4. Effects of non-steady-state operating conditions on algal metabolism (Benemann *et al.*, 1982b).

Important Algal Taxonomic Groups for Biofuel Production

Microalgae are defined as microscopic organisms that can grow via photosynthesis. Many microalgae grow quite rapidly, and are considerably more productive than land plants and macroalgae (seaweed). Microalgae reproduction occurs primarily by vegetative (asexual) cell division, although sexual reproduction can occur in many species under appropriate growth conditions. There are several main groups of microalgae, which differ primarily in pigment composition, biochemical constituents, ultrastructure, and life cycle. There are several main groups of microalgae, which differ primarily in pigment composition, biochemical constituents, ultrastructure, and life cycle. Five groups were of primary importance for biofuel productions are: diatoms (Class: Bacillariophyceae), green algae (Class: Chlorophyceae), golden brown algae (Class: Chrysophyceae), prymnesiophytes (Class Prymnesiophyceae), and the eustigmatophytes (Class: Eustigmatophyceae).The blue-green algae or cyanobacteria (Class: Cyanophyceae). The algal species which are of potentially important for the biofuel production are shown in Figure 3.2.

Green Algae

Green algae, often referred to as chlorophytes, are also abundant; approximately 8,000 species are estimated to be in existence. This group has chlorophyll *a* and chlorophyll *b*. These algae use starch as their primary storage component. However, N-deficiency promotes the accumulation of lipids in certain species. Green algae are the evolutionary progenitors of higher plants, and, as such, have received more attention than other groups of algae.

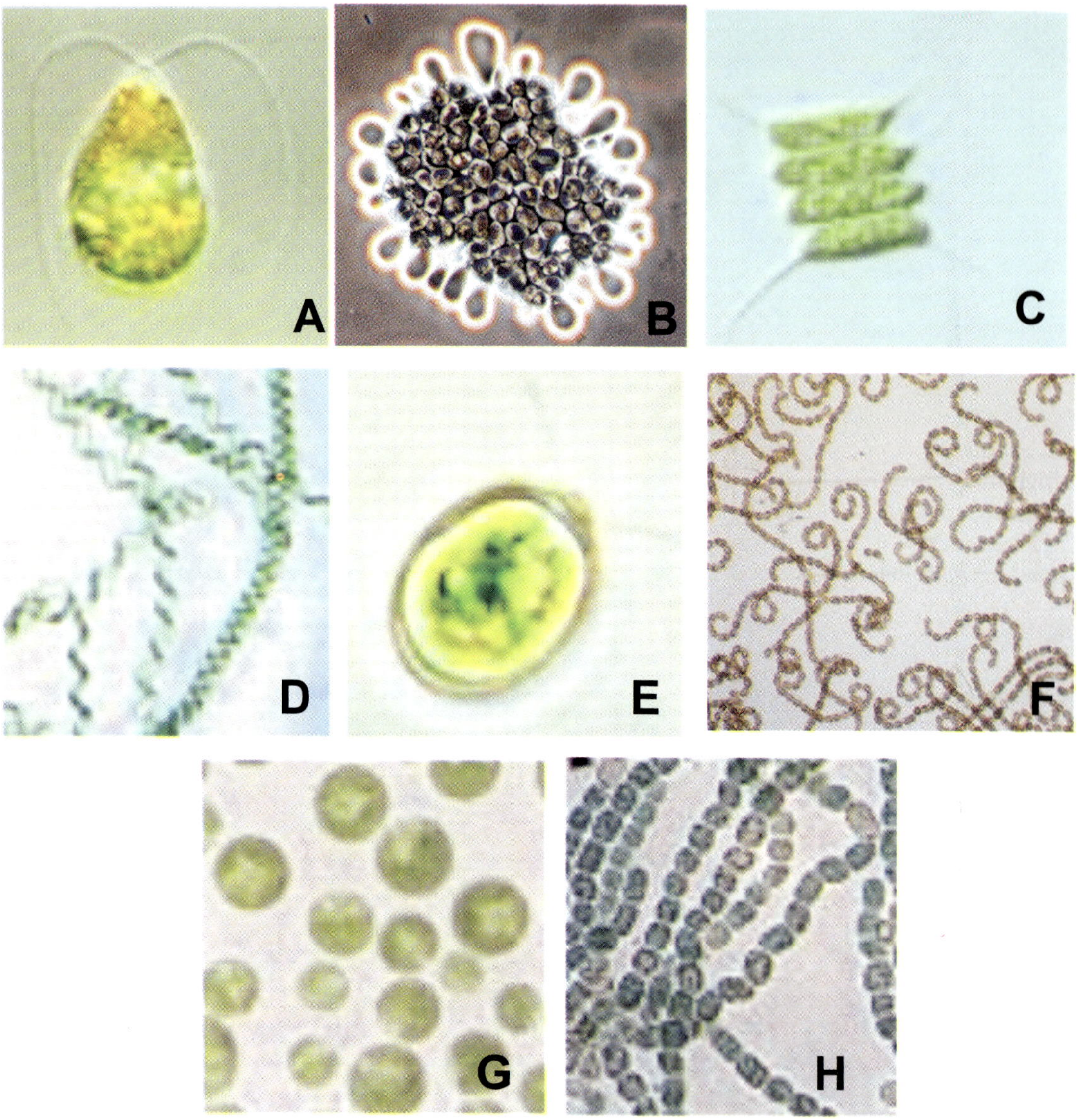

Figure 3.2: Important Algal Species for the Production of Biofuel

A: *Dunaliella*, B: *Botryococcus braunii*, C: *Scenedesmus*, D: *Spirulina*, E: *Chlamydomonas*, F: *Anabaena*, G: *Chlorella* and H: *Nostoc*

Chlamydomonas

A member of chlorophyceae, *Chlamydomonas reinhardtii* (and closely related species) has been studied very extensively, in part because of its ability to control sexual reproduction, thus allowing detailed genetic analysis. Indeed, *Chlamydomonas* was the first alga to be genetically transformed.

However, it does not accumulate lipids, and thus was not considered. But it has been reported for the production of hydrogen.

Chlorella

Another common genus of chlorophyta that has been studied fairly extensively is *Chlorella*. *Chlorella* species are unicellular with a cell size in the range of 3–10m which is ideal for combustion in a diesel engine. The production of a liquid fuel consisting of an emulsion of biodiesel (transesterified rapeseed oil), a surfactant and cells of *Chlorella vulgaris* that will fuel an unmodified stationary diesel engine for the supply of electricity has been reported (Scragg *et al.*, 2003).The fat content of *Chlorella* differs with strains and conditions and it has been found that, the fat content of the alga varies inn the range of 11-63 per cent (w/w). The reduction in nitrogen in the medium increases the lipid content in *Chlorella* strains and in the case of *C.emersonii* and *C. minutissima* values of 63 and 56 per cent were reported. 57.9 per cent of fat for *C. vulgaris*, *C. luteoviridis* 28.8 per cent and *C. capsulata* 11.4 per cent, and *C.pyrenoidosa* 29.2 per cent were reported (Illman *et al.*, 2000). The higher yield of bio-oil (57.9 per cent) production was reported from heterotrophic *Chlorella protothecoides* cells than from autotrophic cells by fast pyrolysis (Miao and Wu, 2004a).

Botryococcus braunii

Botryococcus braunii is a green colonial microalga widespread in freshwater and brackish lakes, reservoirs, ponds and it is recognized as one of the potent renewable resource for production of liquid hydrocarbons. *B. braunii* is classified into A, B and L races depending on the type of hydrocarbons synthesized (Race–A produces C_{23} to C_{33} odd numbered n-alkadienes, mono-, tri-, tetra-, and pentaenes and are derived from fatty acids (Dayananda *et al.*, 2006). These linear olefins can constitute upto 61 per cent of the dry cell mass of the green active state colonies. The L race produces a single hydrocarbon C_{40}-$C_{78,}$ a tetraterpene known as lycopadiene and it constitute upto 2-8 per cent of the dry biomass (Banerjee *et al.*, 2002; Metzger *et al.*, 2005; Dayananda *et al.*, 2005; Dayananda *et al.*, 2006). B race produce triterpenoid hydrocarbons, C_{30}–C_{37} terpenoid hydrocarbons referred to as botryococcenes and C_{31}–C_{34} methylated squalenes (Huang and Poulter, 1989; *Achitouv et al., 2004) and it can accumulate high levels of hydrocarbons ranges from 20–86* per cent on dry weight basis. In natural population, botryococcenes can constitute from 27 to 86 per cent of the dry cell mass (Brown and Knights 1969). However, under cultured conditions the hydrocarbon content varies with the species and physiological conditions from 2-80 per cent. Hydrocarbons oils extracted from *B. braunii*, when hydrocracked, produce a distillate comprising of 67 per cent gasoline fraction, 15 per cent aviation turbine fuel, 15 per cent diesel fuel fraction and 3 per cent residual oil (Hillen *et al.*,1982), these fuels were free from oxides of sulphur and nitrogen (SOX and NOX) after their combustion. It being a photosynthetic organism, it can reduce CO_2 emissions by 1.5 X 10^5 tons/yr (Sawayama *et al.*, 1999). There by it can offers an eco friendly process. In view of the changing energy scenario for renewable energy sources, *Botryococcus* is identified as an untapped resource for production of hydrocarbons. In light of this, the biotechnological exploitation for hydrocarbons needs to be explored. Apart from this, this alga is also useful in bioremediation of arsenic, lead and chromium.

Dunaliella

The unicellular microalga *Dunaliella* naturally occurs in saline habitats and is well known to produce useful compounds, such as b-carotene (3–5 per cent (w/w), glycerol and lipids. *Dunaliella* contains 20.5 per cent of fat, is easier to cultivate on a large scale than *B. braunii.* (Ben-Amotz and Avron 1981; Ben-Amotz *et al.*, 1983), and outdoor mass production systems for *Dunaliella* have already been developed (Borowitzka *et al.*, 1988). Nitric oxide (NO), a major constituent of NO*x* in fossil fuel flue gas, can be removed by the microalga, *Dunaliella tertiolecta*, cultured in a bubble-column-type bioreactor.

Golden-Brown Algae

This group of algae, commonly referred to as chrysophytes, is similar to diatoms with respect to pigments and biochemical composition. Approximately 1,000 species are known, which are found primarily in freshwater habitats. Lipids and chrysolaminarin are considered to be the major carbon storage form in this group. Some chysophytes have lightly silicified cell walls.

Prymnesiophytes

This group of algae, also known as the haptophytes, consists of approximately 500 species. They are primarily marine organisms, and can account for a substantial proportion of the primary productivity of tropical oceans. As with the diatoms and chrysophytes, fucoxanthin imparts a brown color to the cells, and lipids and chrysolaminarin are the major storage products. This group includes the coccolithophorids, which are distinguished by calcareous scales surrounding the cell wall.

Eustigmatophytes

This group represents an important component of the "picoplankton", which are very small cells (2-4 µm in diameter). The genus *Nannochloropsis* is one of the few marine species in this class, and is common in the world's oceans. Chlorophyll *a* is the only chlorophyll present in the cells, although several xanthophylls serve as accessory photosynthetic pigments.

Cyanobacteria

This group is prokaryotic, and therefore very different from all other groups of microalgae. They contain no nucleus, no chloroplasts, and have a different gene structure. There are approximately 2,000 species of cyanobacteria, which occur in many habitats. Although this group is distinguished by having members that can assimilate atmospheric N (thus eliminating the need to provide fixed N to the cells), no member of this class produces significant quantities of storage lipids; therefore, this group was not useful for the production of crude oil. But, cultures of *Anabaena, Spirulina, Synechocystis, Oscilatoria, Chlorogloea, Nostoc* and *Mestigocladus* were reported for the production of hydrogen (Benemann and Wissmann, 1976; Halleenbeck *et al.*, 1978; Mitsui, 1980; Kumazawa and Mitsui, 1982; Karube *et al.*, 1986; Benemann *et al.*, 1986; Lindblad *et al.*, Hall, 1988; Robinson *et al.*, 1986). In 1977 Keenan reported the production of methane gas by *Anabaena flos–aquae*.

The first reported reaction was photohydrogen production of green algae *Scenedesmus obliquusw* which, in the dark, evolved H_2 by the anaerobic degradation of a reserve substance. Heterocystous cyanobaeerium *Anabaena cylindrical* was used for the demonstration of simultaneous production of H_2 and 0_2. Outdoor experiments to study solar energy conversion with this strain resulted in the specific H_2 evolution rate of 0.244mmol/(g.h) which corresponded to the H_2 evolution rate of 0.175mmol/(dm^3.h) (Hallen beck *et al.*,1978). For application purposes, H_2 production was investigated with the filamentous, non-heterocystous marine cyanobactedum *Oscillatoria* sp. strain Miami BG7 of which H_2 production rate reached 0.364mmol/(g.h). Cyanobaeterium, *Spirulina platensis*, was described to evolve H_2 in the dark and under amerubic or microaerobie condition. An extensive study on cyanobacterial H_2 metabolism under a microaerobic condition has shown the capability of H_2 production by *Synechococcus, Microcystis, Gloeobacter, Synechocystis, Aphanocapsa and Gloeocapsa alpicola, Aphanocapsa* sp., and *Apanocapsa mentona*. The condition for H_2 production has been variously reported, however, these observations seem to stimulate further experimental investigations for biofuel productions.

Study on the fixation and utilization of CO_2 through the cultivation of cyanobacterium *Synechococcus leopoliensis* was initiated as a baseline evaluation of above technologies, since this is one of the fastest-growing unicellular photofithotroph which requires radiant energy, using water as electron donor, and exhibits the simplest nutritional requirements of any known organisms. The maximum use of biomass has been argued with multiple biofuel production schemes (Howarth and Codd, 1985: Gaffron and Rubin, 1942; Asada and Kawamura, 1986).

Diatoms

Diatoms are among the most common and widely distributed groups of algae in existence; about 100,000 species are known. This group tends to dominate the phytoplankton of the oceans, but is commonly found in fresh–and brackish-water habitats as well. The cells are golden-brown because of the presence of high levels of fucoxanthin, a photosynthetic accessory pigment. Several other xanthophylls are present at lower levels, as well as a-carotene, chlorophyll *a* and chlorophyll *c*. The main storage compounds of diatoms are lipids (TAGs) and a a-1,3-linked carbohydrate known as chrysolaminarin. A distinguishing feature of diatoms is the presence of a cell wall that contains substantial quantities of polymerized Si. This has implications for media costs in a commercial production facility, because silicate is a relatively expensive chemical. On the other hand, Si deficiency is known to promote storage lipid accumulation in diatoms, and thus could provide a controllable means to induce lipid synthesis in a two-stage production process. Another characteristic of diatoms that distinguishes them from most other algal groups is that they are diploid (having two copies of each chromosome) during vegetative growth; most algae are haploid (with one copy of each chromosome) except for brief periods when the cells are reproducing sexually. The main ramification of this from a strain development perspective is that it makes producing improved strains via classical mutagenesis and selection/screening substantially more difficult. As a consequence, diatom strain development programs must rely heavily on genetic engineering approaches.

Scale Up of Microalgae for Biofuel Production

Photosynthetic organisms, including plants, algae, and some photosynthetic bacteria, efficiently utilize the energy from the sun to convert water and CO_2 from the air into biomass. Microalgae and their ability to make lipids (as a feedstock for liquid fuel or chemical production) or carbohydrates (for fermentation into ethanol or anaerobic digestion for methane production) are of great interest for fuel production. The studies on the growth and chemical composition of microalgae have revealed that, many microalgae (microscopic, photosynthetic organisms that live in saline or freshwater environments), produce lipids as the primary storage molecule. The use of microalgal lipids for the production of fuels and other energy products is well established. Microalgae, like higher plants, produce storage lipids in the form of triacyglycerols (TAGs). Although TAGs could be used to produce of a wide variety of chemicals, many researchers focused on the production of fatty acid methyl esters (FAMEs), which can be used as a substitute for fossil-derived diesel fuel. This fuel, known as biodiesel, can be synthesized from TAGs via a simple transesterification reaction in the presence of acid or base and methanol. Biodiesel can be used in unmodified diesel engines, and has advantages over conventional diesel fuel in that it is renewable, biodegradable, and produces less SO_x and particulate emissions when burned. The technology is available to produce biodiesel from TAGs, and there are growing biodiesel industries both in the United States and Europe that use soybean or rapeseed oil as the biodiesel feedstock. However, the potential market for biodiesel far surpasses the availability of plant oils not designated for other markets. Thus, there was significant interest in the development of microalgal lipids for biodiesel production. Microalgae exhibit properties that make them well suited for use in a commercial-scale biodiesel production facility. Many species exhibit rapid growth and high productivity, and many microalgal species can be induced to accumulate substantial quantities of lipids, often greater than 60 per cent of their biomass. Microalgae can also grow in saline waters that are not suitable for agricultural irrigation or consumption by humans or animals. The growth requirements are very simple, primarily carbon dioxide (CO_2) and water, although the growth rates can be accelerated by sufficient aeration and the addition of inorganic and organic nutrients.

Candidate microalgal species that exhibited characteristics desirable for a commercial production could be used. The desert regions were attractive areas in which to locate microalgal-based biofuel production facilities. The commercial production strain should have certain qualities for its mass production, and are as follows. These characteristics included the ability of the strains to grow rapidly and have high lipid productivity when growing under high light intensity, high temperature, and in saline waters indigenous to the area in which the commercial production facility is located. In addition, because it is not possible to control the weather in the area of the ponds, the best strains should have good productivity under fluctuating light intensity, temperature, and salinity.

For the commercial scale up, the strain should be able to grow in varied range of cultural conditions (pH, temperature, light intensity, salinity, and different media) with a considerable production of biocrude. And it has to be examined the effects of environmental variables and cultural conditions on the growth and lipid composition of the selected strains. No one algal strain was identified that exhibited the optimal properties of rapid growth and high constitutive lipid production. Many microalgae can be induced to accumulate lipids under conditions of nutrient deprivation. If this process could be understood, it might be possible to manipulate either the culture conditions, or to manipulate the organisms themselves, to increase lipid accumulation in a particular strain. Therefore, studies were initiated to study the biochemistry and physiology of lipid production in oleaginous (oil-producing) microalgae. Work performed by several researchers to understand the mechanism of lipid accumulation. In particular, to determine whether there is a specific "lipid trigger" that is induced by factors such as nitrogen (N) starvation and also they studied ultrastructural changes induced in microalgae during lipid accumulation. Efforts are being continuing to produce improved algae strains by looking for genetic variability between algal isolates, attempting to use flow cytometry to screen for naturally-occurring high lipid individuals, and exploring algal viruses as potential genetic vectors. There are many reports on the higher levels of lipid accumulation in microalgae induced by N starvation and silica (Si) depletion in diatoms. Unlike N, Si is not a major component of cellular molecules; therefore it was thought that the Si effect on lipid production might be less complex than the N effect, and thus easier to understand. This initiated a major research effort to understand the biochemistry and molecular biology of lipid accumulation in Si-depleted diatoms. This work led to the isolation and characterization of several enzymes involved in lipid and carbohydrate synthesis pathways, as well as the cloning of the genes that code for these enzymes. One goal was to genetically manipulate these genes in order to optimize lipid accumulation in the algae. Therefore, research is going on simultaneously to develop a genetic transformation system for oleaginous microalgal strains. The successful development of a method to genetically engineer diatoms was used in attempts to manipulate microalgal lipid levels by over expressing or down-regulating key genes in the lipid or carbohydrate synthetic pathways.

Cost-effective production of biodiesel requires not only the development of microalgal strains with optimal properties of growth and lipid production, but also an optimized pond design and a clear understanding of the available resources (land, water, power, etc.) required. On outdoor microalgae mass culture for production of biodiesel, as well as supporting engineering, economic and resource analyses. There are works on utilization of CO_2 from power plant flue gases in mass cultivation of microalga. From 1976 to 1979, researchers at the University of California-Berkeley used shallow, paddle wheel mixed, raceway-type (high-rate) ponds to demonstrate a process for the simultaneous treatment of wastewater and production of energy (specifically methane). Starting in 1980, the Aquatic Species Programme supported outdoor microalgal cultivation projects in Hawaii and California, using fresh and seawater supplies, respectively, in conjunction with agricultural fertilizers and CO_2. From 1987 to 1990, an "Outdoor Test Facility" was designed, constructed and operated in Roswell, New Mexico, including two 1,000 m^2 high-rate ponds. The conclusion from these extensive outdoor mass culture studies was that the use of microalgae for the low-cost production of biodiesel is technically feasible, but still requires considerable long-term R&D to achieve the high productivities required. These studies demonstrated the potential availability of large brackish and saline water resources suitable for microalgae mass cultures, large land and CO_2 resources. Power plant flue gas utilization for greenhouse gas (CO_2) mitigation by microalgae will be of great application. The overall conclusion of these studies was that in principle and practice large-scale microalgae production is not limited by design, engineering, or net energy considerations and could be economically competitive with other renewable energy sources. However, long-term R&D would be required to actually achieve the very high productivities and other assumptions made in such cost analyses.

The most R&D goals for the development of mass cultivation technology are related to the algal cultures themselves (productivity, species control, and harvestability), rather than the engineering aspects, such as the ponds, CO_2 transfer, or biomass processing. In 1960 Oswald and Golueke evaluated for the first time regarding engineering and cost analysis for large-scale microalgae production of fuels. They have projected the costs of electricity generated from biogas (methane) obtained from the anaerobic fermentation of algal biomass. The algae were to be cultivated in very large (40-ha) raceway type ponds, mixed with pumps, and supplied with CO_2 from a power plant. Other nutrients would come from the digesters. Municipal wastewaters would be used as make up for water and nutrients (C, N, P, etc.). The ponds were to be of earthen construction, with a depth of about 30-cm. Harvesting was assumed to be by simple settling. Although few details were provided, the general concept outlined in this early publication has remained essentially unchanged. Perhaps the greatest change is that biomass productivities thought to be achievable at that time were less than 50 mt/ha/yr of biomass, while current projections are roughly two to five times higher. During the mid-1970s, additional engineering and cost analyses were conducted by Benemann *et al.* (1977). The early studies were based on large (8–20-ha) ponds, with multiple channels and mixing by recirculation pumps (the required deep concrete sumps and splash pads were a major cost factor). Both the settling pond for harvesting algae by sedimentation and a covered anaerobic lagoon were part of this initial design. They have reported a yield of about 500 GJ/ha/y (10 GJ/t of algal biomass) of biogas. This study served as a starting point for more detailed later studies.

A relatively detailed analysis of an algal waste water treatment-energy production process was carried out by Benemann *et al.* (1977) as part of a larger study that examined a system integrating waste water algal ponds with tree biomass production. The so-called "Photosynthetic Energy Factory" (Inter Technology Solar Corporation 1978) was to use the effluents of a waste treatment pond system to fertilize short-rotation trees for fuel farming. In turn, the power plant burning the woody biomass would provide CO_2 for the algal ponds. A design of the algal pond subsystem was carried out by Benemann *et al.* (1978) for a typical municipal community of 50,000 people, generating approximately 18,000 m^3 of municipal wastewater per day. The assumption was that algal biomass would be grown up to the N growth potential of the wastewater, containing 65 mg/L of useable N (as organic N and ammonia). This required recycling about 5 to 7 tons of CO_2 per day from the power plant to the algal ponds. A temperate site with an average insolation of about 15 GJ/m2/d was assumed, with a solar conversion efficiency averaging only 2.7 per cent of visible light (about 1.35 per cent of total solar), somewhat higher in winter than summer. This is considerably lower than current assumptions. This study, for the first time, took into consideration monthly variations in temperature, insolation and other parameters. Algal harvesting was assumed to be with microstrainers. This report also carried out the first, though preliminary, analysis of the mixing power required for such large algal ponds and of the transfer requirements for CO_2 to the algal culture. A 160-ha algal pond system was required to treat this wastewater flow year-round. This was about three times larger than a conventional oxidation pond system. Costs were projected to be competitive with conventional wastewater treatment systems. Energy outputs were twice the energy inputs, based on digester gas production and requirements for pumping the wastewater, mixing the ponds, etc. The overall economics were very favorable because of the wastewater treatment credits. Although this concept appeared favorable, in practice the relatively small scale of the locally available municipal wastes could supply only a small fraction of fertilizer needs for the very large (>10,000 ha) energy plantations being projected. It does, however, point to the potential of this technology in wastewater treatment. This kind of large scale algal (*B. braunii*) hydrocarbon production was carried out by Regan (1980).

Biosorption of Heavy Metals

Microalgae exhibit mechanisms for the removal of free metal ions from waters, thereby both detoxifying and remediating the water. A number of microalgal strains the green algae species *Chlorella vulgaris, Scenedesmus* sp., and *Chlorococcum* sp., and the blue-green algae species, *Phormidium*,, and *Spirulina*), which are potentially suitable for Hg(II), Cd(II) and Pb(II) removal in aqueous solution. Among these, unicellular alga, *Chlamydomonas reinhardtii* has only recently gained greater attention for the treatment of heavy metal contamination in aqueous solutions. Because of its wide range of heavy metal tolerance, this species can sequester an array of trace metals including: copper, zinc, lead, cadmium, cobalt, nickel, mercury, silver, and gold. In addition, it can be genetically engineered so as to enhance its ability to selectively sequester toxic heavy metals (Perales-Vela *et al.*, 2006; Tuzun *et al.*, 2005). All these organisms are reported for the hydrogen production with the ability to sequester the above said heavy metals. Another hydrocarbon rich green microalga, *Botryococcus braunii* is also reported for the biosorption of heavy metals like chromium, arsenic, cadmium and lead in considerable amounts (Metzer *et al.*, 2005).

Conversion of Biomass to Energy

Biomass is considered as renewable energy source and it can be converted into either direct energy or energy carrier compounds by:

1. Direct combustion
2. Anaerobic digestion systems
3. Destructive distillation
4. Gasification
5. Chemical or biochemical hydrolysis

Biomass supplies may be achieved by three ways:

1. Cultivation of energy crops (sugar cane, cassava, jatropha, Soya, honge and other special trees).

2. Harvesting of natural vegetation (grass, foliage, timber and algae).
3. Utilization of agricultural and other organic wastes.

The conversion of this resulting biomass in to usable fuels can be accomplished by biological or chemical means or by a combination of both (Figure 3.3). Currently using methods of conversion of microalgal biomass in to fuels are given below:

Hydrocracking of Hydrocarbons

The crude oil (at the rate of 4 ml min^{-1}) and preheated hydrogen gas are passed in to the stain less steel reactor (6.5m x 6.35mm) containing cobalt–molybdate catalyst. Surplus hydrogen is vented out through the gas vent. The pressure inside the reactor chamber is maintained at 20 MPa at 400°C. After the process, 80 per cent of the product is recovered and cooled. The cooled product is subjected to conventional distillation to get combustiable fuels such as gasoline, diesel and aviation turbine fuel (Figure 3.4) (Hillen *et al.*, 1982).

Thermo-chemical Liquefaction

Another instant method used for the conversion of biomass into fuel hydrocarbons is called thermochemical liquefaction. The process involves subjecting the dry or wet biomass of *B.braunii* for

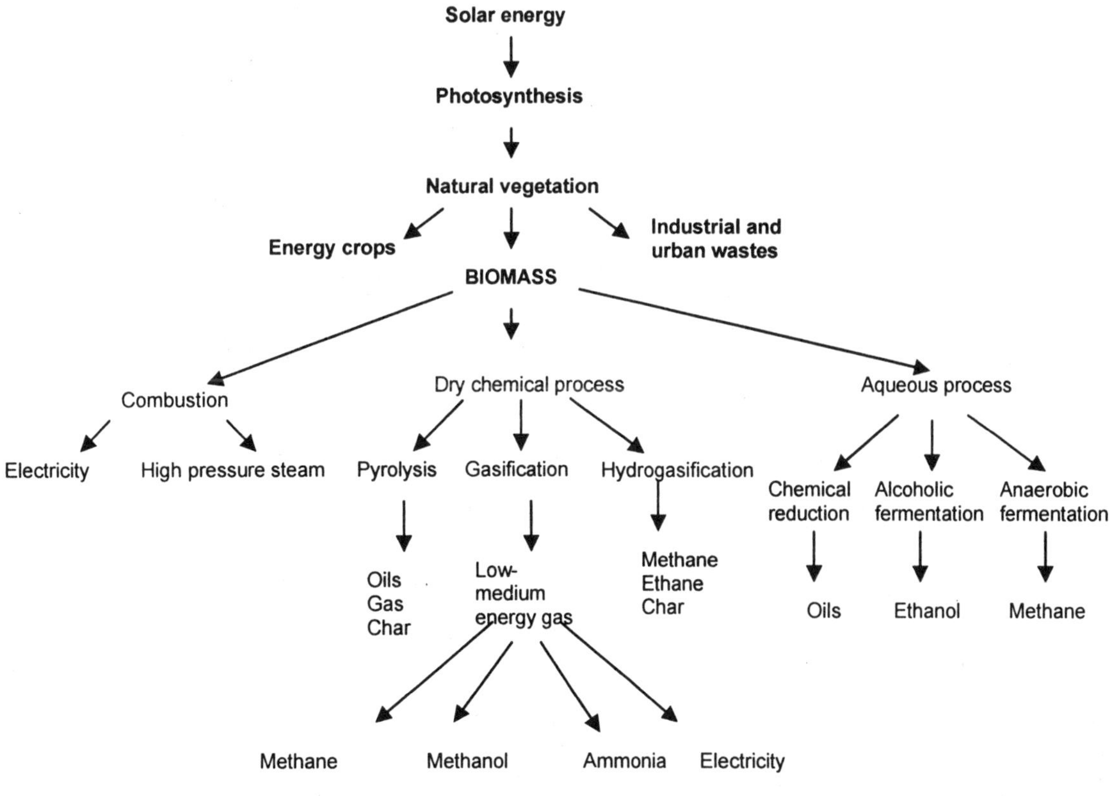

Figure 3.3: Options for the Conversion of Biomass to Energy (Adopted from A. K. Das, Chemical weekly, October 10, 2000)

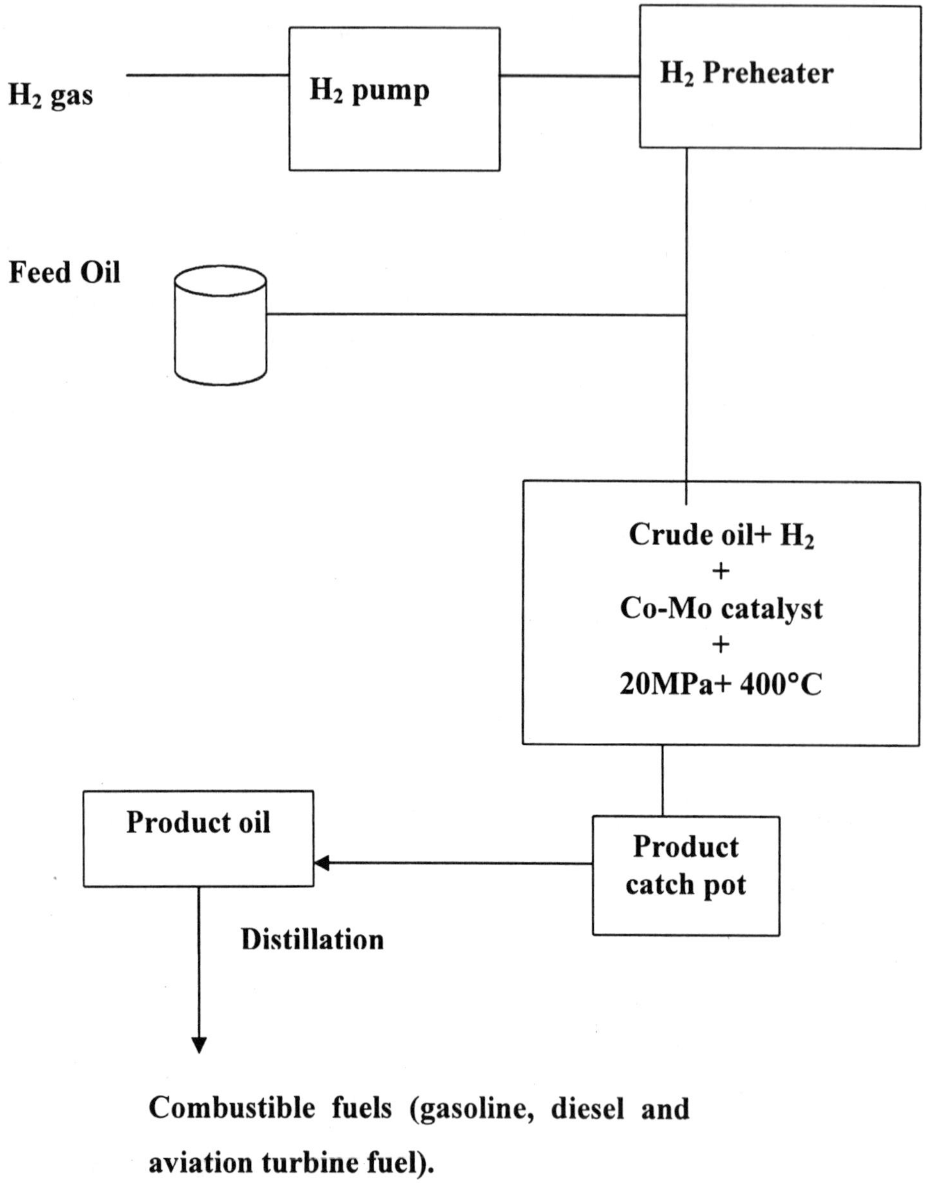

Figure 3.4: Flow Chart of Hydrocraker

thermochemical liquefaction for 1 hour at 300°C in presence of 5 per cent $Na_2 Co_3$ as a catalyst at 10 MPa pressure. The oil obtained after liquefaction would be fractionated by silica gel column chromatography in to three fractions (1) a low molecular weight (197–281 Da), (2) high molecular weight (438–572 Da) and (3) polar substances (867–2209 Da).

Thermo-chemical liquefaction eliminates the need of initial solvent extraction for hydrocarbons. Although overall yield of liquefied biomass is less compared to the yield that obtained by hexane extraction (Sawayama *et al.*, 1999; Inoue *et al.*, 1994; Dote *et al.*, 1992). However, thermo-chemical liquefaction is not applicable to continuous culture operations, where all the biomass is retained in the culture flask or bioreactor and hydrocarbon extraction is done continuously *in situ* using biocompatible solvents (Frenz *et al.*, 1989a and 1989b).

Crude hydrocarbons of *B. braunii* can be converted in to gasoline (60-70 per cent), light cycle oil (10-15 per cent), heavy cycle oil (2-8 per cent) and coke (5-10 per cent) by subjecting to catalytic cracking. Yield of gasoline and other fractions depend on the reaction temperature and the catalysts (Cobalt –Molybdenum or Zeolites etc.) (Matsui *et al.*, 1997).

Fast Pyrolysis

Fast pyrolysis is a high temperature process in which biomass is rapidly heated in the absence of oxygen. The essential features of a fast pyrolysis process are very high heating and heat transfer rates, carefully controlled pyrolysis reaction temperature of around 500°C, short vapour residence times of less than two seconds and rapid cooling of the pyrolysis vapour.

The algal cell samples were screened to pass through a screen of 0.18mm aperture to give small particles to ensure rapid heat transfer rates in the reactor. Two hundred grams of sample was subjected to pyrolysis in the fluid bed reactor (Figure 3.5). The combustor was used for the further pyrolysis of some biomass that was not completed pyrolyzed in the reactor. The biomass-feeding rate was 4 g min^{-1}. The experiments were carried out at temperature of 500 C with a heating rate of 600°C s^{-1} and a sweep gas (N_2) flow rate of 0.4 m^3 h^{-1} and a vapour residence time of 2 to 3 seconds. The total liquid products were comprised of an aqueous and an oil phase, which were separated and weighed. After pyrolysis, the solid char was removed and weighed, and then the gas yield was determined by the difference. All the yields were expressed on the basis of the dry weight of samples. However, the higher yield (about 21 per cent and could be expected to be higher) of flowing bio-oil products were directly produced from microalgae while in the fast pyrolytic process. The bio-oil from fast pyrolysis of microalgae has a higher heating value of 29 MJ/kg, which is about 1.4 times of that of wood. The

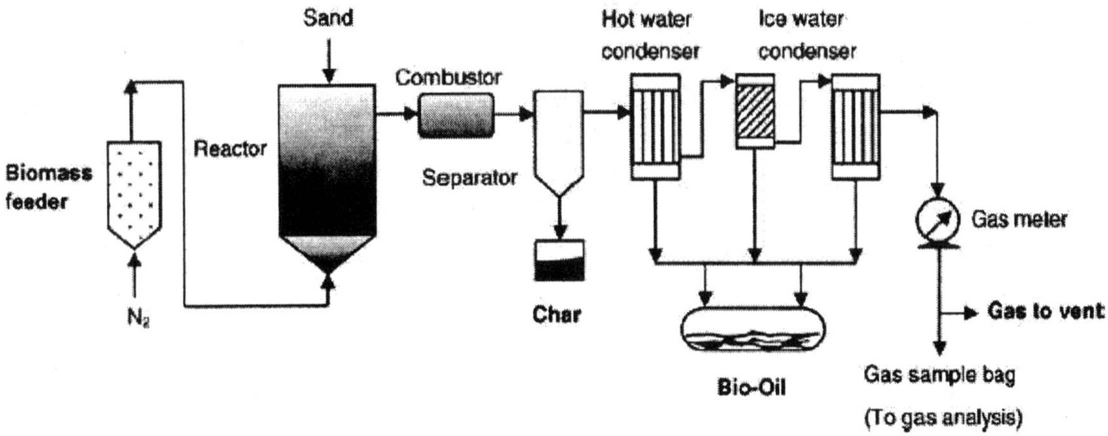

Figure 3.5: Fluid Bed Fast Pyrolysis System

lower oxygen contents of microalgae bio-oil will make them have better storage stability than that of wood. The liquid fuels from fast pyrolysis of microalgae can be used in many applications as direct substitutes for conventional fuels or as a source of chemicals (Miao *et al.*, 2004b).

Future Possibilities

Fossil fuels are fast depleting energy resources is well known. Due to very limited domestic deposits, our import dependence on petro-fuels is projected to rise from the current 70 per cent to 90 per cent by the end of this decade. Fuel extenders such as ethanol or methanol can be used as partial solution to the fuel crisis. Gasohol is a mixture of ethanol and gasoline, was used in the II World War and has been use in Brazil from 1970s and many American automobiles are now equipped to run on E85, a mixture of 80 per cent ethanol and 15 per cent gasoline. In India and in other developing countries, the use of gasohol can be controversial. Ethanol is usually obtained by fermentation and distillation of maize, wheat, potatoes or sugarcane and in India these crops are more valued as food. For any developing countries, biodiesel could be the best alternative transport fuel from renewable resources. Soya oil is the main source of biodiesel in United States, and in India experiments are being carried out on non edible oils from jatropha and honge (*Pongemia pinnata*) plants. India can overcome from the oil crisis by further exploiting its existing coal deposits using clean coal technology, speed up the railway electrification programme and popularizing biodiesel and also by minimizing the fossil fuel usages.

Trees or plants grown for energy are known as "energy crops" some of these include willow, aspen poplar, miscanthus, oil seeds and maize, sugarcane, honge, jatropha etc. Alternative energy sources such as wind, hydro and solar have already achieved stable market. Energy crops are quite expensive to plant and maintain as well as to harvest and transport.

Coal, tar sand and other fossil fuels should remain plentiful for at least another century. But these dirtier fuels carry a steep environmental cost: Generating electricity from coal instead of natural gas, for example releases twice as much carbon dioxide. And in order to power vehicles, they must be converted to a liquid or gas, which requires energy and therefore raises their cost. Even if plenty of fossil fuels available, it's doubtful we'll want to use them all. Burning fossil fuels has already increased the concentration of CO_2 in the atmosphere from 280 to 370 parts per million (ppm) over the past 150years, which is likely responsible for the 0.6°C raise in the average global surface temperature over the past century. Unchecked, it is expected to pass 500 ppm this century, according to New York Physicist Martin Hoffert and colleagues (Hoffert *et al.*, 2002; Service,2004). As populations explode and economies surge, global energy use is expected to raise by 70 per cent by 2020, according to a European commission report (2003), much of it to be met by fossil fuels. To limit the amount of CO_2 pumped in to the air, many scientists have urged for capturing a sizable fraction of that CO_2 from electric plants, chemical factories, and the like and piping it deep underground. Instead of piping that emitted CO_2 in to deep underground the same can be used for algal cultivation, which in turn reduces the cost of algal production.

On the face of it, hydrogen seems like the perfect alternative. When burned or oxidized in a fuel cell, it emits no pollution, including no greenhouse gases. Gram per gram, it releases more energy than any other fuel. No wonder it's being touched as a clean fuel of the future and the answer to modern society's addiction to fossil fuels. So, hydrogen producing microalgae, *Anabaena, Spirulina, Oscilatoria, Chlorogloea, Nostoc, Mestigocladus* and *Chlamydomonas reinharditii* could be the suitable candidates for the future fuel. Apart from these most of the "energy algae" (algae cultivated for energy purposes) also found there application in bioremediation. And commercial microalgal production (Figure 3.6)

Figure 3.6: Typical Commercial Microalgae Production Facility (Cyanotech Corp., Kona, Hawaii. Note green ponds culturing Spirulina and red ponds with *Haematococcus pluvialis*)

methodologies were also developed for some of the taxa like *Spirulina, Chlorella, Haematococcus* and *Dunaliella.*

Conclusion

Reducing the build-up of atmospheric CO_2, the major driving force in projected global warming, can be accomplished by three conceptually different methods: 1. reducing the use of fossil fuels; 2. removing CO_2 from the atmosphere; and 3. capturing and sequestering or utilizing the CO_2 emitted by fossil fuel combustion before it enters the atmosphere. Microalgae systems can use land, such as hard-pan and high clay soils, and water resources, such as waste or brackish waters, not suitable for conventional agriculture or forestry, minimizing the competition with food and fiber production (Benemann, 1977).

The biotechnology of microalgae has gained considerable importance in recent decades. Applications range from simple biomass production for fuel, food and feed to valuable products for ecological applications. For most of these applications, the market is still developing and the biotechnological use of microalgae will extend into new areas. Considering the enormous biodiversity of microalgae and recent developments in genetic engineering, this group of organisms represents one of the most promising sources for new products and applications. With the development of sophisticated culture and screening techniques, microalgal biotechnology could meet the high demands of fuel, food and pharmaceutical industries (Pulz and Gross, 2004)

In conclusion, microalgae will have an industrial future, but it is necessary to have a cooperative work with industry and scientific institutes to avoid disappointments and failures, which may condemn microalgal biotechnology for a long time. It is a difficult challenge but it can be succeeded if both scientific and industrial communities work together.

References

Achitouv E, Metzger P, Rager M-N and Largeau C (2004) C_{31}–C_{34} methylated squalenes from a Bolivian strain of *Botryococcus braunii*. *Phytochemistry* 65:3159–3165

Asada Y and Kawamura S (1986) Screening for cyanobacteria that evolve molecular hydrogen under dark and anaerobic conditions. J.*Ferment.Technol* 64(6):553-556.

Banerjee A, Sharma R, Chisti Y and Banerjee UC (2002) *Botryococcus braunii*: A renewable Source of Hydrocarbons and other Chemicals. *Critical Reviews in Biotechnology* 22(3): 245-279.

Ben-Amotz A and Avron M (1981) Glycerol and b-carotene metabolism in the halotolerant alga *Dunaliella*: a model system for biosolar energy conversion. *Trends Biochem. Sci.* 6: 297–299.

Ben-Amotz A and Avron M (1983) Accumulation of metabolites by halotolerant algae and its industrial potential. *Ann. Rev. Microbiol.* 37: 95–119.

Benemann JR, Augenstein DC and Weissman JC (1982a) "Microalgae as a source of liquid fuels, appendix: technical feasibility analysis." *Final Report,* U.S. Department of Energy, unpublished, pp 126.

Benemann JR, Baker D, Koopman BL and Oswald WJ (1977) "A systems analysis of bioconversion with microalgae." *Proc. Symposium Clean Fuels from Biomass and Wastes,* (Klass D, ed.) Institute of Gas Technology, Chicago, pp. 101-126.

Benemann JR, Goebel RP, Weissman JC and Augenstein DC (1982b) "Microalgae as a source of liquid fuels." *Final Report,* U.S. Department of Energy, pp 202.

Benemann JR, Koopman BL, Baker D, Goebel R and Oswald WJ (1977) "Preliminary design of the algae pond subsystem of the photosynthesis energy factory." *Final Report to Inter-Technology Solar Corp.,* Sanitary Eng. Res. Lab., Univ. of Calif.-Berkeley.

Benemann JR, Koopman BL, Weissman JC, Eisenberg DM and Oswald WJ (1978) "An integrated system for the conversion of solar energy with sewage-grown microalgae." *Report, Contract D (0-3)-34,* U.S. Dept. of Energy, SAN-003-4-2, June 1978.

Benemann JR, Persoff P and Oswald WJ (1978) "Cost analysis of microalgae biomass systems." *Final Report prepared for the U.S. Dept. of Energy,* HCP/t1-605-01 Under Contract EX-78-X-01-1 605.

Benemann JR, Weissman JC, Goebel RP and Augenstein DC (1986) Microalgae fuel economics. In: *Algal Biomass Technologies.* Barclay WR and McIntosh RP (eds), Nova Hedwigia Beih, 83, J. Cramer, Berlin, pp. 176-191.

Borowitzka MA, Borowitzka LJ (1988) *Dunaliella* in: M.A. Borowitzka LJ (ed.) New York, 27–58.

Cohen Z (1986) Products from microalgae. In: *Hand book of Microalgal Mass culture.* Richmond A (ed), CRC Press, Boca Raton, FL,421-454.

Dante RC (2005) Hypotheses for direct PEM fuel cells applications of photobioproduced hydrogen by Chlamydomonas reinhardtii. *International Journal of Hydrogen Energy* 30: 421– 424.

Dayananda C, Sarada R, Bhattacharya S and Ravishankar GA (2005) Effect of media and culture conditions on growth and hydrocarbon production from *Botryococcus braunii*. *Process Biochemistry* 40: 3125-3131.

Dayananda C, Sarada R, Srinivas P, Shamala TR and Ravishankar G A (2006). Presence of methyl branched fatty acids and saturated hydrocarbons in botryococcene producing strain of *Botryococcus braunii*. *Acta Physiologiae Plantarum* 28(3):523-528.

Dote Y, Sawayama S and Yokayama S (1992) Liquefaction of hydrocarbon –rich microalga. *Am. Chem. Soc. Div. Fuel. Chem.* 37:1836.

Gaffron R and Rubin J (1942) Fermentative and photochemical production of hydrogen in algae'" J. Gen. *Physiol.*, 26:209.

Gavrilescu M and Chisti Y (2005) Biotechnology a sustainable alternative for chemical industry. *Biotechnology Advances* 23: 471–499.

Graboski MS and McCormick RL (998) Combustion of fat and vegetable oil derived fuels in diesel engines. *Prog Energy Combus Sci* 24:125– 64.

Hall DO (1998) Immobilized photosynthetic system for the production of fuels and chemicals. In: 8Th International Biotechnology Symposium, Paris, July 17-22, Abstrak Book 34.

Hallen back PC, Kochain LV, Weissman JC and Benemann JR (1978) Solar energy conversion with hydrogen producing cultures of the blue green algae Anabaena cylindrical. *Biotechnol. Bioeng. Symp.* 8: 283-297.

Hillen LW, Pollard G, Wake LV and White (1982) Hydrocracking of the Oils of *Botryococcus braunii* to Transport Fuels. *Biotechnology and Bioengineering* 24:193-205.

Howarth DC and. Codd GA (1985) The uptake and production of molecular hydrogen by unicellular cyanobacteria. *J. Gen. Microbiol.* 131:1561-1569.

Huang Z and Poulter CD (1989) Tetramethylsqualene, a triterpene from *Botryococcus braunii* var. Showa. *Phytochemistry* 28:1467–1470.

Hoffert M I, Caldeira K, Benford G, Criswell DR, Green C, Herzog H, Jain AK, Kheshgi HS, Lackner KS, Lewis JS, Lightfoot H D, Manheimer W, Mankins JC, Mauel ME, Perkins LJ, Schlesinger ME, Volk T and Wigley TM L (2002) Advanced Technology Paths to Global Climate Stability: Energy for a Greenhouse Planet. *Science* 298 (5595):981–987

Illman AM, Scragg AH and Shales SW (2000) Increase in *Chlorella* strains calorific values when grown in low nitrogen medium. *Enzyme Microb Technol* 27:631–635.

Inoue S, Dote Y, Sawayama S, Minowa T, Ogi T and Yokayama S (1994) Analysis of oil from liquefaction of *Botryococcus braunii*. *Biomass Bioenergy* 6:269.

Karube I, Ikemoto H, Kajiwawa K, Tamiya E and Matsuoka H (1986) photochemical energy conversion using immobilized blue-green algae. *Journal of Biotechnology* 4:73.

Keenan J K (1977) Bioconversion of solar energy to methane. *Energy* 2: 365-369.

Kumazawa S and Mitsui A (1982) Hydrogen metabolism of photosynthetic bacteria and algae. In: *CRC Handbook of Biosolar Resources*. (ed) Zaborsky, Vol.1. Part1, CRC Press Inc., Boca Raton, Florida.

Lind blad P, Oxelfelt F, Tamagini P and Troshino O (1998) Cyanobacterial Biotechnology Nostoc PCC 73102 and H_2: knowledge, research and biotechnological challenges. In: Cyanobacterial Biotechnology. (eds) Subramanian, G., Kaushik, B. D. and Venkataraman, G.S., *Oxford and IBH Publishing Co. Pvt. Ltd.*, Borowitzka (eds.), *Micro-Algal Biotechnology*, Cambridge University Press, New Delhi, 1-5.

Metzger P and Largeau C (2005) Botryococcus braunii: a rich source for hydrocarbons and related ether lipids. *Appl Microbiol Biotechnol* 66: 486-496.

Miao X and Wu Q (2004b) High yield bio-oil production from fast pyrolysis by metabolic controlling of *Chlorella protothecoides. Journal of Biotechnology* 110: 85–93.

Miao X, Wu Q and Yang C (2004a) Fast pyrolysis of microalgae to produce renewable fuels *J. Anal. Appl. Pyrolysis* 71: 855–863

Mitsui A (1980) Saltwater based biological solar energy conversion for fuel, chemicals, fertilizer, food and feed. In: *Proc. Bioenergy'*80. Bioenergy Council, Washington, D.C., 486-491.

Matsui T, Nishihara A, Ueda C, Ohtsuki M, Ikenaga N and Suzuki T (1997) Liquefaction of micro-algae with iron catalyst. *Fuel* 76 (11):1043-1048

Nandi R and Sengupta S (1998) Microbial production of hydrogen: an overview. *Crit Rev Microbiol* 24:61–84.

Neenan B, Feinberg D, Hill A, McIntosh R and Terry K (1986) Fuels from microalgae: Technology status, potential and research requirements, *Solar Energy Research institute, Golden Co.*, SERI/SP, pp: 31-255.

Perales-Vela HV, Pena-Castro JM and Canizares-Villanueva RO (2006) Heavy metal detoxification in eukaryotic microalgae. *Chemosphere* (in press).

Pulz O and Gross W (2004) Valuable products from biotechnology of microalgae. *Appl Microbiol Biotechnol* 65: 635–648.

Regan DL (1980) Marine biotechnology and the use of arid zones. *Search* 111:377-381.

Robinson PK, Mak AL and Trevan MD (1986) Immobilized Algae: A Review. *Process Biochemistry* 21:122-127.

Sawayama S, Minowa T and Yokayama S (1999) Possibility of renewable energy production and carbon dioxide mitigation by thermochemical liquefaction of microalgae. *Biomass Bioenergy* 17: 33.

Sawayama S, Minowa T, Dota Y and Yokayama S (1999) possibility of renewable energy production and carbon dioxide mitigation by thermochemical liquification of microalgae. *Biomass Bioenergy* 17: 33-39.

Scragg AH, Morrison J and Shales SW (2003) The use of a fuel containing *Chlorella vulgaris* in a diesel engine. *Enzyme and Microbial Technology* 33: 884–889.

Service RF (2004) The Hydrogen Backlash. *Science* 305(5686)958–961.

Tuzun I, Bayramoglu G, Yalcyn E, Basaran G, Celik G and Aryca MY (2005) Equilibrium and kinetic studies on biosorption of Hg(II), Cd(II)and Pb(II) ions onto microalgae Chlamydomonas reinhardtii. *Journal of Environmental Management* 77:85–92.

© R.K. Gupta & V.D. Pandey (*Editors*)
Advances in Applied Phycology, 2007
Daya Publishing House, New Delhi

75–84

Chapter 4

Factors Influencing Algicide Production by *Microcystis* sp and its Effect on Selected Cyanobacteria

P. Jaiswal*, R. Prasanna and P.K. Singh

Centre for Conservation and Utilization of Blue Green Algae (CCUBGA), Indian Agricultural Research Institute (IARI), New Delhi – 110 012, India

ABSTRACT

Five *Microcystis* strains were isolated from diverse habitats and screened for algicidal activity. Intrestingly, only one of them (*Microcystis* sp.: M1) showed algicidal activity. Time course studies on the influence of different environmental and physico-chemical factors revealed that culture age had a significant impact on algicide production. Addition of phosphate was observed to enhance the biocidal activity. The addition of culture extract of M1 brought about a reduction in protein and chlorophyll content of test organisms (*Synechococcus* sp. and *Synechocystis* sp.) on the 6th day of incubation. Crude extract of freshly collected samples of M1 showed toxicity against mice and the mice died within 1hr of injection, while lab grown sample did not show any toxicity, indicating that repeated laboratory-level culturing leads to reduction in toxicity of the strain.

Keywords: Algicide; Cyanobacterium; Inhibition zone; Microcystis sp.

Introduction

In freshwater ecosystems phytoplankton/suspended algae are the key source of organic matter production in the food web. Phytoplankton productivity is dependent on adequate nutrient supplies,

* Corresponding Author: Tel: 91-011-25848431; Fax: 91-011-25741648; E-mail: pranitajaiswal1@hotmail.com.

when the nutrient concentration increases, (which is mainly due to human intervention) it leads to the acceleration of primary production/eutrophication, an event which is known as bloom formation. Blooms are the primary agent involved in water quality deterioration, including foul odor and taste, deoxygenation of bottom water, toxicity, fish kill and food-web alterations. Although in freshwater ecosystems, phytoplankton belonging to diverse phyla are capable of forming blooms, blue green algae (cyanobacteria) are the most notorious bloom formers.

Cyanobacteria are simple primitive photosynthetic microorganisms exhibiting wide occurrence in fresh, brackish and marine environments. Their success as a group in a wide range of aquatic habitats has been attributed to their unique physiological characters and high adaptive ability under a wide range of environmental conditions. They are capable of growing into big biomass, often dominating other aquatic flora and under certain circumstances, can accumulate near the water surface producing scums. Cyanobacterial blooms are hazardous due to production of secondary metabolites which can have severe physiological effects on other organisms and human beings and many of them have pharmaceutical importance (Sivonen *et al.*, 1990; Carmichael, 1992). These biotoxins may reach humans, through their domestic animals and wild life consuming such waters, through recreational purposes, or through food chain/direct contact with water bodies/algae. Also, the ecological and economic impacts of toxin production by these blooms is of serious concern.

Some of the bloom forming cyanobacteria are *Microcystis, Anabaena, Nodularia, Oscillatoria, Aphanizomenon* etc. Among them *Microcystis* is most common, as it forms dense blooms in eutrophic lakes. It is known to produce a family of related cyclic hepatopeptides that pose a considerable threat to aquatic ecosystems.

These bioactive compounds can also find use as biocidal agents against mosquitoes, algae, fungi, bacteria and other plants (Tyagi and Kaushik, 1998; Singh *et al.*, 2001; Metcalf *et al.*, 2004). Such investigations bring to light the other facet of cyanobacterial toxins–which may prove beneficial to mankind. Also, a number of interesting algicidal compounds have been extracted from cyanobacteria (Schlegel *et al.*, 199; Patterson and Bolis, 1993), which show a common mode of action similar to DCMU i.e inactivation of PSII mediated electron transport. (Srivastava *et al.*, 2001), which may be indicative of evolutionary significance and basis for their dominance in aquatic ecosystem. The present investigation is focused at comparing five *Microcystis* strains for their biocidal activity and evaluating the effect of physical factors on algicide production.

Materials and Methods

Isolation of Cyanobacteria

Standard microbiological techniques were followed for isolation and purification of *Microcystis* sp. from different water and soil samples using Parker's medium (Kaushik, 1987). Incubation was done at 27±1°C, 3000 lux light intensity and 16:8 L/D cycles. The flasks were regularly monitored for cyanobacterial growth and examined under microscope. Cultures were made unialgal by repeated streaking. The cultures were made axenic by UV/antibiotic treatment. The length and doses of UV/antibiotic treatment varied with cyanobacterial strains (Kaushik, 1987).

Algicidal Activity

Algicidal activity of different *Microcystis* strains were tested on the lawn of *Synechococcus* and *Synechocystis* (prepared using double layer technique), and the diameter of inhibition zones were measured (Jaiswal *et al.*, 2005).

Effect of Different Environmental and Physicochemical Factors

In order to study the effect of different physicochemical factors on algicidal activity the medium was supplemented with different levels of P and N (½ and 2 times as in Parker's medium). Effect of light intensities(1500-2000; 3000-4000; 5000-6000), temperature (20; 30; 40) and pH (7; 9; 11) was evaluated by growing the cyanobacterium in respective conditions.

Extraction of Secondary Metabolites

Both cell free extracts and lyophilized cells were used for extraction of secondary metabolites. Culture extract was partitioned with non-polar, polar and highly polar solvents followed by concentration of extracts by rotary evaporation. Extracts were finally dissolved in desired volume of acetone and loaded on TLC plates. Extraction from lyophilized cells was done using methanol as solvent.

Mouse Bioassay

Toxicity was tested by intraperitoneal injection of samples (toxins extracted from the lyophilized cells) in 18-22g male mice (Albino Parks strains).Three mice per dose level were used. The injected mice were observed closely over a period of 7h.

Results and Discussion

Blooms of toxic cyanobacteria in water bodies have drawn attention of scientists world-wide. They produce a wide range of secondary metabolites, which may be toxic to animals and plants. Many freshwater cyanobacterial blooms include sp. of toxigenic genera *Microcystis, Anabaena, Planktothrix* etc. Among them, *Microcystis* is the most common bloom forming cyanobacterium.

During the course of the present investigation, five *Microcystis* strains were isolated from diverse habitats (Table 4.1), and tested for algicidal activity by placing discs impregnated with culture extracts/ filtrates on the lawn of test organisms *Synechococcus* and *Synechocystis* sp. Among these strains only *Microcystis aeruginosa* (M1), isolated from Lakshmikund showed algicidal activity against both the test organisms (Figure 4.1), emphasizing that all *Microcystis* strains are not toxic. Production of algicidal compounds by cyanobacteria has already been reported by many workers (Bagchi and Ray, 2001; Hu *et al*,. 2004; Jaiswal *et al*,. 2005).

Table 4.1: Source of Five *Microcystis* Strains

Strains	*Source*
M1	Laxmikund,Varanasi
M2	Kandawa,Varanasi
M3	Khojwa,Varanasi
M4	Rice field, IARI
M5	Boat club pond, India Gate

Time course study were taken up to study the effect of different environmental and physicochemical factors and measured in the form of diameter of zone of inhibition (ZOI). Algicidal activity of the cyanobacterial extracts was checked at fixed time intervals starting from log phase to decline phase (5-25) days (Figure 4.2). It was found that cultures of all stages (log-decline phase) showed algicidal activity but maximum effect was recorded in 15-20 days culture. Hence, age of the culture had a significant effect on algicide production, indicating that algicide production is correlated with particular stage of the growth cycle (late exponential to decline phase). Ray and Bagchi (2001) also reported that algicide started appearing in the medium after mid exponential phase and paralleled the biomass yield.

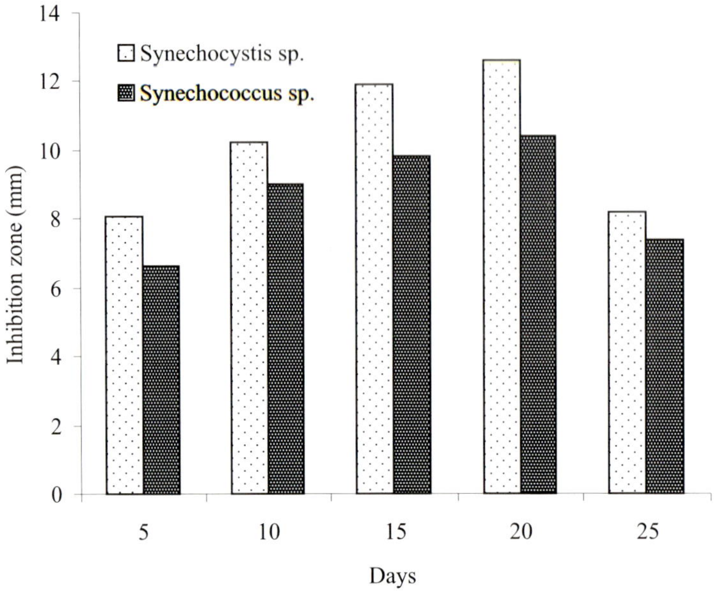

Figure 4.1: Algicidal Activity by *Microcystis aeruginosa* (M1) on the Lawn of *Synechococcus* sp. and *Synechocystis* sp.

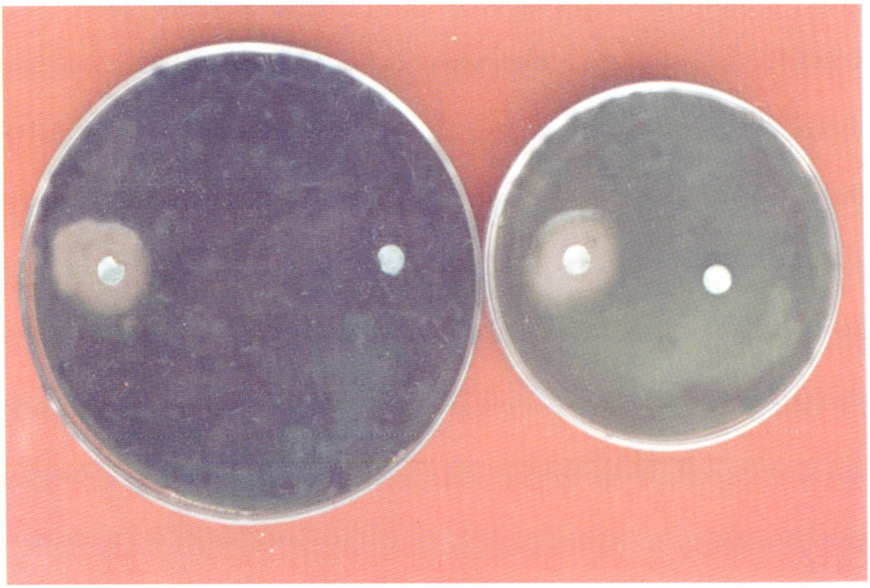

***Synechococcus* sp.** ***Synechocystis* sp.**

Figure 4.2: Diameter of Inhibition Zone (mm) Produced by M1 on the Lawn of *Synechococcus* sp. and *Synechocystis* sp.

In order to study the role of phosphate on algicide production, the cyanobacterium was grown with different phosphate concentrations. Maximum activity was recorded (Figure 4.3), when the phosphate concentration was doubled (2P) in comparison to basal medium. However, no activity was recorded, when phosphate concentration was reduced to half (1/2P) of basal medium (P). So it may be concluded that increase in phosphate concentration by 2 folds, leads to enhancement in algicidal activity. This can be correlated with *M. aeruginosa* growing as blooms in water bodies. The reduction in P in eutrophic waters may lower microcystin producing rate of *M. aeruginosa*. There have been several reports regarding effect of phosphate on secondary metabolite production (Sivonen, 1990; Utkilen and Gjolme, 1995). Utkilen and Gjolme (i995) reported that phosphate limiting conditions have no effect on toxin production by *M. aruginosa*, On the other hand, Oh *et al.* (2000) showed more P in the culture medium stimulates growth and toxin production by *M. aeruginosa*. Recently, Ray and Bagchi (2001) reported algicide production by *Oscillatoria latevirens* was negetively regulated with phosphate and pH. However, we did not find any effect of change in pH. Change in light intensity, temperature, and nitrate concentration also did not show any significant effect on algicide production. Then algicidal activity was recorded under optimum growth conditions for the cyanobacterium. Any variation in conditions, which is affecting growth of the organisms was observed to affect algicide production by the cyanobacterium (Table 4.2). Wicks and Thiel (1990) also reported that there is no strong releationship between total toxin concentration and organic and inorganic nutrients in surface water.

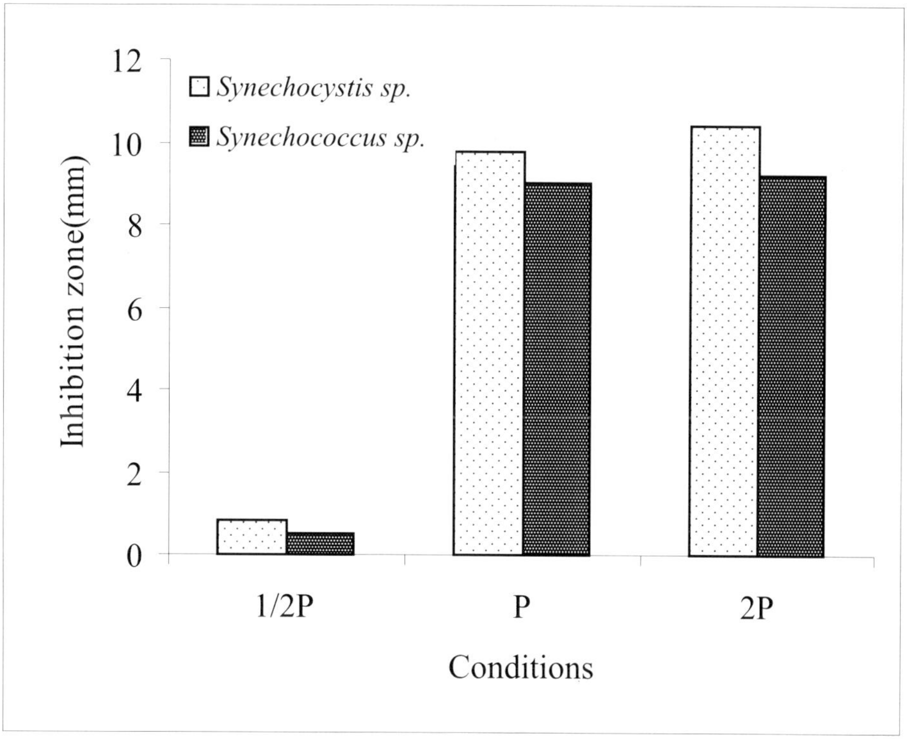

Figure 4.3: Effect of Phosphate Concentration on the Diameter of Inhibition Zone (mm) Produced by M1 on the Lawn of *Synechococcus* sp. and *Synechocystis* sp.

Table 4.2: Effect of Environmental Factors on Algicidal Activity of *Microcystis aeruginosa*

Condition	*Synechocystis sp.*	*Synechococcus sp.*
Light Intensity		
1500-2000	-	-
3000-4000	+	+
5000-6000	-	-
Temperature		
20	-	-
30	+	+
40	-	-
pH		
7.0	-	-
9.0	+	+
11.0	-	-

Figure 4.4: TLC Picture Showing Fractionation of Compounds Using Solvents Hexane : Benzene (1:1)

Secondary metabolites from cyanobacterial filtrate was extracted by using different solvent systems and loaded on TLC plates. Maximum separation of extracts was recorded, when hexane and benzene were used as solvent system in ratio of 1:1. In this solvent system, hexane extract showed 3 spots (Figure 4.4), while dichloromethane and ethyl acetate extracts of M1 showed the presence of 5 spots.

In order to see the effect of cyanobacterial filtrate extract on the physiology of test organisms, culture filtrate of *Microcystis* was added (10 per cent) to *Synechococcus* and *Synechocystis* culture and kept in culture room conditions for regular observations. The culture extract of *Microcystis* sp lowered the protein content in *Synechococcus* sp.as compared to control on the fourth day of incubation. Similarly in case of *Synechcystis* sp a decline in protein content was recorded (Figure 4.6), when supplemented with culture extract. Further incubation till the 8th day caused further decrease in the protein content of test organisms.

The effect of culture extract was also observed on chlorophyll content of test organisms (*Synechococcus* and *Synechocystis*), and it was observed that till the 6th day of incubation (Figures 4.7 and 4.8), there was no significant effect. Only on the 8th day of incubation a reduction of 6.5 per cent and 1.5 per cent in the chlorophyll content of *Synechococcus* and *Synechocystis* was observed, respectively Further incubation resulted in further reduction in chlorophyll content of the test organisms.

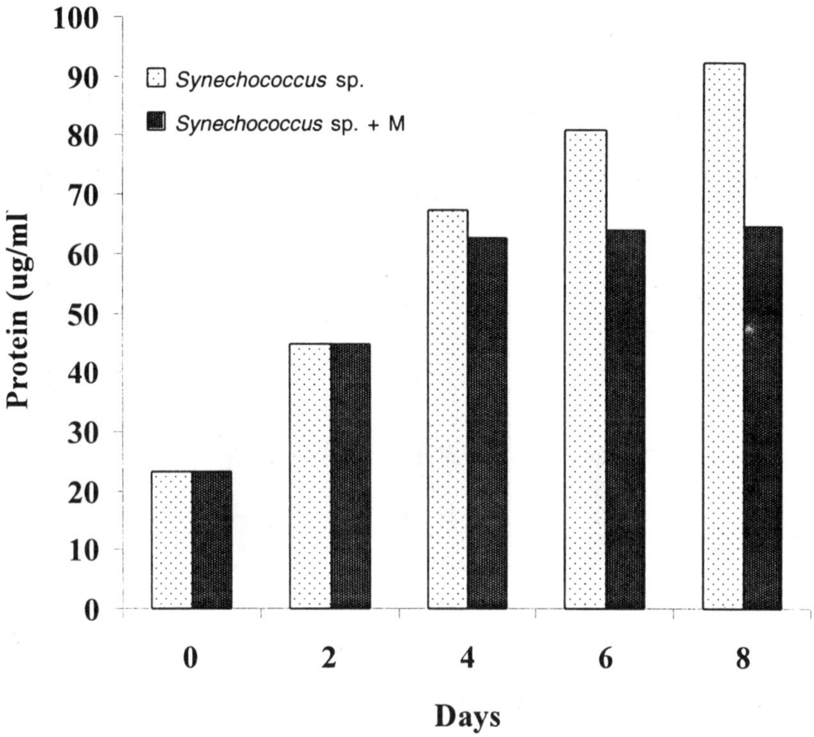

Figure 4.5: Effect of M1 Culture Filtrate on Protein Content of *Synechococcus* sp.

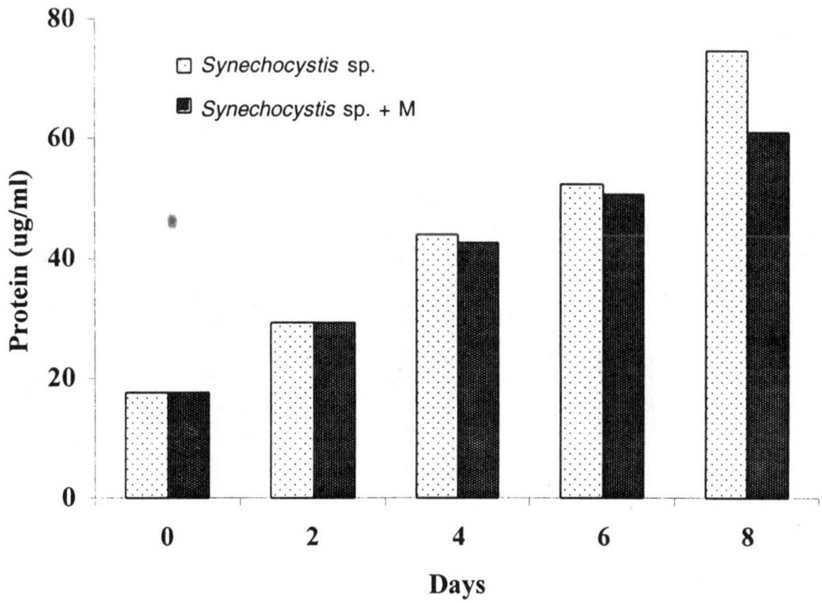

Figure 4.6: Effect of M1 Culture Filtrate on Protein Content of *Synechocystis* sp.

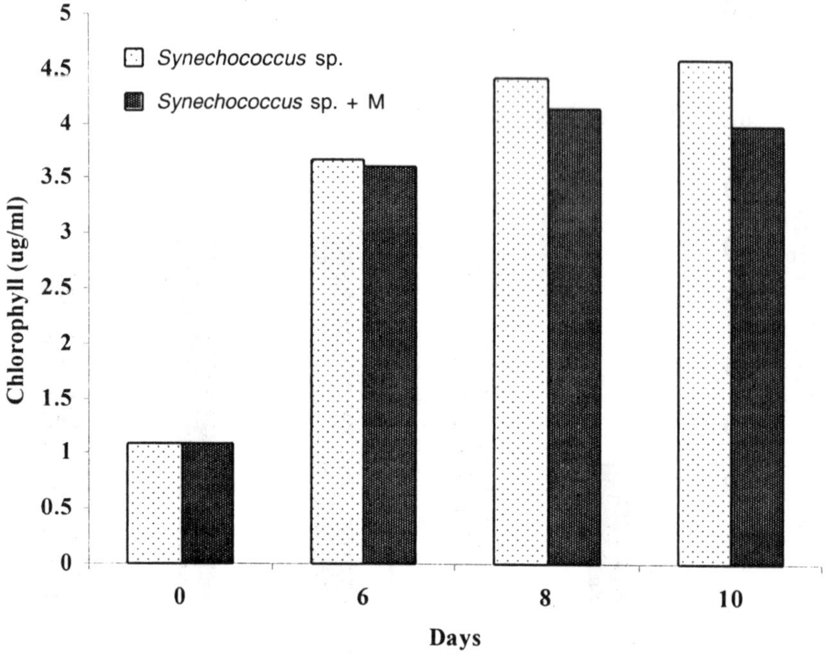

Figure 4.7: Effect of M1 Culture Filtrate on Chlorophyll Content of *Synechococcus* sp.

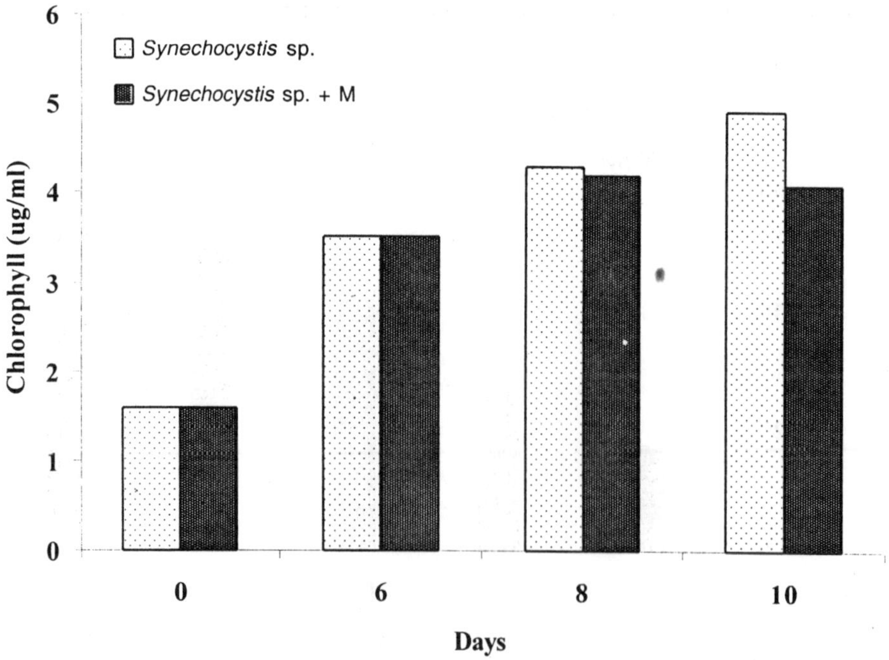

Figure 4.8: Effect of M1 Culture Filtrate on chlorophyll Content of *Synechocystis* sp.

Toxicity of the five *Microcystis* strain were also tested by performing mouse bioassay experiments and it was observed that none of the lab grown samples showed toxicity against mice. However, crude extract of freshly collected samples of *Microcystis aerugonosa* showed toxicity and the mice died within one hour of injection, indicating that the potency of toxicity decreases with repeated culturing in laboratory conditions.

The salient results of our investigation are

1. The genus *Microcystis* consist both toxic and non toxic species/genera
2. There is a need to simulate the natural conditions of the habitat of the organism for obtaining high levels of toxicity as indicated by our experiment on mice.
3. Concentration of phosphate is critical for production of biocidal compounds
4. Chlorophyll/Protein accumulation of target organism is more severely affected i.e mode of action may be towards inhibition of protein synthesis.

Concerted efforts need to be undertaken to characterize the nature of biocidal compound and its mode of action. This will provide pointers for alternative and novel uses of these compounds in agriculture and a better understanding of the biological interactions occurring in aquatic ecosystems.

Acknowledgements

Authors are thankful to Dr. Ashok Kumar, School of Biotechnology, B.H.U. for providing laboratory facilities to carry out some part of the experiments. One of the authors is thankful to Council of Scientific and Industrial Research (CSIR), New Delhi, India for providing financial assistance. The facilities required for carrying out this investigation provided by the Centre for Conservation and Utilization of Blue Green Algae, IARI, New Delhi is gratefully acknowledged.

References

Bagchi, SN and Ray S (2001) Extraction and purification of an algicide metabolite from a cyanobacterium, *Oscillatoria latevirens*. *Ind. J. Microbiol.* 41: 163-167.

Bagchi, SN and Ray S (2001) Nutrient and pH regulate algicide production in cultures of cyanobacterium *Oscillatoria latevirens*. *New Phytologist* 149: 455-460

Carmichael, W. W. 1992. Cyanobacterial secondary metabolites: the cyanotoxins. *J. Appl. Bacteriol.* 72: 445-459.

Hu ZQ, Liu YD and Li DH (2004) Physiological and biochemical analyses of microcystin-RR toxicity to the cyanobacterium *Synechococcus elongatus Environ. Toxicol.* 19: 571-577.

Jaiswal P, Prasanna, R. and Singh, PK (2005) Isolation and Screening of rice field cyanobacteria exhibiting algicidal activity. *Asian J. Microbiol. Biotech. Env. Sci.* 7: 767-770.

Kaushik BD (1987) *Laboratory methods for blue green algae.* Associated Publishing Company, New Delhi.

Oh HM, Lee SJ, Jang MH and Yoon BD (2000) Microcystin production by *Microcystis aeruginosa* in a phosphorus limited chemostat. *Appl. Environ. Microbiol.* 66: 176-179.

Metcalf JS, Barakate A and Codd GA (2004) Inhibition of plant protein synthesis by the cyanobacterial hepatotoxin cylindrospermopsin. *FEMS Microbiol. Lett.* 235: 125-129.

Patterson GML and Bolis CM (1993) Regulation of scytophycin accumulation in cultures of *Scytonema ocellatum*. I. Physical factors. *Appl. Microbiol. Biotechnol.* 40: 375-381.

Ray S, and Bagchi, SN (2001) Nutrients and pH regulate algicide accumulation in cultures of cyanobacterium *Oscillatoria latevirens*. *New Phytologist* 149: 455-460.

Rippka R, Deruelles J, Waterbury JR, Herdman M and Stanier RY (1979) Generic assignments, strain histories and properties of pure culture of cyanobacteria. *J. Gen. Microbiol.* 111: 1-61

Schlegel I, Doan NT, Chazal ND and Smith, GD (1998) Antibiotic activity of new cyanobacterial isolates from Australia and Asia against green algae and cyanobacteia. *J. Appl. Phycol.* 10: 1-9.

Singh DP, Tyagi MB, hakur JK and Kumar A (2001) Antialgal activity of a hepatotoxin –producing cyanobacterium, *Microcystis aeruginosa*. *World J. Micro. Biotech.*, 17: 15-22.

Sivonen K, Niemela SI, Niemi RM, Lepisto L, Luoma TH and Rasanen LA (1990) Toxic cyanobacteria (blue green algae) in Finnish fresh and coastal waters. *Hydrobiologia*. 190: 267-275.

Srivastava R, Ray S and Bagchi SN (2001) Action of algicide from cyanobacterium, *Oscillatoria latevirens*, on photosystem II. *Indian J. Exptl. Biol.* 39:1268-1273.

Tyagi R and Kaushik B D (1999) Cyanobacterial toxins and their effect on some microorganisms and *Anopheles* mosquito larvae. In: Vohra S B and Agarwal VP (ed) *Toxicology and Environmental Health*. Society of Biosciences/Jamia Hamdard/Asiatech Publ. Tnc. New Delhi. 285-290.

Utkilen H and Gjolme N (1992) Toxin production by *Microcystis aeruginosa* as a function of light in continuous cultures and its ecological significance. *Appl. Environ. Microbiol.* 58: 1321-1325.

Wicks RJ and Thiel PG (1990) Environmental factors affecting tha production of peptide toxins in floating scums of the cyanobacterium *Microcystis aeruginosa* in hypertrophic African reservoir. *Environ. Sci. Technol.* 24: 1413-1418.

© R.K. Gupta & V.D. Pandey (*Editors*)
Advances in Applied Phycology, 2007
Daya Publishing House, New Delhi

Chapter 5

Seaweeds as a Human Diet: An Emerging Trend in the New Millennium

P.V. Subba Rao*, Vaibhav A. Mantri, K. Ganesan and K. Suresh Kumar

Marine Algae and Marine Environment Discipline,
Central Salt and Marine Chemicals Research Institute (CSIR),
Gijubhai Badheka Marg, Bhavnagar – 364 002, Gujarat, INDIA

ABSTRACT

In the Far East and Pacific, there has been a long tradition of consuming seaweeds as sea vegetables, while in Western countries the principal use of seaweeds has been as source of phycocolloids having their applications, some times in food preparations. Edible seaweeds have played a significant role in economy of some Nations such as Japan, Korea and China. In world market, there are 13 algae authorized as vegetable and condiments, although 152 seaweed species have been utilized for food preparations. The nutritive value of the seaweeds is mainly due to the presence of rich protein, amino acids, minerals, dietary fibers and antioxidants. The protein content in seaweeds varies from 3 to 47 per cent of dry wt. Aspartic and glutamic acid constitute large part of the amino acid fraction in the seaweed. Mineral content (including macro and micro nutrients) recorded in seaweeds is found in the range of 8 to 40 per cent. In addition, seaweeds constitute an interesting source of dietary fibers and antioxidant compounds with health protective effects. This paper is an overview of edible seaweeds and their nutritional values.

Introduction

SEAWEEDS are a fascinating and diverse group of organisms living in the oceans that occupy almost ¾ of the globe. They are seen attached to rocks in the intertidal zone, washed up on the beach,

* Corresponding Author: E-mail: pvsubbarao@csmcri.org.

and also are found as giant underwater forests, and as well floating on the ocean's surface. They can be very tiny, or quite large, growing up to 30 meters long. Seaweeds refer to any large marine benthic algae that are multicellular, macrothallic, and thus differentiated from most algae that are of microscopic size (Smith, 1944). Although they have many plant-like features seaweeds are not true vascular plants; they are algae. Algae are part of the Kingdom Protista, which means that they are neither plants nor animals. Seaweeds are not grouped with the true plants because they lack a specialized vascular system (an internal conducting system for fluids and nutrients), roots, stems, leaves, and enclosed reproductive structures like flowers and cones. As all the parts of a seaweed are in contact either fully or partially with the seawater, they are able to take up fluids, nutrients, and gases directly from the seawater, and as a result they do not need an internal conducting system. Like true plants, seaweeds are photosynthetic; they convert energy from sunlight into the materials needed for growth. Within their cells seaweeds have the green pigment chlorophyll, which absorbs the sunlight they need for photosynthesis. Chlorophyll is also responsible for the green colouration of many types of seaweed. In addition to chlorophyll some seaweeds contain other light absorbing pigments. These pigments are red, blue, brown, or golden, and are responsible for the beautiful colouration of red and brown algae. Seaweeds are classified into three major groups; the green algae (Chlorophyta), the brown algae (Phaeophyta), and the red algae (Rhodophyta). They are placed into one of these groups based on their pigments and colouration. Other features used to classify seaweeds include cell wall composition, reproductive characteristics, and the chemical nature of their photosynthetic products (oil and starch). Within each of the three major groups of seaweeds, further classification is done based on characteristics such as plant structure, form, and shape (Druehl, 2000). These seaweeds form a very important renewable resource in marine environment and have been a part of human civilization from time immemorial. The reports on uses of seaweeds have been cited as early as 2,500 years ago in Chinese literature (c. f. Tseng, 2004). The long history of the seaweed utilization for a variety of purposes has led to the gradual realization that some of their constituents are more superior and valuable in comparison to their counterparts on land. Seaweeds synthesize a wide range of chemicals, some of which stand the only natural resource *e.g.* agar, carrageenan and alginates. Every year about 7.5 to 8 million tons of wet seaweeds have been produced along the coastal regions world–wide (McHugh, 2003). Seaweeds form as a valuable source of food containing protein, carbohydrate, lipid, iodine and vitamin etc. (Table 5.1).

Seaweed as Food: Global Vs. Indian Scenario

Seaweeds referred as Seavegetables are daily food items among the Celtic cultures of Europe, the Island nations of Oceania and the countries of East Asia especially Japan, Korea and China where they form the basis for a multi-billion USD business employing seaplant farmers and processors. Seaweeds have been consumed as a vegetable since the beginning of fourth century in Japanese and six century in China. On average, the Japanese eat 1.4 kg seaweed per person per year. This ancient tradition and everyday habit have made possible a large number of epidemiological researches showing the health benefits linked to seaweed consumption (Teas 1981, Hiqashi *et al.*, 1999, Funahashi *et al.*, 1999). The spread of Japanese and Chinese cuisine and of health foods throughout the world has brought new attention to seavegetables. There is immense potential for increasing the consumption of seavegetables as food items, dietary supplements, food texturisers, flavouring, colouring and condiments especially among "mainstream" markets. France is the first european country to establish a specific regulation concerning the use of seaweeds for human consumption as non–traditional food substances. Currently, 13 algae (5 brown seaweeds, 4 red seaweeds, 2 green seaweeds and 2 micro algae) are authorized as vegetables and condiments (Burtin, 2003) (Table 5.2).

Table 5.1: Nutritional Composition of Edible Common Seaweeds

Parameters	*Alaria*	*Dulse*	*Kelp*	*Nori*
Protein g/100 g	17.7	21.5	16.1	28.4
Fat g/100 g	3.6	1.7	2.4	4.5
Carbohydrate g/100 g	39.8	44.6	39.3	45.1
Calories/100 g	262	264	261	318
Iodine mg/100 g	16.6	5.2	144	1.4
Vit A.I.U	8,487	663	561	4,286
Vit B_1 mg/100 mg	0.558	0.073	0.549	0.577
Vit B_2 mg/100 mg	2.73	1.91	2.48	2.93
Vit B_3 mg/100 mg	10.5	1.89	3.62	5.92
Vit B_6 mg/100 mg	6.23	8.99	8.63	11.21
Vit B_{12} mg/100 mg	5.03	6.6	2.6	17.5
Vit C mg/100 mg	5.9	6.34	4.16	12.03
Vit E. I. U.	4.92	1.71	2.71	5.09

Source: www.seaveg.com/chart.html as accessed on 15th February 2006.

Table 5.2: Seaweeds Authorized as Vegetables and Condiments in World Market

	Phylum	*Seaweed Species*
Macroalgae	Brown seaweeds	*Ascophyllum nodosum*
		Fucus vesiculosus
		Fucus serratus
		Himanthalia elongata
		Undaria pinnatifida
	Red seaweeds	*Porphyra umbilicalis*
		Palmaria palmata
		Gracilaria verrucosa
		Chondrus crispus
	Green seaweeds	*Ulva* spp.
		Enteromorpha spp
Microalgae		*Spirulina* sp.
		Odontella aurita

Source: Burtin P. 2003. *Elect J Env Agri Food Che.*, 2, 498–503.

Global food security will remain a worldwide concern for the next 50 years, due to growing differences in food-to-population ratio (Rosegrant and Cline, 2004). This ever-increasing population is compelling the human race to opt for non-conventional and alternative food resources. The World annual production of edible seaweeds through aquaculture is 6 million tons (Fresh wt) (Fleurence, 1999). In all 152 seaweed species have been utilized for food purpose globally (Table 5.3), of which 83 are from Rhodophyta, 40 from Phaeophyta and 29 from Chlorophyta. In India though seaweeds are not much popular as human food there are stray reports of their use in food formulations. In some coastal places of Tamil Nadu (Tiruchandur–Kaniyakumari sector) people used to take seaweed gangi (water in which seaweed-*Hypnea* species–is boiled) for getting rid of stomach disorders especially from worms as well as seaweed extracts from species of *Ulva* in the preparation of sweets (senior author personal observation). By keeping in mind the seaweeds as alternate vegetarian diet, it is worthwhile here to overview the nutritional value of seaweeds for human food in the new millennium.

Table 5.3: Seaweeds being Used as a Food in Different Counties

Seaweed	*Country*
Capspsiphon fulvescens	Korea
Caulerpa spp	Malaysia, Thailand
Caulerpa lentillifera	Philippines
Caulerpa peltata	Philippines
Caulerpa racemosa	Bangladesh, Japan, Philippines, South Pacific Islands, Vietnam
Caulerpa sertularioides	Philippines
Caulerpa taxifolia	Philippines
Codium spp	Argentina
Codium bartletti	Philippines
Codium edule	Philippines
Codium fragile	Korea, Philippines
Codium muelleri	Hawaii
Codium taylori	Israel
Codium tenue	Indonesia
Codium tomentosum	Indonesia
Colpomenia sinuosa	Philippines
Enteromorpha spp	Bangladesh, France, Hawaii, Myanmar
Enteromorpha compressa	Korea, Indonesia
Enteromorpha clathrata	Korea
Enteromorpha grevillei	Korea
Enteromorpha intestinalis	Indonesia, Japan, Korea
Enteromorpha linza	Korea
Enteromorpha nitidum	Korea
Enteromorpha prolifera	Indonesia, Japan, Korea, Philippines
Monostroma nitidum	Japan
Scytosiphon lomentaria	Korea, France
Ulva spp.	Argentina, Canada, Chile, Hawaii, Japan, Malaysia
Ulva lactuca	Vietnam Indonesia
Ulva reticulata	Vietnam

Contd...

Table 5.3–Contd...

Seaweed	Country
Acanthophora spicifera	Philippines, Vietnam
Asparagopsis taxiformis	Hawaii, Indonesia
Betaphycus gelatinum	Vietnam
Calaglossa adnata	Indonesia
Catenella spp	Myanmar
Chondria crassicaulis	Korea
Chondrus crispus	Ireland, France
Chondrus ocellatus	Japan
Eucheuma cartilagineum	Japan
Eucheuma gelatinae	Indonesia, Japan, Philippines, Caribbean
Eucheuma muricatum	Indonesia
Gelidiella acerosa	Philippines
Gelidiella tenuissima	Bangladesh
Gelidium spp	Hawaii
Gelidium anansii	Korea, Indonesia
Gelidium latifolium	Indonesia
Gelidium pusillum	Bangladesh
Gloiopeltis spp	Vietnam
Gracilaria changii	Thailand
Gracilaria cornea	Caribbean
Gracilaria coronopifera	Hawaii, Vietnam
Gracilaria crassissima	Caribbean
Gracilaria domingensis	Brazil, Caribbean, Chile
Gracilaria eucheumoides	Indonesia, Vietnam
Gracilaria firma	Vietnam
Gracilaria fisheri	Thailand
Gracilaria gracilis	Vietnam
Gracilaria lemaneiformis	Japan
Gracilaria parvispora	Hawaii
Gracilaria salicornia	Thailand, Vietnam
Gracilaria tenuistipitata var. liui.	Thailand, Vietnam
Gracilaria verrucosa	France, Indonesia, Japan, Korea
Grateloupia filicina	Indonesia, Japan
Halymenia spp.	Myanmar
Halymenia discoidea	Bangladesh
Halymenia durvillaei	Philippines

Contd...

Table 5.3–Contd...

Seaweed	Country
Hypnea spp	Myanmar
Hypnea muscoides	Vietnam
Hypnea nidifica	Hawaii
Hypnea pannosa	Bangladesh, Philippines
Hypnea valentiae	Vietnam
Iridaea edulis	Iceland
Kappaphycus alvarezii	Philippines
Kappaphycus cottonii	Vietnam
Laurencia obtusa	Indonesia
Laurencia pinnitifida	Portugal
Mastocarpus stellatus	Ireland
Mazzaella splendens	Canada
Meristotheca papulosa	Japan
Meristotheca procumbens	South Pacific Islands
Nemalion vericulare	Korea
Palmaria hecatensis	Canada
Palmaria mollis	Canada
Palmaria palmata	Canada, France, Iceland Ireland, UK, US
Porphyra spp	Israel, New Zealand, UK
Porphyra abbottae	Alaska, Canada
Porphyra acanthophora	Brazil
Porphyra atropurpurae	Indonesia
Porphyra columbina	Argentina, Chile, Peru
Porphyra crispata	Thailand, Vietnam
Porphyra fallax	Canada
Porphyra haitanensis	China
Porphyra kuniedae	Korea
Porphyra leucostica	Portugal
Porphyra perforata	Canada
Porphyra psuedolanceolata	Canada
Porphyra seriata	Korea
Porphyra spiralis	Brazil
Porphyra suborbiculata	Korea, Vietnam
Porphyra tenera	Japan, Korea
Porphyra torta	Alaska, Canada
Porphyra umbilicalis	France, US

Contd...

Table 5.3–Contd...

Seaweed	Country
Porphyra vietnamensis	Thailand
Porphyra yezoensis	China, Japan, Korea
Pterocladia capillacea	Korea
Scinaia moniliformis	Philippines
Solieria spp	Myanmar
Alaria crassifolia	Japan
Alaria fitulosa	Alaska
Alaria marginata	Canada
Alaria esculenta	Iceland, Ireland, US
Cladosiphon okamuranus	Japan
Durvillaea antarctica	Chile
Ecklonia cava	Japan
Ecklonia stolonifera	Korea
Egregia menziesii	Canada
Fucus serratus	France
Fucus vesiculosus	France, Portugal
Hizikia fusiformis	Japan, Korea
Hydroclathrus clathratus	Bangladesh, Philippines
Laminaria angustata	Japan
Laminaria bongardiana	RoK Alaska
Laminaria diabolica	Japan
Laminaria digitata	Ireland
Laminaria groenlandica	Canada
Laminaria japonica	China, Japan, Korea
Laminaria longicruris	US
Laminaria longissima	Japan
Laminaria octotensis	Japan
Laminaria religiosa	Japan, Korea
Laminaria saccharina	Alaska, Canada, Ireland, Rok Alaska
Laminaria setchelli	Canada
Macrocystis pyrifera	Argentina
Nemacystis decipiens	Japan
Nereocystis luetkaena	US
Pelvetia siliquosa	Korea
Postelsia spp	US

Contd...

Table 5.3–Contd...

Seaweed	*Country*
Sargassum aquifolium	Indonesia
Sargassum crassifolium	Thailand
Sargassum spp.	Bangladesh, Hawaii, Malaysia, Myanmar, Philippines, Thailand, Vietnam
Sargassum filipendula	Egypt
Sargassum horneri	Korea
Sargassum oligosystum	Thailand
Sargassum polycystum	Indonesia, Thailand
Sargassum siliquosum	Indonesia
Undaria pinnitifida	Australia, China, Japan, Korea
Undaria peterseniana	Korea

Source: Zemke-White W. and Ohno M. 1999, *J. Appl. Phyco.* 11, 369–376

Seaweed Proteins and Amino Acids

The protein content of seaweeds differs according to the species. Generally, protein fraction of brown seaweeds is low (3 to 15 per cent of dry wt) as compared with that of green or red seaweeds (10 to 47 per cent of dry wt) [Arasaki and Arasaki, 1983]. Except for the species *Undaria pinnatifida* (Wakame) that has a protein level between 11 to 24 per cent (dry wt), most brown seaweeds industrially exploited (*Laminaria digitata, Ascophylum nudosum, Fucus vesiculosus and Himanthalia elongata*) have a protein content lower than 15 per cent (dry wt). In some green seaweeds like species of *Ulva*, the protein content varies between 10 to 26 per cent (dry wt) of the plant. For instance, *U. pertusa*, which is frequently consumed under the name of '*Ao-nori*' by the Japanese people, has a high protein level between 20 to 26 per cent (dry wt.) [Fujiwara–Arasaki *et al.*, 1984]. Higher protein levels are recorded for the red seaweeds such as *Porphyra tenera* (47 per cent dry wt) or *Palmaria palmata* (35 per cent dry wt) [Morgan, *et al.*, 1980]. These seaweeds known under the names of '*Nori*' and '*Dulse*' respectively have protein levels higher than those found in protein rich pulses such as Soybean.

In case of most of the seaweeds, aspartic and glutamic acid together constitute a large part of the amino acid fraction. In the *Fucus* spp. these two amino acids are present between 22 to 44 per cent (Munda, 1977), while in *Ulva rigida* and *U. rotundata* they are in the range of 26 to 32 per cent (Fleurence, *et al.*, 1995) of the total amino acids. But these amino acids are present in less quantity of 14 to 19 per cent in edible red seaweeds *viz. Porphyra tenera* and *Palmaria palmata* (Fujiwara–Arasaki *et al.*, 1984, Indegaard and Minsaas, 1991). In *Palmaria palmata*, leucine, valine and methionine are well represented in the essential amino acid fraction. The average levels of these amino acids are close to those generally reported for ovalbumin. On the other hand, isolucine and threonine contents are similar to those recorded for leguminous protein. Except for histidine content, the essential amino acids profile of *Porphyra tenera* seems to be relatively close to those of leguminous plants. The amino acid composition of *Ulva pertusa* reveals the presence of the main essential amino acids *viz.* valine, leucine and lysine. Histidine, which is essential amino acid for children, is present at a similar level to leguminous and egg protein (cf Fleurence, 1999).

The first record of nitrogen content of the Indian seaweeds is reported by Chidambaram and Unny (1952). Subsequently Lewis (1962, 1963, 1964, 1966) has reported protein and amino acid composition from different Indian seaweeds. It has been observed that protein and amino acid composition have varied greatly in different seaweeds and the range of protein content is found to be between 15 to 30 per cent (dry wt) [Lewis, 1967]. Green seaweeds are found to be rich in protein content as compared with red and brown seaweeds. However *Grateloupia lithophila* has been found to record maximum protein.

Indian seaweeds are rich in amino acid composition and in the species of *Champia viz. C. compressa, C. compressa* var. *scindica, C. indica, C. parvula,* 25 different amino acids are recorded (Lewis, 1973). Among the amino acids estimated from Indian seaweeds leucine, isoleucine, lysine, methionine, phenylalanine, thereonine, valine and tryptophan are found to be comparable with those of common vegetables, nuts, seeds and cereals consumed as food except for leucine, isoleucine.

Mineral Constituents of Seaweeds

Seaweeds provide all necessary 56 minerals and trace elements. They are rich in macro and microelements, which are essential for human nutrition (Mabeau and Fleurence 1993). Seaweed mineral content, 8–40 per cent is higher than that of land plants and animal products (Ito and Hori 1989; Ortega-Calvo *et al.*, 1993). Seaweeds provide these essential nutrients to the human body in a chelated, colloidal optimally balanced form so that they are made bio-available. Calcium is necessary for skeletal health and nervous system functioning. Magnesium activates enzymatic action, healthy heartbeats. Potassium naturally prevents high blood pressure and provides cellular energy. Sodium is essential for the correct balance of body fluids. Iron is essential for hemoglobin transport and distributes oxygen to all the cells. Trace elements are especially essential to the enormous enzymatic functions constantly occurring in human body. The composition of major minerals and trace metals in six commonly consumed edible seaweeds is furnished in Table 5.4.

Table 4: Mineral composition of common edible seaweeds (mg/100g)

Minerals	*Fucus*	*Laminaria*	*Wakamae*	*Chondrous*	*Nori*
Na	5,469	3,818	7,064	4,270	3,627
K	4,322	11,579	8,699	3,184	3,500
Ca	938	1,005	931	420	390
Mg	994	659	1181	732	565
Fe	4.20	3.29	7.56	3.97	10.3
Zn	3.71	1.77	1.74	7.14	2.21
Mn	5.50	<0.5	0.87	1.32	2.72
Cu	< 0.5	< 0.5	<0.5	<0.5	<0.5
Total cations	11,736	17,066	17,885	8618	8097

Source: Rupérez, R., 2002, *Food Chemistry*. 79,23-26

Seaweeds as a Source of Dietary Fibers and Antioxidants

Apart from proteins, amino acids, carbohydrates and macro and microelements seaweeds are a good source of dietary fibers and antioxidants.

They contain different type of dietary fibers depending on the seaweed phyla. For brown seaweeds (Phaeophyta), the soluble fibres are alginates, fucans, and laminarans. However, for red seaweeds (Rhodophyta), the soluble fibers are sulphated galactans (agar and carageenans) or soluble xylans. Insoluble fibers are essentially made of cellulose, except for the red alga Nori, which contain insoluble mannan and xylan (Lahaye and Kaeffer 1997). The dietary fibers range from 33–50 per cent of dry wt of seaweed depending up on the species. Ruperez (2001) has reported total dietary fibers of 33.78 per cent (dry wt) in nori, 34.29 per cent (dry wt) in *Chondrus*, 33.58 per cent (dry wt) in Wakame, 36.12 per cent (dry wt) in *Laminaria* and 50.09 per cent (dry wt) in *Fucus*. In the soluble and insoluble dietary fibers following sugars have been detected *viz.* Rhamnose, Fucose, Arabinose, 3,6 anhydro-Gal, Xylose, Mannose, 6-O-Me-Gal, Galactose, Glucose.

Oxidative modification of DNA, proteins, lipids and small cellular molecules by reactive oxygen species (ROS) plays a role in wide range of common diseases and age-related degenerative conditions. Antioxidants play a very crucial role in treating such disorders. Though there is less research in this area some seaweeds are reported to have antioxidant properties *viz. Scytosiphon lomentaria* (kayamo-nori), *Papenfussiella kuromo* (kuromo), *Nemacystus decipiens* (mozuku) and *Porphyra* spp. (nori) [Kuda *et al.*, 2005] and *Kappaphycus alvarezii* (eucheuma) [Mohamed Fayaz *et al.*, 2005].

Acknowledgements

Authors would like to thank Dr. Pushpito Kumar Ghosh, Director for encouragement and to allow this article to be published. Thanks are also due to Prof. B. Jha, Head Marine Algae and Marine Environment Discipline for facilities.

References

Arasaki A. and Arasaki T.(1983). Low calories, High Nutrition. Vegetables from the Sea to help you Look and Feel Better, pp.39-42, Japan publications Inc.

Burtin P. (2003) Nutritional value of seaweeds. *Elect J Env Agri Food Che.*, 2, 498–503.

Chidambaram K and Unny M. J. (1952) First Intl. Seaw. Symp. Edinburgh, 67–68.

Druehl L. (2000) *Pacific Seaweeds.* Harbor Publishing, Madeira Park, B.C., Canada.

Fleurence J., C. Le Coeur, S. Mabeau, M. Maurice & A. Landrein (1995) Comparison of different extractive procedures for proteins from the edible seaweeds *Ulva rigida* and *Ulva rotundata. J. Appl. Phycol.*, 7(2): 577–528.

Fujiwara-Arasaki T., Mino N and Kuroda M (1984). The protein value in human nutrition of edible marine algae in Japan. *Hydrobiol*, 116/117, 513-516.

Funahashi H., Imai T., Tanaka Y., Tsukamura K., Hayakawa Y., Kikumori T., Mase T., Itoh T., Nishikawa M., Hayashi H., Shibata A., Hibi Y., Takahashi M. and Narita, T. (1999) Wakame seaweed suppresses the proliferation of 7,12-dimethylbenz(a)-anthracene-induced mammary tumors in rats. *Jpn J Cancer Res.*, 90, 922-7.

Hiqashi Okaj K., Otani S. and Okai Y (1999) Potent suppressive effect of a Japanese edible seaweed, *Enteromorpha prolifera* (Sujiao-nori) on initiation and promotion phases of chemically induced mouse skin tumorigenesis. *Cancer Lett.*, 140, 21-5.

Indegaard M. and Minsaas J. (1991). In Animal and Human Nutrition. Seaweed Resources in Europe. Uses and potential (Gury, M. D. and Blunden, G. eds.) pp.21-64., John wiley and Sons.

Ito K. and Hori K. (1989). Seaweed: chemical composition and potential uses. *Food review International*, 5, 101-144.

Kuda T., Tsunekawa M., Goto H. and Araki Y. (2005) Antioxidant properties of four edible algae harvested in the Noto Peninsula, Japan. *J. Food Comp. Anal.*, 18, 625–633.

Lahaya M. and Kaeffer B. (1997) *Sci. Aliments*. 17, 619–639.

Lewis E. J. (1962) Studies on the proteins, peptides and free amino acid contents in some species of brown algae from Southeastern coast of India. Revue. Algologique. Sr. 2, 6: 209–216.

Lewis E. J. (1963) Studies on the proteins, peptides and free amino acid contents in some species of red algae from Southeastern coast of India. Proceedings of the National Institute of Science of India. Part B–Biological Science. 29: 137–145.

Lewis E. J. (1964) Proteins, peptides and free amino acid contents of some species of *Gracilaria* from Southeastern coast of India. *Fish. Tech.*, 1, 158–163.

Lewis E. J. (1966) Protein concentrate from *Ulva fasciata*. *Salt Res. Indu.*, 3, 19.

Lewis E. J. (1967) A review on protein, peptides and free amino acid contents of Indian marine algae. In V. Krishnamurthy (Ed.), Proceedings of the seminar on Sea, Salt and Plants. Bhavnagar, India., 296–308.

Lewis E. J. (1973) The protein, peptides and free amino acid composition in species of Champia from Saurashtra coast, India, *Bot. Mar.*, 15, 145–147.

Mabeau S. and Fleurence J. (1993). Seaweed in food products: Biochemical and nutrional aspects. *Trends in Food Science and Technology*, 4, 103-107.

McHugh D. J. (2003) FAO Fisheries Technical Paper. 441, 105 pp.

Mohamed Fayaz, Namita K. K., Chidambra Murthy K. N., Mahadevasamy, M., Sarada, R., Khanam. S., Subba Rao P. V. and Ravisankar G. A. (2005) Chemical composition Iron bioavailablity and antioxidant activity of *Kappaphycus alvarezii* (Doty). *J. Agri. Food. Chem.*, 53, 792-797.

Morgan K.C., Wright J. L. C. and Simpson F. J. (1980). Review of the chemical constituents of the red alga *Palmaria palmata* (Dulse). *Econ. Bot.* 34, 27-50.

Munda I. M. (1977) Difference in amino acid composition of estuarine and marine fucoids. *Aqua Bot.*, 3, 273–280.

Ortega-Calvo, J. J., Mazuelos C., Hermosín, B, and Sáiz–Jiménez, C. (1993). Chemical composition of Spirulina and Eukaryotic algae food products marketed in Spain. *Journal of Applied Phycology*, 5, 425-435.

Rosegrant M. W. and Cline S. A. (2004) Global Food Security: Challenges and Policies. *Science*, 302, 1917–1919.

Rupérez, R. (2002). Mineral content of edible marine seaweeds. *Food Chemistry*. 79,23-26.

Smith G. M. (1944) Marine Algae of the Monterey Peninsula, California. Stanford Univ., 2nd Edition.

Teas J. (1981) The consumption of seaweed as a protective factor in the etiology of breast cancer. *Med. Hypotheses*, 7, 601-13.

Tseng C. K. (2004) Past, present and future of phycology in China. *Hydrobiol.* 512, 11–20.

www.seaveg.com/chart.html as accessed on 15th February 2006.

Zemke-White W. and Ohno M. (1999) World seaweed utilization: An end–of–century summary. *J. Appl. Phyco.* 11, 369–376.

©R.K. Gupta & V.D. Pandey (*Editors*)
Advances in Applied Phycology, 2007
Daya Publishing House, New Delhi

97–103

Chapter 6
Potential Biotechnological Applications of Cyanobacteria

G. Abraham*

Centre for Conservation and Utilization of Blue Green Algae (CCUBGA), Indian Agricultural Research Institute, New Delhi – 110 012, India

ABSTRACT

Cyanobacteria are the scientific name for "blue-green algae" or "pond scum". The first recognized species were blue-green in color and hence the name. They are a group of gram negative photosynthetic prokaryotes which possess oxygen evolving photosynthetic apparatus similar to higher plant chloroplasts. A variety of them are able to fix atmospheric nitrogen. They have been living on earth for over 2.5 billion years since Pre-Cambrian era. They occupy a variety of ecological habitats ranging from freshwater to saline and psychrophilic, thermophilic, acidophilic and alkalophilic forms have been reported. In the recent times cyanobacteria received a greater attention owing to their importance as food, feed, colorant, vitamins, toxins, enzymes, pharmaceuticals etc. The focus of the article is mainly on the potential biotechnological applications of cyanobacteria.

Keywords: *Biotechnology, Biofertilizer, Cyanobacteria, Neutraceuticals, Pharmaceuticals.*

Introduction

The cyanobacteria or blue-green algae are a large and morphologically diverse group of prokaryotes, which are found to occupy almost every possible ecological habitat on earth. They possess a photosynthetic apparatus similar to that of higher plant chloroplasts. During the course of evolution, these organisms were the first to derive electrons for CO_2 fixation from water, evolving O_2 as a by-

* Corresponding Author: Tel: 011-2584831; E-mail: gabraham1@rediffmail.com.

product and were instrumental in changing the prevalent reducing conditions to an oxidizing one (Haselkorn, 1978; Wolk, 1996). In a way, the cyanobacteria are credited with providing an oxidizing environment for the development of many other microbes and higher eukaryotic life forms with complex metabolic pathways. Many of these forms are able to convert molecular nitrogen to ammonia with the help of the nitrogen fixing enzyme nitrogenase which is present in a highly specialized cell known as heterocyst. Some cyanobacteria, although do not form heterocysts fix atmospheric nitrogen efficiently (Stal, 2000). A survey of rice fields has shown that the population of non-heterocystous cyanobacteria is much higher that of the heterocystous forms (Desikachary, 1959; Anand, 1989). According to an estimate, cyanobacteria constitute about 15 per cent of the total algal flora in the tropics about 2 per cent in the temperate regions (Thajuddinn and Subramanian, 2005).

Some cyanobacteria are of nuisance value as they grow luxuriantly and pollute a variety of water bodies by forming water–blooms. *Microcystis*, an abundantly occurring nuisance cyanobacterium in many eutrophic ponds and reservoirs of India and other tropical countries, is responsible for unpleasant odor of water bodies (Rai *et al.*, 1998). The cyanobacteria are important primary producers in many ecosystems including unfriendly extreme environments. Some of the cyanobacterial strains are extremophiles and are able to grow in extreme habitats with low and high pH, brines, low and high temperature, metal polluted soils and in environments where no other microalgae can exist (Broady *et al.*, 1996; Grant and Tindall, 1986; Gaur and Rai, 2001; Thajuddin *et al.*, 2002).

Cyanobacteria also form symbiotic associations with plants. Compatible cyanobacteria are recognized by respective host plants which influence their behavior. Symbiotic relationship exists with, for example bryophytes, fungi, pteridophytes, gymnosperms and angiosperms (Rai, 1990). In addition to this, they are potential source of food and medicine which make them potential organisms useful to mankind.

Commercial Applications

Because of the potential economic implications much interest is currently being devoted to bioprospecting of cyanobacteria. Cyanobacteria offer immense potential as biofertilizer, food, feed, medicine etc. Since cyanobacteria contain a variety of fine chemicals such as pigments, vitamins and enzymes and have been widely used in scale-up technologies for the production of such chemicals. Centre for conservation and utilization of blue green algae (CCUBGA) at Indian Agriculture Research Institute, New Delhi maintain collection of several fresh water cyanobacterial forms and is a prestigious National facility. Similarly a National facility for marine cyanobacteria (NFMC) was established at the Department of Microbiology, Bharatidasan University, Tiruchirapalli, Tamilnadu. Many private companies have now established Research and Development wing focusing on the commercial exploitation of cyanobacteria. For example M/S ABL Biotechnologies, Chennai has a very active R&D set up for the commercial exploitation of cyanobacteria.

Biofertilizer

The cyanobacteria provide valuable nutrients to plants in terms of fixed nitrogen, amino acids and plant growth promoting substances which improve the soil health and texture. The role of cyanobacteria in maintaining the fertility of Indian rice fields was studied by De (1939) and Singh (1961). Algalization of paddy fields is beneficial and cyanobacterial contribution to the total nitrogen in paddy fields may vary. However, it was estimated that cyanobacteria contribute 25-30 Kg nitrogen per hectare per season to the rice crop. Significant increase in yield with the application of cyanobacteria in paddy fields of Uttar Pradesh was observed (Dwivedi *et al.*, 2000). Algalization also improves phosphate solubilization (Singh *et al.*, 1981; Bisoy and Singh 1988). The application of cyanobacteria was shown to improve yield due to the production of growth promoting substances (Marsalek *et al.*, 1992). Extra cellular growth promoting substances containing amino acids have been reported in certain strains of cyanobacteria such as *Nostoc muscorum* and *Haplosiphon fontinalis* (Wikstrom *et al.*, 1997). Because of these properties strain improvement was attempted in cyanobacteria with a view to augment their nitrogen fixing potential (Tiwari *et al.*, 1991; 2001).

Food and Nutraceuticals

Some strains of cyanobacteria such as *Nostoc* and *Anabaena* have already been reported to be good for human consumption. *Spirulina* and *Nostoc* have been used as a source of protein and vitamin for humans and animals (Ciferri, 1983; Kay, 1991; Gao, 1998; Takanaka *et al.*, 1998). The protein content of *Spirulina* is unusually high (up to 70 per cent of the dry weight) and is used as a dietary supplement. It is also a rich source of beta–carotene, thiamine and riboflavin and is commercially available in the market. M/s. Dabur India Private Limited is marketing *Spirulina* tablets under the trade name "Sunova". In Philippines and China *Nostoc flagelliforme* and *Nostoc commune* are used as delicacy (Martinez, 1988; Gao, 1998; Takanaka *et al.*, 1998). The planktonic cyanobacterium *Spirulina platensis* is gaining increasing attention due to its nutritional and medicinal properties. Protein extracts of *S. platensis* fractions showed antioxidant activities and inhibited microsomal lipid peroxidation (Pinero *et al.*, 2001). Cyanobacteria contain significant quantities of lipids with a composition similar to vegetable oils. Many of these cyanobacterial fatty acids are essential components of human and animal diet as well as important feed additives in aquaculture (Borowitzka, 1988). The lipids of some cyanobacteria are rich in essential fatty acids (Singh *et al.*, 2002). Carotenoids are widely used as natural colorant for food, drug and cosmetic products. Shukla and Kashyap (2003) assessed the biopotential of few Antarctic cyanobacterial strains with respect to carotenoid production and suggested that these strains have a potential for commercial production of carotenoids. In India Murugappa Chettiar Research Centre, Chennai and Thapar Institute of Science and Technology, Patiala are involved in the production of *Spirulina* biomass on a large scale using open pond system of cultivation.

Pharmaceuticals and Fine Chemicals

The medicinal value of cyanobacteria was appreciated as early as 1500 BC, when strains of *Nostoc* were used to treat gout, fistula and several forms of cancer (Pietra, 1990). People in Aztec civilization used to consume *Spirulina* pills four hundred years ago. In traditional as well as in modern medicine there are few appreciated algal preparations and specified bioactive compounds. However, with the availability of powerful biotechnological tools and techniques cyanobacterial biotechnology reached newer heights in terms of process and product development. Phycobiliproteins and carotenoids are effective natural blue colorants used commercially (Emodi, 1978). *Spirulina platensis* is a blue green microalga which can produce large quantities of high value products such as phycocyanin (Chen *et al.*, 1996). Biologically active compounds from *Lyngbya majuscula* showing molluscicidal activity were isolated and characterized (Orjala *et al.*, 1995; Orjala and Gerwick 1996). A study conducted by Moore (1996) revealed novel biologically active compounds from cyanobacteria which could be exploited as potentially rich sources of natural products. Novel extra cellular diterpenoids with biological activity from *Nostoc commune* have been isolated (Jaki *et al.*, 2000). Over 40 different Nostacales species, mainly *Anabaena* and *Nostoc* sp. having activities such as anti-HIV, anticancer, antifungal, antimalarial and antimicrobial have been reported (Burja *et al.*, 2001). Among the several strains with bioactivity *Nostoc ellipsoporum* was found to show potential activity against several immunodeficiency viruses such as HIV-1, M–and T-tropic strains of HIV-1, HIV-2, SIV and FIV due to a 101 amino acid protein Cyanovirin (CV-N, cyanovirin-N). Many peptide metabolites with bioactivity have been reported. Examples are cyanopeptolin, microviridin or micropeptin produced by bloom forming freshwater cyanobacteria (Neilan *et al.*, 1999; Tillett *et al.*, 2000; Kurmayer *et al.*, 2004). Marine cyanobacterial forms are richest sources of novel bioactive compounds including toxins with pharmaceutical applications (Gustafson *et al.*, 1989; Raghavan *et al.*, 2002).

An important enzyme in the molecular biology, the restriction endonucleases have been isolated from cyanobacteria (Murray *et al.*, 1976). Isotopically labeled cyanobacterial metabolites are commercially available (Patterson, 1996). More than 200 cyanobacterial restriction endonucleases have been reported (Roberts and Macelis, 2001). Saravanan *et al.* (2003) identified and characterized a new new type II restriction endonuclease from a nonheterocystous cyanobacterium *Oscillatoria foreaui*. These results have prompted scientists to exploit cost effective cultivation of cyanobacterial strains in mass culture conditions for the production of such compounds. In addition, cyanobacteria are a rich source of several polyols, polysaccharides, lipids, fatty acids, halogenated compounds and have properties to be employed as flocculants, surfactants and others (Becker, 1994).

Cyanobacteria have also been used in the control of certain bacterial diseases on plants. Recently Pandey and Pandey (2002) reported antibacterial property of three strains of cyanobacteria such as *Lyngbya majuscula*, *Microcystis aeruginosa* and *Plectonema boryanum* and it's efficacy in controlling leaf spot disease of chilli due to *Xanthomonas vesicatoria*. This has opened new vistas for eco-friendly control of diseases using a broad spectrum of biologicals isolated from cyanobacteria against the conventional pesticides which are otherwise hazardous and unsafe to plants as well as animals.

Most of the cyanobacteria contain a wide range of products which are important industrially and pharmacologically. However, a full documentation of such potential of cyanobacteria is yet to be done. With the recent advancement in biotechnology and renewed interest in natural product chemistry of cyanobacteria it is expected that many more products will be identified and synthesized from them. Bioprospecting of cyanobacteria is a fast emerging discipline with potential economic implications. These organisms are undoubtedly important in the Indian context because of the economic wealth they are expected to generate in coming days in the biotechnological sphere.

Acknowledgements

The author is grateful to Council of Scientific and Industrial Research, New Delhi for financial support.

References

Anand N (1989) *Hand book of Blue Green Algae*. Bishen Singh Mahendra Pal Singh, Dehradun, 1-74.

Becker EW (1994) *Microalgae: Biotechnology and Microbiology*, Cambridge University Press, United Kingdom.

Bisoy RN and Singh PK (1988) Effect of phosphorous fertilization in blue-green algal inoculum production and nitrogen yield under field conditions. *Biol Fert Soils* 5: 338-343.

Borowitzka MA (1988) Fats, oils and hydrocarbons. In: Borowitzka MA Borowitzka, LJ (eds), *Microalgal Biotechnology*. Cambridge University, Press, Cambridge, 257-287.

Broady PA, Garrick R and Anderson GM (1996) Diversity, distribution and dispersal of Antarctic terrestrial algae. *Biodiversity Conservation* 5: 1307-1335.

Burja AM, Banaigs B, Abou-Mansour E and Burgess JG (2001) Marine cyanobacteria–a prolific source of natural products. *Tetrahedron* 57: 9347-9377.

Chen F, Zhang Y and Guo S. (1996) Growth and phycocyanin formation of *Spirulina platensis* in photoheterotrophic culture. *Biotechnol Lett* 18: 603-608.

Ciferri O (1983) *Spirulina*, the edible microorganism. *Microbiol Rev.* 47: 551-578.

Desikacachary TV (1959). *Cyanophyta*. Indian Council of Agricultural Research, New Delhi.

De PK (1939) The role of blue green algae in nitrogen fixation in rice fields. *Proc Roy Soc* 127: 121-139.

Dwivedi CP, Sinha SK, Ashraf SS, Verma DC and Dubey AK (2000) Application of blue green algae (BGA) fertilizer in paddy cultivation under farm conditions in Uttar Pradesh. In: *Biofertilizers, Blue green algae and Azolla*, Singh PK et al. (eds), Venus Printers and Publishers, New Delhi, 133-141.

Emodi A (1978) Carotenoids: Properties and Applications. *Food Technol* 32: 38-42.

Gao K (1998) Chinese studies on the edible blue-green algae *Nostoc flagelliforme*: a review. *J Appl Phycol* 10: 37-39.

Gaur JP and Rai LC (2001) In: *Algal adaptation to environmental stresses*, Rai LC and Gaur JP (eds), Springer, Berlin, 363-388.

Grant WD and Tindall WJ (1986) In: *Microbes in extreme environments*, Herbert TA and Codd GA (eds), Academic Press, London, 25-54

Gustafson KR, Cardellina JH, Fuller RW, Weislon OS, Kiser RF and Snader KM. (1989) Antiviral sulfolipids from cyaqnobacteria (blue-green algae). *J Nat Cancer Inst* 81: 1254.

Haselkorn R (1978) Heterocysts. *Ann Rev Plant Physiol* 29: 319-344.

Jaki B, Orjala J, Heilmann H, Linden A, Vogler B and Sticher O (2000) Extracellular diterpenoids with biological activity from the cyanobacterium *Nostoc commune*. *J Nat Prod* 63: 339-343.

Kay RA (1991) Microalgae as food and supplement. *Crit Rev Food Sci Nutr* 30: 555-573.

Kurmayer R, Christiansen G, Fastner J and Borner T (2004) Abundance of active and inactive microcystin genes in populations of the toxic cyanobacterium *Plankthrix* sp. *Environ Microbiol* 6: 831-841.

Marsalek B, Zahradnickova H and Hronkova M (1992). Extracellular abscisic acid production by cyanobacteria under salt stress. *J Plant Physiol* 139: 506-508.

Martinez MR. (1988) *Nostoc commune* Vauch. A nitrogen fixing blue-green algae, as source of food in the Philippines. *Philippine Naturalist*. 71: 295-307.

Moore RE (1996) Cyclic peptides and depsipeptides from cyanobacteria: a review. *Ind J Microbiol* 16: 134-143.

Murray K, Gughes SG, Brown JS and Bruce SA (1976). Isolation and characterization of two sequence specific endonucleases from *Anabaena variabilis*. *Biochem J* 138: 123-127.

Neilan BA, Dittman E, Rouhianinen L, Bass RA, Sivonen K and Borner T (1999) Non ribosomal peptide synthesis and the toxigenecity of cyanobacterium. *J Bacteriol* 181: 4089-4097.

Orjala J, Nagle D and Gerwick WH (1995) Antillatoxin, an exceptionally ichthyotoxic cyclic lipopeptide from the marine cyanobacterium *Lyngbya majuscula*. *J Am Chem Soc* 117: 8281-8282.

Orjala J and Gerwick WH (1996) Barbamide, a chlorinated metabolite with molluscoidal activity from the carribean cyanobacterium *Lyngbya majuscula*. *J Nat Prod* 59: 427-430.

Pandey U and Pandey J (2002) Antibacterial properties of cyanobacteria: A cost effective and eco-friendly approach to control bacteria leaf spot disease of chilli. *Curr Sci* 82 (3): 262-264.

Patterson GML (1996) Biotechnological applications of cyanobacteria. *J Sci Ind Res* 55: 669-684.

Pietra F (1990) A Secret World: Natural Products of Marine Life. Ist ed. Birkhauser, Basel, Switzerland.

Pinero Estrada JE, Bermejo Bescos P and Villar del Fresner AM (2001) Antioxidant activity of different fractions of *Spirulina platensis* protean extract. *Farmaco* 56(5-7): 497-500.

Raghavan C, Kadlamani B, Thirunalasundari T, Subramahnian G and ASkbarsha M A (2002) A study of the male antifertility properties of the marine cyanobacterium *Oscillatoria willei* BDU 135011–a preliminary report. In: *Proceedings of the XX symposium*. Bharathidasan University, Thiruchirapalli, 137-138.

Rai AN (1990) Cyanobacteria in symbiosis. In: AN Rai (ed), *Hand book of Symbiotic cyanobacteria* CRC Press, Bocca Raton.

Rai LC, Singh S and Pradhan S (1998) Biotechnoloogical potential of naturally occurring and laboratory grown *Microcystis* in biosorption of Ni^{2+} and Cd^{2+}. *Curr Sci* 74 (5): 461-464.

Roberts RJ and Macelis D (2001) REBASE-restriction enzymes and methylases. *Nucliec Acids Research*. 29: 268-269.

Shukla SP and Kashyap AK (2003) An assessment of biopotential of three cyanobacterial isolates from Antarctic for carotenoid production. *Ind J Biochem Biophys* 40: 362-366.

Saravanan M, Elango K, Chandrashekaran S, Anand N and Nagaraja V (2003) A new type II restriction endonuclease, *Ofo* I from nonheterocystous cyanobacterium *Oscillatoria foreaui*. *Curr Sci* 85(2): 188-191.

Singh RN (1961) *Role of blue green algae in nitrogen economy of Indian Agriculture*. Indian Council of Agricultural Research, New Delhi, 175.

Singh PK, Panigrahi BC and Satpathy KB (1981). Comparative efficiency of *Azolla*, blue-green algae and other organic manures in relation to N and P availability in flooded rice soil. *Plant and Soil* 62: 35-44.

Singh SC, Sinha RP and Hader DP (2002) Role of lipids and fatty acids in stress tolerance in cyanobacteria. *Acta Protozool* 41: 297-308.

Stal LJ (2000) Cyanobacterial mats and stromatolites. In: Whitton BA and Potts M (eds), *The ecology of cyanobacteria*. Kluwer Academic Publishers, Netherlands, 61-120.

Takanaka H, Yamaguchi Y, Sakaki S, Watarai K, Tanaka N, Hori M, Seki H, Tsuchida M, Yamada A, Nishimori T and Mrinaga T (1998). Safety evaluation of *Nostoc flagelliforme* (Nostocales, Cyanophyceae) as a potential food. *Food Chem Toxicol* 36: 1073-1077.

Thajuddin N and Subramanian N (2005). Cyanobacterial biodiversity and potential applications in biotechnology. *Curr Sci*. 89(1): 47-57.

Thajuddin N, Nagasthya A, Chelladevi R and Saravanan P (2002) Biodiversity of cyanobacteria in different saltpans of Pudukottai District, Tamilnadu. *Sea weed Res Utiln* 24: 1-11.

Tillet D, Dittman E, Erhard M, Von Dohren H, Borner T and Neilan BA (2000) Structural organization of microcystin biosynthesis in *Microcystis aeruginosa* PCC 7806: An integrated peptide-polypeptide synthetase system. *Chem Biol* 7: 753-764.

Tiwari DN, Kumar A and Mishra AK (1991) Use of cyanobacterial diazotrophic technology in rice agriculture. *Appl Biochem Biotech* 28: 387-396.

Tiwari DN, Mishra AK and Singh AP (2001) Transposan mutagenesis and other molecular tools for strain improvement in cyanobacteria. In: *Recent advances in the exploitation of blue green algae*, PK Singh *et al.* (eds), Venus Printers and Publishers, New Delhi, 85-94.

Wikstrom P, Szwajeer E, Brodelius P, Nilsson KN and Mosbach K (1997) Formation of alphaketo acids from amino acids using immobilized bacteria and algae. *Biotechnol Lett* 4: 153.

Wolk CP (1996) Heterocyst formation. *Ann Rev Gen* 30: 59-78.

© R.K. Gupta & V.D. Pandey (*Editors*)
Advances in Applied Phycology, 2007
Daya Publishing House, New Delhi

Chapter 7

Chlorophyll Fluorescence Analysis: A Potential Tool for Rapid Measurement of Photosynthesis

Jagat Narayan Maurya*

Department of Plant Sciences, Faculty of Applied Sciences, M.J.P. Rohilkhand University, Bareilly – 243 006, India

ABSTRACT

Photosynthesis is the process by which living organisms exploit the inexhaustible solar energy and converts it into the chemical energy of organic molecules. Chlorophyll fluorescence analysis allows instantaneous measurements of the key aspects of photosynthetic light capture and its subsequent transformation to chemical forms through electron transports. Fluorescence analysis can be used to investigate the functioning of PS II. Fluorescence signals can generate information about functional state of electron carriers especially QA, pigment concentration under different growth conditions, O_2 - evolution and CO_2 - fixation under various environmental stresses.

Keywords: Chlorophyll fluorescence, Modulated fluorometer, Photosynthesis, Stress.

Abbreviations

Chl *a* –Chlorophyll *a*, PS I–Photosystem I, PS II–Photosystem II, DCMU-3 (3,4-dichlorophenyl)-1,1-dimethylurea,

Introduction

Photosynthesis is vital to the maintenance of life on earth. It is the process by which living organisms exploits the inexhaustible solar energy and convert it into the chemical energy of organic molecules for their complex physico–chemical reactions.

*Tel: 9897108929

Key to the photosynthesis is capture of the energy of light. The pigment–protein beds on the thylakoid membranes serves as the site of energy harvest and its transformation to chemical form. The thylakoid membranes possesses two distinct pigment–protein beds namely, photosystem I (PS I) and photosystem II (PS II). Two specific chlorophyll molecules, designated as P_{700} and P_{680}, each capable of undergoing light—induced charge separation and subsequent electron transfer, constitute the reaction center of PS I and PS II, respectively. Both the photosystems (I and II) act in series. After excitation by the photon, an electron leaves P_{700} of PS I and via several intermediate electron carriers finally reach to the terminal electron acceptor $NADP^+$. Similarly, P680 of PS II when excited by photon passes an electron via mobile plastoquinone (PQ), cytochrome b-f complex and plastocyanine (PC) to P_{700} of PS I. This fills the existing electron hole on P_{700}. The electron hole left on P_{680} in turn is filled by electrons coming from photolysis of water. Figure 7.1 shows schematic representation of thylakoid membrane organization with respect to electron transport and ATP synthesis.

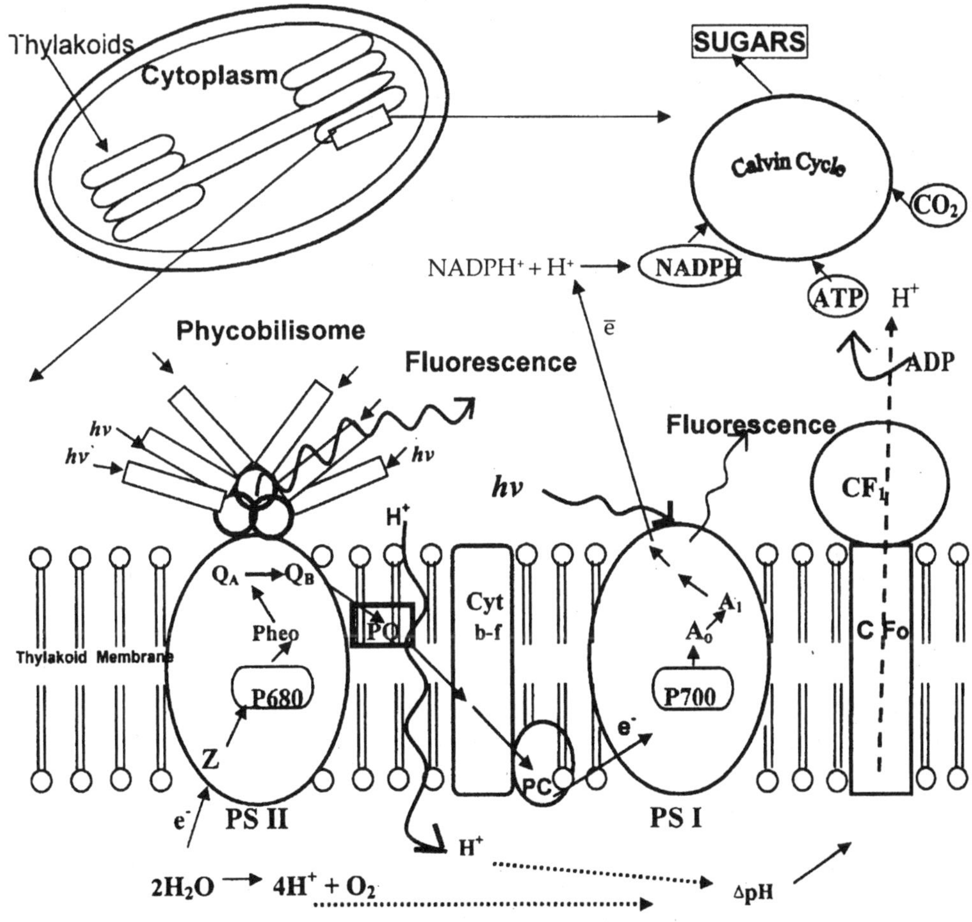

Figure 7.1: Schematic Representation of Photosynthetic Apparatus Showing the Components Involved in Conversion of Light Energy to Redox Energy (NADPH and ATP) and their Utilization in Fixation of CO_2 (Calvin cycle) in a Cyanobacterium. In higher plants the components related with electron transport and NADPH and ATP synthesis are also organized in same basic plan. However in higher plants distinct light harvesting complexes namely LHC –I and LHC–II are found in association with PS–I and PS–II respectively.

Chlorophyll fluorescence analysis allows nearinstantaneous measurements of the key aspects of the photosynthetic light capture and electron transports. Fluorescence signals can be used for extracting physiological and ecological useful information about plants in both field and laboratory settings.

Chlorophyll Fluorescence and Principle of Measurement

Chlorophyll fluorescence is emission of radiation of a wavelength longer than that of the initially absorbed by the chlorophyll molecules. When a pigment absorbs the energy of a photon and enters an excited electronic state, there are essentially three routes for the return to the ground state:

1. *Photochemistry*–in which the excited electron leaves the pigment molecules and enters an electron transport chain;
2. *Non-radiative decay* (heat dissipation)–in which the excited electron return to ground state by releasing heat (conversion to kinetic energy); and
3. *Fluorescence*–emission of a radiation of a wavelength longer than that initially absorbed.

These three processes are in direct competition for every exciton generated. Decrease in dissipation of excitation energy in one route results an increase in dissipating energy of other route (Figure 7.2).

In plants and cyanobacteria the most chlorophyll fluorescence is emitted by Chl *a* of the photosystem II. Although PS II fluorescence is a minor pathway for excitation dissipation, it competes with the quantitatively more important energy dissipation routes such as of PS II photochemistry, exciton transfer to other pigment systems (such as PS I), and heat dissipation. Therefore, changes in photochemistry or in the two nonphotochemical routes (energy transfer and heat) cause changes in the fluorescence yield from PS II. When the potentials for photochemistry and nonphotochemical dissipation are minimal, the fluorescence yield is maximal. Quenching or lowering of the fluorescence yield below its maximum occurs when excitation flow increases to the competing photochemical or nonphotochemical pathways. Further the level of fluorescence emission depends on the pigments concentration and the excitation light intensity. For comparision and quenching analysis, the excitation intensity and pigment concentration must be constant, so that changes in fluorescence reflect the

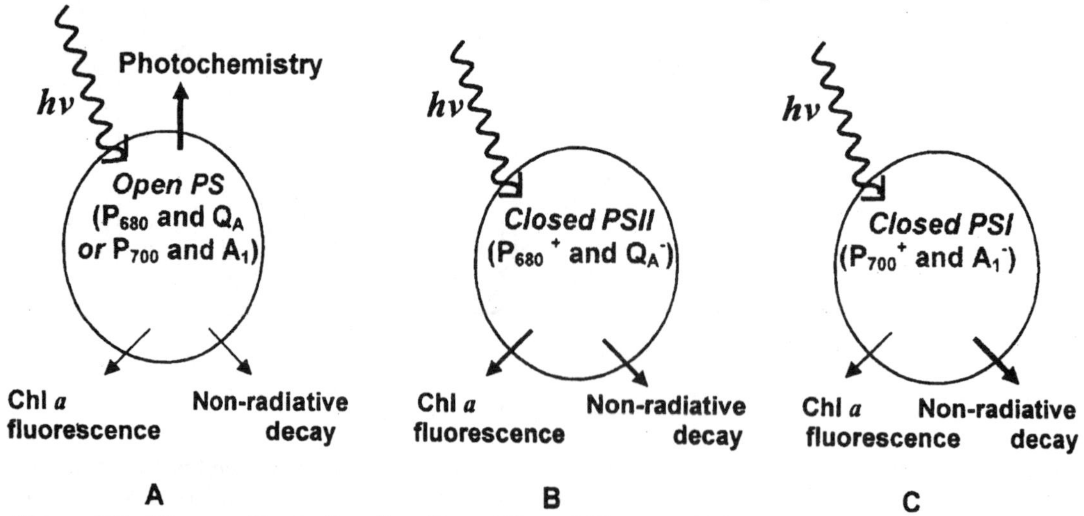

Figure 7.2: Diagram Illustrating the Absorption of Photons (*hv*) and Dissipation of Excitons Subsequently Generated within the Pigment Matrix of Photosystem Complex (light–harvesting system plus reaction centre). The thickness of arrow reflects the relative yield of the corresponding dissipative process, in a non-proportional manner.

changes in fluorescence yield which result from the competing photochemical and nonphotochemical de-excitation pathways.

Measurement of Chlorophyll Fluorescence

Various Chl *a* fluorescence imaging systems have been developed by a number of research groups (Fenton and Crofts 1990; Genty and Meyer 1995; Scholes and Rolfe 1996; Osmond *et al.*, 1999; Nedbal *et al.*, 2000). In addition, commercial Chl *a* fluorescence imaging systems have been developed by PSI (Brno, Czech Republic), Walz Systems (Effeltrich, Germany), Technologica Ltd. (Colchester, UK) and Hansatech (Great Britain,UK). These imagimg systems called Modulated fluorometers specifically detect and amplify only the fluorescence excited by a weak, constant measuring beam consisting of a train of low light pulses at a frequency of 1 to 100 kHz. Therefore, the excitation intensity is constant and changes in the fluorometer signal reflect changes in fluorescence yield. The modulated measuring beams are too weak to drives photosynthetic processes and so allow the determination of the fluorescence yield of dark–adapted samples. Furthermore, the detection system in modulated fluorometers ignores fluorescence excited by other light.

A typical modulated fluorometer as shown in Figure.7.3 has a cuvette with apertures for lights and fluorescence detectors. Detectors are connected to a recorder which record fluorescence emitted from the samples under study. The recorder can be connected to computer to get amplified fluorescence signals. Imaging system has three different sources of illumination: (*i*) to excite Chl *a* fluorescence

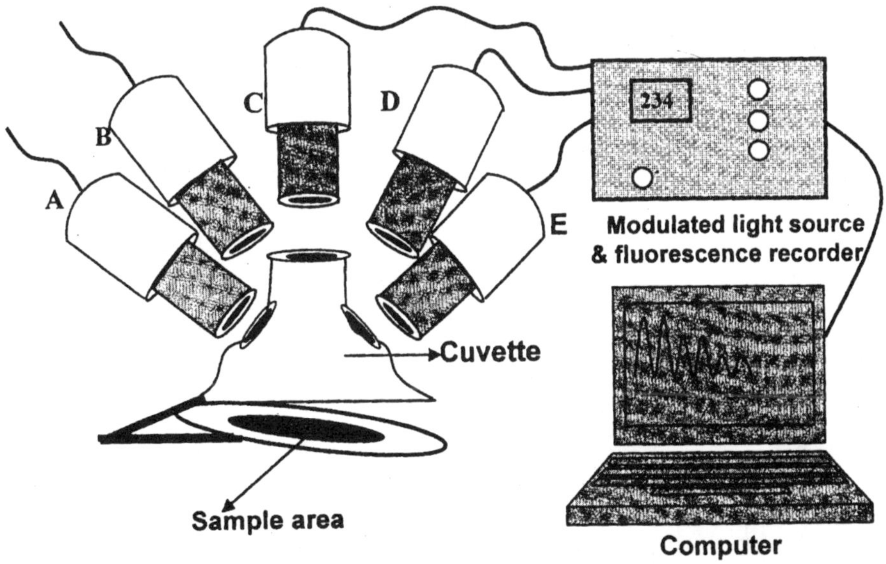

Figure 7.3: Schematic Representation of a Modulated Fluorometer Showing Arrangement of Actinic, (A) Saturating Light Pulse (B) and Modulated Light (C) Source and Continuous (D) and Modulated (E) Fluorescence Detectors. Actinic light was provided from a source positioned at 90° to the sample. The modulated light source and the continuous and modulated fluorescence detectors are placed at an angle of 45o to the sample surface in the position shown such that they should not shade the sample from actinic light. Further it is essential to ensure that the modulated and continuous fluorescence detectors should receive emission from the same area of the sample, which must completely and evenly irradiated with the modulated light source. Detectors are connected to a recorder. Recorder can be connected to computer which generates amplified signals.

(called modulated light), (*ii*) to provide constant actinic illumination (usually growth light), and (*iii*) to provide the multiple turnover of saturating light pulses. A fluorometer–compatible system of cuvette, magnetic stirrer and oxygen electrode can be used for the simultaneous measurement of fluorescence and oxygen evolution in cultures. Figure 7.4 represent procedure of measurement as described by Campbell *et al.*, 1998. The samples are dark adapted for 15 to 30 minutes prior to the measurement. The minimal fluorescence level (Fo) with all PS II reaction centres open is measured by measuring modulated light which is sufficiently low (0.1–1.0 $m^{-2} s^{-1}$) and unable to induce any significant variable fluorescence (essentially no photosynthesis). The maximal fluorescence level (Fm_{dark}) with PS II reaction

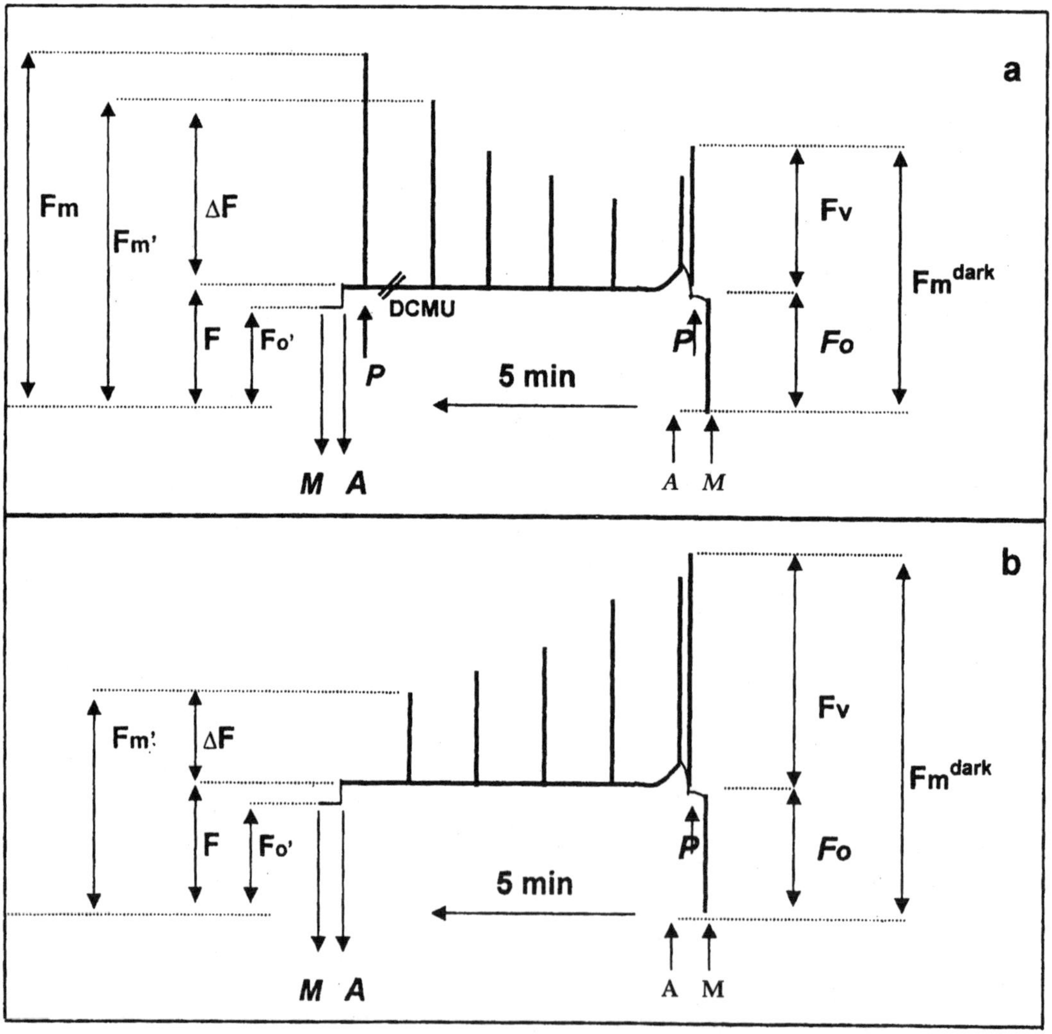

Figure 7.4: Typical Kinetics of Chlorophyll Fluorescence in a Cyanobacterium *Scytonema geitleri* (a) and in a Higher Plant *Mangifera indica* (b). Upward–pointing arrows on the abscissa shows: 'switching on' of the modulated light (M), saturating light pulse (P) and actinic light (A) followed by saturating light pulse at regular intervals. Downward–pointing arrows on abscissa shows: 'switching off' of actinic light (A), modulated light (M).

centres closed is determined by applying 1 second saturating light pulse of photon flux density (PFD) 8000–10,000 μmole m^{-2} s^{-1} on dark–adapted samples. Then, the samples are continuously illuminated with a white actinic light at an irradiance of PFD 75–200 μmole m^{-2} s^{-1} which is equivalent to the actual growth light of the cyanobacteria and higher plants. Within about 10 minutes the steady–state level of fluorescence (Fs) is achieved. The steady–state fluorescence (Fs) is recorded and a second saturating light pulse of 8000–10,000 μmole m^{-2} s^{-1} PFD is imposed on samples to determine the maximal fluorescence level in light–adapted state (Fm^{I}). The actinic light is then removed and the minimal fluorescence level in light–adapted state (Fo^{I}) is determined by illuminating the samples for 3 seconds with far–red light. The maximal fluorescence (Fm) with all PS II reaction centres closed in case of cyanobacteria is determined by addition of DCMU (10 μM).

Interpretation of Fluorescence Parameters

Parameters nomenclature is according to van Kooten and Snel (1990).

Fo

The initial fluorescence (minimal fluorescence) Fo in dark–adapted samples is believed to come from excited Chl *a* molecules in the antennae of the PS II before the exciton migration to the reaction centres. It represents the fluorescence level when all reaction centres are open *i.e.* Q_A is fully oxidized.

Fm and Fm^{DCMU}

It is the maximal fluorescence yield in dark-adapted samples and represents fluorescence level when all reaction centres are close *i.e.* Q_A is fully reduced (no photochemistry). Rise of emission from Fo F_m reflects the reduction of Q_A.

Fv

It is variable fluorescence yield in dark-adapted sample and calculated as Fv = Fm–Fo. Fv reflects reduced state of Q_A, however, it is not directly proportional to the redox state of Q_A. It is usually lowered by environmental stresses such as heat, freezing and high light intensity, which cause damage to thylakoid membranes. This decrease may be due to increase in Fo and decrease in Fm.

Fo′

It is minimal fluorescence level in light adapted state. It is measured with samples (cells or tissues) under a given light acclimation status. Prior to the measurement the samples are transferred briefly to darkness or far-red light. The determination of Fo′ in cyanobacteria is a problematic aspect of quenching analysis, since under moderate light intensities it is often very close to the steady-state Fs fluorescence level. Furthermore, unlike in higher plants, Fo′ in cyanobacteria is usually higher than the Fo fluorescence as a result of the dark–to–light increase in PS II fluorescence yield, *i.e.*, the state transition.

Fm′

It is maximal fluorescence yield in light adapted state. In cyanobacteria Fm′ is higher than Fm *i.e.*, the fluorescence of the light-adapted samples at the end of the time–courses is larger than fluorescence during the first light-saturating pulse after a period of 15-30 min of predarkening. However this phenomenon is not universal. The reasons for this phenomenon are not yet clear. It appears that they are related to photosynthetic state shifts.

Fv′

Variable fluorescence yield in light adapted state (Fm′–Fo′)

Fv/Fm

It is an index of maximal photochemical efficiency of PS II but this interpretation depends on both Fo and Fv originating predominantly from PS II. In higher plants under ideal conditions, Fv/Fm is near 0.8 and lower values reflect inhibition of PS II function. In cyanobacteria, changes in Fv/Fm under conditions of constant pigment content correlated well with changes in independent measurements of PS II function such as oxygen evolution, but the absolute level of Fv/Fm is not a reliable indicator of PS II function as phycobiliprotein fluorescence contributes to Fo and PS II accounts for only a small proportion of total chlorophyll. The distortion of Fo fluorescence in cyanobacteria is pronounced only at high cellular concentration of phycocyanin, which may be achieved primarily under nutrient-rich artificial culture conditions. However, when interpreted with cautions, same sample is monitored repeatedly over time and if the cellular pigment content is constant, Fv/Fm is a useful parameter to assess the photosynthetic performance under varied environmental conditions.

Fv′/Fm′

In higher plants, Fv′/Fm′, defined as (Fm′–Fo′)/Fm′ reflects the photochemical efficiency of open PS II centers under a given light acclimation status. Fv′/Fm′ generally varies inversely with qN, since nonphotochemical energy dissipation lowers the photochemical efficiency of PS II below the maximum levels reflected by Fv/Fm. A drop in Fv/Fm as occurs during photoinhibition of PS II activity, also feeds through and results in a drop in Fv′/Fm′. Thus in higher plants changes in the Fv′/Fm′ parameter reflect the combined regulation of PS II through both reversible nonphotochemical quenching and photoinhibitory inactivation of PS II.

In cyanobacteria, changes in Fv′/Fm′ also combine nonphotochemical influences on PS II function and photoinhibitory inactivation of PS II. In cyanobacteria, nonphotochemical quenching results primarily from changes in excitation distribution between the two photosystems rather than from excitation dissipation. Therefore, in cyanobacteria a drop in Fv′/Fm′ can result from a photoinhibitory drop in Fv/Fm or from a regulatory redistribution of excitation from PS II to PS I. In higher plant fluorescence analysis, a common implicit assumption is that down-regulation of PS II reflects overall down-regulation of photosynthetic electron transport. This assumption is not applicable to cyanobacteria, which have more flexible excitation distribution and electron transport systems. Furthermore, in cyanobacteria Fv′/Fm′ suffers the same limitations as described above for Fv/Fm, which is further compounded by the difficulty of measuring the Fo′ fluorescence level. Fv′/Fm′ in cyanobacteria is a useful integrated measure of PS II activity, even though various mechanisms may underlie the changes in PS II function.

ΔF/Fm′

It is quantified as Fm′–Fs/Fm′–Fo′ and reflects the actual quantum yield of PS II electron transport in the light adapted state which is equal to the product of qP and Fv′/Fm′. Therefore it depends on the degree of the closed PS II reaction centres and the efficiency of excitation energy capture in PS II.

qP (Photochemical Fluorescence Quenching)

It reflects a lowering of fluorescence below maximal levels through photochemical competition with fluorescence emission. Thus, when all PS II reaction centers are open, the potentials for photochemistry is maximal, photochemical quenching of fluorescence is maximal and fluorescence yield is low (Fo or Fo'). Conversely, when all PS II centers are closed and no photochemistry can occur, photochemical quenching is zero and fluorescence yield is maximal (Fm or Fm'). qP reflects the balance between excitation of PS II centers, which closes them, and removal of electrons from PS II by the electron transport chain, which reopens the centers. In practice, photochemical quenching is quantified by photochemical quenching coefficient which is calculated by formula given below:

$$qP = (Fm' - Fs)/(Fm' - Fo')$$

This parameter gives the position of steady-state fluorescence (Fs), on the scale from Fo' (all PS II centers open) to Fm' (all PS II centers closed). Thus, if steady–state fluorescence is equal to Fo', qP =1, indicating that all PS II centers are open. If steady-state fluorescence equals Fm', qP =0, indicating that all PS II centers are closed. Between these extremes, progressive reaction center closure is reflected by a decline in qP, although the relationship is not strictly linear.

In cyanobacteria qP typically stays high over a broad range of light intensity, up to 10 times higher than the growth light intensity. This contrasts sharply with the typical pattern of higher plants, where qP falls progressively as the light intensity exceeds the growth light. This cyanobacterial capacity to maintain PS II centers open under excess light reflects a complex and flexible electron transport system, as well as a generally high PS I/PS II ratio.

qN (Nonphotochemical Fluorescence Quenching)

Nonphotochemical fluorescence quenching reflects processes other than photochemistry which lowers the yield of variable fluorescence. It can be quantified by using formula given below:

$$qN = 1 - (Fm' - Fo')/(Fm - Fo) = 1 - (Fv'/Fv)$$

Quenchings which account for nonphotochemical quenching are:

1. *Energy dependent quenching (qE)*: It is due to creation of transthylakoid pH gradient by light.
2. *State transition quenching (qT)*: Quenching due to energy loss in state–I to state–II transition.
3. Spillover quenching: Quenching due to energy transfer to weakly fluorescent PS I by spillover.
4. *Photoinhibitory quenching (ql)*: Quenching due to inhibition of photosynthesis by illumination with high light intensity.
5. *Quenching by pigment radicals*: Sometimes excitation energy captured by radicals or excited states of photosynthetic pigments is converted to heat and fluorescence is lowred. Such molecular species are Pheo+, P680*, Chl* and carotenoid* but contribution is much less.

qN compares the span of variable fluorescence (Fv') under a given light condition with the maximal potential variable fluorescence (Fv). In a cyanobacterium, qN is high in the dark and drops to a minimum near the growth light intensity. In higher plant, qN climbs steadly as the light intensity surpasses the growth level. These differing patterns reflect a fundamental difference in the predominant processes contributing to nonphotochemical quenching. In plants nonphotochemical quenching is dominated by a mechanism(s) for excitation-dependent thermal dissipation of energy from PS II and its antennae, in competition with fluorescence and photochemistry. In contrast, nonphotochemical quenching in cyanobacteria largely reflects changes in the PS II fluorescence yield as a result of the state transition mechanisms, which regulates the distribution of excitation energy between PS II and PS I.

Prediction of Photosynthesis from Chlorophyll Fluorescence

For field measurements, a fluorescence-based estimate of electron transport and carbon dioxide fixation is very valuable. The fluorescence transients arise largely from PS II, and so calculations based on fluorescence reflect PS II activity and electron transport through PS II. Sundberg *et al.* (1997) have given the empirical relation between CO_2 fixation and quenching parameters.

$$P = \phi PS\ II \times I_i \times 1\ CO_2\ fixed/10\ photons$$

where,

P = micromoles of CO_2 fixed per square meter per hour, ϕPS II = (Fm'-Fs)/Fm' = (Fv'/Fm') x qP, I_i = number of incident photons per square meter per hour, and 1 CO_2 fixed/10 photons is an empirical conversion factor. This predictor gave good estimates of actual CO_2 fixation near the acclimated growth light intensity. Under high light, the predictor, which is based on light-driven electron flow through PS II, progressively overestimated actual CO_2 fixation. The overestimation probably reflects electron flow back to O_2 under excess light, which maintains PS II centers open but does not contribute to CO_2 fixation.

Similar empirical relation is developed by Campbell *et al.* (1998) to predict gross oxygen evolution from PS II.

$$Oxygen\ evolution = \phi PS\ II \times I_i \times 1\ O_2/12\ photons \times 1/Chl$$

where,

Oxygen evolution is expressed as micromoles of O_2 per milligram of chlorophyll per hour, ϕPS II δ = (Fm'–Fs)/Fm' = (Fv'/Fm') x qP, I_i = number of micromoles of photons incident per hour, 1 O_2/12 photons is empirical conversion factor, and Chl is the chlorophyll content in milligrams.

At higher light intensities, like CO_2 fixation predictor, the O_2 evolution predictor also increasingly overestimates measured oxygen evolution, again because of pseudocyclic electron flow, with electrons extracted from water by PS II ultimately reaching oxygen (O_2) as a terminal acceptor.

Similarly multiplication of ΔF/Fm' by the PPFD (photosynthetic photon flux density) under the constant conditions gives an empirical approximation of the relative electron transport rates (ETR).

Application of Fluorescence Analysis

Fluorescence analysis is an integral part of studies of photosynthesis in plants and cyanobacteria. In recent years, advances in instrumentation and interpretation have greatly expanded the application of fluorescence to ecophysiological and molecular studies. For cyanobacteria, estimating the acclimated light in a population from the light response of nonphotochemical quenching will allow rapid tracking of acclimation in laboratory or field studies. Cells must integrate changing environmental signals to regulate the expression of abundant proteins such as ribulose-1, 5–bisphosphate carboxylase/oxygenase or phycobiliproteins, in order to produce appropriate long-term levels of proteins.

Fluorescence imaging systems can be used to study heterogenous patterns of photosynthetic performance in leaves. For example, photosynthesis performance after a prolonged dark period (Bro *et al.*, 1996), after fungal infection (Scholes and Rolfe 1996), during source to sink transition (Meng *et al.*, 2001), after exposure to ozone (Leipner *et al.*, 2001) and from heterogenous populations of algal cells within intact biofilms (Oxborough *et al.*, 2000) has been studied using fluorescence imaging systems. Fluorescence analysis can be used to detect herbicide effects on the efficiency of PS II photochemistry several days before any visible effects on the plants are observed (Barbagallo *et al.*, 2003). Fluorescence signals can be used for rapid screening of mutant or transgenic colonies (Bennoun and Beal 1997) and for tracking physiological processes during gene regulation experiments. Rapid screening has now become increasingly important with the advent of genomic sequencing and saturation mutagenesis.

Fluorescence analysis can be used to assess the impacts of various environmental stresses such as light (Hong and Xu 1999; Luttge *et al.*, 1995), high temperature (Inoue *et al.*, 2000), water stress (Luttge *et al.*, 1995; Sanchez-Rodriguez *et al.*, 1999) and salinity (Netondo *et al.*, 2004). For example environmental stresses that cause structural alterations at the PS II pigment level affect minimal fluorescence Fo. Thermal damage is characterized by a drastic increase in Fo. By comparision, photoinhibition may lead to a slight increase in Fo. Freezing damage to thylakoid membranes does not affect Fo. Water stress cause decrease in Fo.

Conclusions

Chlorophyll fluorescence imaging systems can be used to investigate the functioning of PS II and may show how cyanobacteria and plants set their targets for gene expression and metabolic acclimation in the face of changing environmental factors and circadian status. Although several aspects of photosynthetic performance under varied environmental stresses can be studied, the results should be confirmed by the independent methods of measuring chlorophyll fluorescence as astonishingly different biological species show little difference between them.

References

Barbagallo RP, Oxborough K, Pallett KE and Baker NR (2003) Rapid, non-invasive screening for perturbations of metabolism and plant growth using chlorophyll fluorescence imaging. Plant Physiol 132: 485-493.

Bennoun P and Beal D (1997) Screening algal mutant colonies with altered thylakoid electrochemical gradient through fluorescence and delayed luminescence digital imaging. Photosyn Res 51: 161-165.

Bro E, Meyer S and Genty B (1996) Heterogeniety of leaf CO_2–assimilation during photosynthetic induction. Plant Cell and Environ 19: 1349-1358.

Campbell D, Hurry V, Clarke AK, Gustafsson P and Oquist G (1998) Chlorophyll fluorescence analysis of cyanobacterial photosynthesis and acclimation. Microbio and Mol Bio Rev 667-683.

Fenton JM and Crafts AR (1990) Computer–aided fluorescence imaging of photosynthetic systems: application of video imaging to the study of fluorescence induction in green plants and photosynthetic bacteria. Photosyn Res 26: 59-66.

Genty B, Briantais J-M and Beker NR (1989) The relationship between the quantum yield of photosynthetic electron transport and quenching of chlorophyll fluorescence. Biochimica et Biophysica Acta 990: 87-92

Hong S-S and Xu D-Q (1999) Light–induced increase in initial chlorophyll fluorescence Fo level and the reversible inactivation of PS II reaction centres in soybean leaves. Photosyn Res 61: 269-280.

Inoue N, Emi T, Yamane Y, Kashino Y, Koike H and Satoh K (2000) Effects of high-temperature treatments on a thermophilic cyanobacterium *Synechococcus vulcanus*. Plant Cell Physiol 41(4): 515-522.

Leipner J, Oxborough K and Baker NR (2001) Primary sites of ozone-induced perturbations of photosynthesis in leaves: identification and characterization in Phaseolus vulgaris using high resolution chlorophyll fluorescence imaging. J Exp Bot 52: 1689-1696.

Luttge U, Budel B, Ball E, Strube F and Weber P (1995) Photosynthesis of terrestrial cyanobacteria under light and desiccation stress as expressed by chlorophyll fluorescence and gas exchange. J Exp Bot 46(284): 309-319.

Meng Q, Siebke K, Lippert P, Baur B, Mukhergee U and Weis E (2001) Sink source transition in tobacco leaves visualized using chlorophyll fluorescence imaging. New Phytologist 151: 585-596.

Nedbal L, Soukupova J, Kaftan D, Whitmarsh H and Trilek M (2000) Kinetic imaging of chlorophyll fluorescence using modulated light. Photosyn Res 66: 25-34.

Netondo GW, Onyango JC and Beck E (2004) Shorgham and Salinity. II. Gas exchange and chlorophyll fluorescence of Shorgham under salt stress. Crop Sci 44: 806-811.

Osmond B, Schwartz O and Gunning B (1999) Photoinhibitory printing on leaves, visualized by chlorophyll fluorescence imaging and confocal microscopy, is due to diminished fluorescence from granna. Australian J Plant Physiol 26: 717-724.

Oxborough K, Hanlon ARM, Underwood GJC and Baker NR (2000) *In vivo* estimation of the photosystem II photochemical efficiency of individual microphytobenthic cells using high-resolution imaging of chlorophyll a fluorescence. Limnology and Oceanography 45: 1420-1425.

Sanchez-Rodriguez J, Perez P and Martinez-Carrasco R (1999) Photosynthesis, carbohydrate level and chlorophyll fluorescence-estimated intercellular CO_2 in water-stressed *Casuarina equisetifolia* Forst. and Forst. Plant Cell and Environ 22: 867-873

Scholes JD and ROlfe SA (1996) Photosynthesis in localized regions of oat leaves infected with crown rust (*Puccinia coronata*): quantitative imaging of chlorophyll fluorescence. Planta 199: 573-582.

Sundberg B, Campbell D, and Palmqvist K (1997) Predicting CO_2 gain and Photosynthetic light acclimation from fluorescence yield and quenching in cyano-lichens. Planta 201: 138-145

van Kooten O and Snel JFH (1990) The use of chlorophyll fluorescence nomenclature in plant stress physiology. Photosyn Res 25: 147-150.

© R.K. Gupta & V.D. Pandey (*Editors*)
Advances in Applied Phycology, 2007
Daya Publishing House, New Delhi

115–130

Chapter 8

Botryococcus: A Hydrocarbon Producing Microalga

Yashveer Singh*

Department of Botany, Dr. S.P.M. Govt. Degree College, Bhadohi – 221 401 Sant Ravidas Nagar (U.P.)

ABSTRACT

Botryococcus, a freshwater biochemically bizarre green microalga, can be grown in both batch and mass cultures. It forms relatively large colonies in the matrix of which there is a massive accumulation of hydrocarbons in the outer walls of cells. *Botryococcus* species exists in three (green exponential, yellow stationary and brown resting) physiological states and are characterized by varying growth rates and production of hydrocarbons. The lipid profiles of two species of this hydrocarbon rich microalga *viz. B. braunii* and *B.protuberans* grown in batch cultures under optimum conditions were studied with emphasis on hydrocarbon production at three different physiological states. Highest biomass (4.2 gl^{-1} in *B. braunii and* 4.0 gl^{-1} in *B. protuberans*) was obtained at the end of the exponential state. The contents of proteins, chlorophylls, carotenoids, carbohydrates and lipids were higher in *B. braunii* at each state. In the resting state, caroteniods frequently exceeded chlorophylls and the colonies being brownish. Apparently, at this state an oily layer of hydrocarbons could be seen on the surface of the culture medium. The highest hydrocarbon content (70 per cent of the total lipid in *B. braunii* and 60 per cent in *B. protuberans*) was found at the resting state whereas its productivity was higher at the exponential and early stationary states. Radioactive carbon labelling studies revealed that the contents of polar lipids declined at resting state whereas nonpolar lipids particularly hydrocarbons increased.

The production of lipids and hydrocarbons in both the species of *Botryococcus* has been also studied in batch cultures with respect to nitrogen limitation under aerobic and anaerobic conditions. Nitrogen deficiency significantly decreased the dry weight, chlorophyll and protein contents but

* Tel: 9450236424

the amounts of carotenoids, carbohydrates and lipids increased in both the species. Nitrogen starvation gave a 1.6 fold increase in lipid content. Anaerobiosis under nitrogen deficient conditions gave greater lipid production than anaerobiosis in nitrogen supplemented medium. Anaerobiosis induced hydrocarbon biosynthesis more significantly than nitrogen deficieney but decreased other nonpolar and polar lipids. Lower dose of ultraviolet-B(280-320 nm) radiation was found conducive to enhance lipids and hydrocarbons production in both the species of *Botryococcus*. Whole-cell immobilization of this microalga in alginate beads under air-lift batch cultures resulted in a significant increase in caroteniod and lipid contents at stationary and resting growth states as compared to free cells.

Keywords: *Hydrocarbons, Lipids, Microalga, Immobilization, Botryococcenes.*

Introduction

The increasing population pressure, the limits the earth's arable lands, and the increasing cost and depletion of petroleum have compelled scientists to develop new sources of agricultural products as alternative sources of energy that will not tax the earth's declining energy resources (Aaronson and Dubinsky, 1982). The enormous dependence of modern industrial society on petroleum and its derivatives for energy and the limited supply of these chemicals for industry have become serious economic problems for the world. A search for alternative sources of energy is thus one of the most important undertaking for ensuring the long term survival of mankind.

In recent years much work has been done on microalgae for producing a wide range of fine chemicals, pigments, oils and polysaccharides. The rapid increase in research and development of algal biomass cultivation technology for commodity bioproducts has made it possible for microalga to play new and different roles in relation to humanity. The mass culture of microalgae has been advocated as a source of pigments, fats and oils (Ben-Amotz and Avron, 1983; Borowitzka, 1998), fuels (Barclay *et al.*, 1988); pharmaceuticals (Aaronson and Dubinsky, 1982) and polysaccharides (Renn, 1988). Fats and oils from microalgae are popularly called as single cell oil (SCO). Microalgae may also be grown on arid lands in the tropics and subtropics in saline or alkaline waters and at relatively high temperature under conditions that are not suitable for conventional agriculture.

Microalgae are still a relatively untapped resource for lipids and hydrocarbons. Algal lipids are highly reduced hydrocarbons, and are similar to those of vegetable oils in composition. Historically, the global energy source has been hydrocarbon compounds. Average lipid (fat and oil) contents in microalgae are in the range of 20-40 per cent of dry weight. Under certain environmental conditions some algae produce upto 85 per cent of the dry weight as lipids (Dubinsky *et al.*, 1978). Algae that produce large quantities of lipid, generally demonstrate the ability to subsist under stress for prolonged period. Algal oils and fats can also be used as vegetable oil substitutes (Rattray, 1984). In addition to nutrition, a much larger diversity of algal lipids is important in modern daily life as food additives, fuels, pharmaceuticals, cosmetics and pigments. Lipid cover a wide range of compounds and includes long-chain hydrocarbons, alcohol, aldehydes, fatty acids and derivatives. Industrial nations like USA and Germany was initiated research on microalgal lipids as a sources of food and fuel oils. In recent years actual commercial production has been started by a few companies. In Belgium, Bioprex S.A. extracts terpeniods and other compounds from *Scenedesmus*, *Chlorella* and other algae for use as ingredients of cosmetics. Work conducted in the Russia led to the industrial production of green algae for essential oils and pigments (Dubinsky and Aaronson, 1982). Detailed studies and development of microalgae for liquid fuel production are being carried out by a number of groups, especially the Solar

Energy Research Institute in Golden, Colorado. This group is placing particular emphasis on algae capable of growing in saline environments at high temperature and light intensity (SERI, 1985).

The production of large quantities of hydrocarbons is rather a rare phenomenon in algae. Small quantities of C_{15} to C_{27} hydrocarbons (less than 5 per cent of dry weight) are produced by various green algae (Gelpi *et al.*, 1968), but one of the most promising alga is *Botryococcus*. This green colonial microalga produces unusually large amounts of hydrocarbons (Wake and Hillen, 1980; Lein, 1981; Singh and Kumar, 1992; Kumar and Singh, 1996). *Botryococcus* has the capacity to divert a very large fraction of its photosynthate into several distinctive hydrocarbons which have been named alkadienes, botryococcenes and lycopadienes, found in three different races of *B. braunii*. The presence of hydrocarbons is conspicuous only as the careteniods in green microalga *Dunaliella* and as a mixture of C_{17}-C_{35} hydrocarbons in the slow growing alga *B. braunii*. Both of these microalgae continue to receive attention as unique biological producers of these hydrocarbons. Metzger and Casadevall (1988) found considerable chemical variability among strains of *B. braunii*, isolated from Bolivia, France, Ivory coast, Morocco, Peru, Thialand and the West Indies. On the basis of hydrocarbon chemical structure, *B. braunii* is sub-divided into three races: Race A: it has hydrocarbons which are straight chain alkadienes and trienes, odd numbered from 23 to 31. In this race hydrocarbons accounts for 50-87 per cent of the oil. Race B: the triterpenoid hydrocarbons, termed botryococcenes are always the major components (75 per cent of the oil), and are acyclic olefin molecules C_{34} H_{58}. In the third race called T race, tetraterpene (lycopadiene) $C_{40}H_{78}$ hydrocarbons make up 2-40 per cent of the oil (Metzger *et al.*, 1990). Large amounts of non-isopreroid hydrocarbons occur in several strains of *B.braunii* and derive from oleic acid via an elongation-decarboxylation process (Templier *et al.*, 1984). Both hydrocarbons and oxygenated compounds may serve as an excellent source of liquid fuels. The properties of a liquid fuel are that it can be converted into a gaseous form that provides greater handling, storage and combustion capabilities. The routes of formation of hydrocarbons and oxygenated compounds are mechanistically related and both are synthesized through fatty acid precursors. Hydrocarbons are the end products of the reduction of organic compounds derived from decorboxylation or decorboxylation–condensation reactions of fatty acids (Tornabene, 1982). *B. braunii* is found widely in fresh water lakes and pools as well as in brackish water on all continents. The thick water blooms which it sometime forms, float on the surface of water and form combustible sediments. Such sediments are regarded as the source of the boghead coals and tar-like deposits found in different locations, known as torbanite, coorongite or balkaschite (Cane, 1977). *Botryococcus* has an interesting role in several areas of petrochemistry. Botryococcenes were reported by Moldowan and Seifert (1980) to make up approximately 1 per cent of a crude oil from Sumatra.

Materials and Methods

Test Organisms

Botryococcus braunii was isolated from a local pond (pH 7.6) near Varanasi and its identity was confirmd as per Philipose (1967). *B.protuberans* was obtained from the Texas University Culture Collection, Austin, U.S.A.

Culture Conditions

Both the species were grown in batch cultures in improved Chu-13 culture medium (Singh and Kumar, 1994). Standard methods were followed for the maintenance of axenic cultures. The cultures were kept in culture room at 27 $\pm$ 1°C under 18h light and 6h dark cycle. Illumination was provided by white fluorescent tube lights giving 16 Wm^{-2} intensity at the surface of the culture vessels. Flasks were plugged with cotton wool. Sterile air, enriched with 0.4 per cent (v/v) CO_2, was bubbled through the cultures for 6-8h daily. Unless otherwise stated, exponentially growing cells were used as inocula.

Dry Weight Determination

The dry weight of algal mass was determined by filtration of appropriate volumes of cultures through a tared Whatman No. 1 filter paper. The filter with algae was dried in a vacuum oven at 80 °c for 24h, cooled and weighed.

Pigments, Protein and Carbohydrate Determination

Pigments (chlorophyll and carotenoid) and protein were determined by the methods of Myers and Kratz (1955) and Lowry, respectively. Total carbohydrate of the algal cells was estimated by the Phenol method of Dubois *et al.* (1956). The chlorophyll, carotenoid, protein and carbohydrate contents were calculated and expressed as percent of dry weight.

Lipid and Hydrocarbon Estimation

Total lipids were assayed by the method of Kates (1972). The lipids were phase separated by adjusting the resolution ratio to 10:10:9 (methanol-chloroform-water, by volume). The chloroform phase was evaporated under a stream of N_2 at 42 °c, and weight of lipids was determined gravimetrically. The total lipid extracts were fractionated on a column (1x20cm) packed with heat-activated Silicic acid (Silica gel 60F_{254} E. Merck) using chromatography grade hexane, benzene, chloroform and methanol in order to isolate hydrocarbons, nonpolar lipids, and polar lipids, respectively. The isolated lipids in each eluate were measured gravimetrically after evaporation of the solvents. The eluates were further identified by thin layer chromatography (TLC) by comparison with standards.

Radioactive Carbon Labelling

Cultures were grown in the medium containing sodium (^{14}C) bicarbonate obtained from Bhabha Atomic Research Centre, Bombay (specific activity 5 μ ci/ml) and harvested at different physiological states for quantitative estimation of major lipid components. ^{14}C counting was done in a liquid scintillation counter (Beckman LS-7000).

Nitrogen Limitation

For nitrogen starvation experiments, cultures were first grown in normal medium until they reached mid growth phases, which was observed around the 23rd to 25th days. They were centrifuged aseptically and resuspended in nitrogen free medium. Controls were run in normal nitrogen containing medium.

Anaerobiosis

Anaerobiosis was aehieved in N-free and N-containing medium by placing the cultures in an anaerobic incubator (Don Whitley Scientific Ltd, England); N_2 and CO_2 (9:1 v/v) were passed into the incubator at a pressure of 2.15 Kg Cm^{-2}. H_2 was supplied at 0.86 Kg Cm^{-2}. During the course of study insignificant quantities of soluble nitrogenous compound were produced because of cell lysis.

Immobilization

A known volume of concentrated homogeneous algal suspension was mixed with 4.0 per cent (w/v) sodium alginate (Fluka AG, Chemische Fabrik, Switzerland) solution with continuous shaking for 20 win. Sodium alginate mixed cultures were taken into a 10ml syringe fitted with a needle of 1.0mm bore. The mixture was extruded dropwise through the syringe from a height of about 10cm into a glass container having excess of 0.2 M $CaCl_2$. Sodium alginate beads were allowed to harden in sterilized $CaCl_2$ solution for 30min. Stabilized beads of about 3mm in diameter were washed with double distilled water and then transferred into basal medium for growth. Immobilized cells were suspended and released by addition of K_2CO_3 Sol (50g/100ml distilled water) for the estimation of lipids and other chemicals.

Features of *Botryococcus*

Microalga *Botryococcus*, a colonial member of chlorophyceae is widely distributed in fresh water lakes, pools as well as in brakish water on all continents (Bachofen, 1982). It is unicellular, but the cells form aggregates of different sizes. *Botryococcus* is a slow growing alga with doubling time of more than two days (Singh and Kumar, 1994). Thallus is a free floating colony that forms 'blooms' and give rise to palmelloid forms. The colonies are irregular with a tough, hyaline, orange or brown and yellowish-green envelope of mucilage during culture (Figure 8.1). The single oval cells are completely embedded in a gelatinous mass or matrix containing oils and carotenoids (Schnepf and Koch 1978; Singh 1992). Cells are radially arranged in one layer at the periphery of young colony but later become scattered at maturity and turn from green to brick red. Cells lie close to each other in aggregates connected by delicate mucilaginous connections of colonial envelope. Compound colonies varied from 800-1000 µm in diameter and simple colonies are upto 70µm in diameter. Single cells are 3.6–5.0 µm broad and 6.7-10.0µm long. The morphological differences in both the species (*B. braunii and B. protuberans*) are that the cells of *B. braunii* are completely enclosed in mucilagenous envelope while those of *B.protuberans* are covered only at basal portion. In fact, *B. braunii* produces more mucilage. The colony density of *B. protuberans* is higher than *B.braunii*. There is a regular vegetative multiplication of colonies by a breaking or dissolution of the connecting strands. *Botryococcus* has long been regarded as aberrant in structure, behaviour and metabolism, and indeed it seem to have no close parallel in any algal groups.

Results and Discussion

Growth States

During the course of growth studies, in batch cultures, *Botryococcus braunii and B. protuberans* existed in three growth or physiological states, which were characterized by differences in chlorophyll-caroteniod ratio and hydrocarbon formation (Singh, 1992):

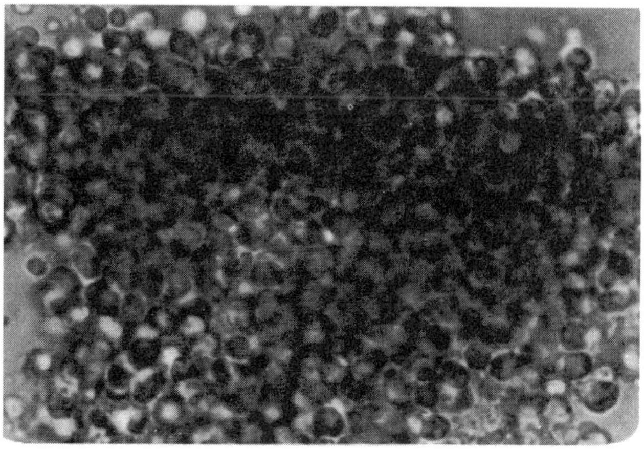

Figure 8.1: Microphotograph of the Colony of *B. braunii*

Green Exponential State

During this state colonies were green and heavier than water and cells were dominated by chlorophyll *a* and *b*, and a hydrocarbon content of around 18 per cent of total lipid.

Yellowish-green Stationary State

Colonies gradually become yellow-orange due to accumulation of carotenoids, and rise on the surface of the culture medium in a layer. Buoyancy was due to accumulation of oil in a colony matrix.

Brownish-orange Resting State

Cells in this state produced orange colored oil often in such quantity and abundance that the cell contents were completely masked and the entire colony appeared orange or brick red. There was also on oily layer on the surface of the medium whose viscosity increases due to exopolysaccharide which was dissolved in the medium.

Under optimum culture conditions *B. braunii* grew exponentially upto 22 days after a lag of 3 days whereas *B.protuberans* grew upto 24 days after a lag of 5 days (Singh and Kumar, 1994). Stationary state lasted in *B.braunii* from the 23rd to the 35th day and in *B. protuberans* from 25th to the 35th day. Maximum biomass was produced during the early stationary state in both the species (Figure 8.2). From 35th to 50th day in both species there was a brownish-orange resting state. In the early part of

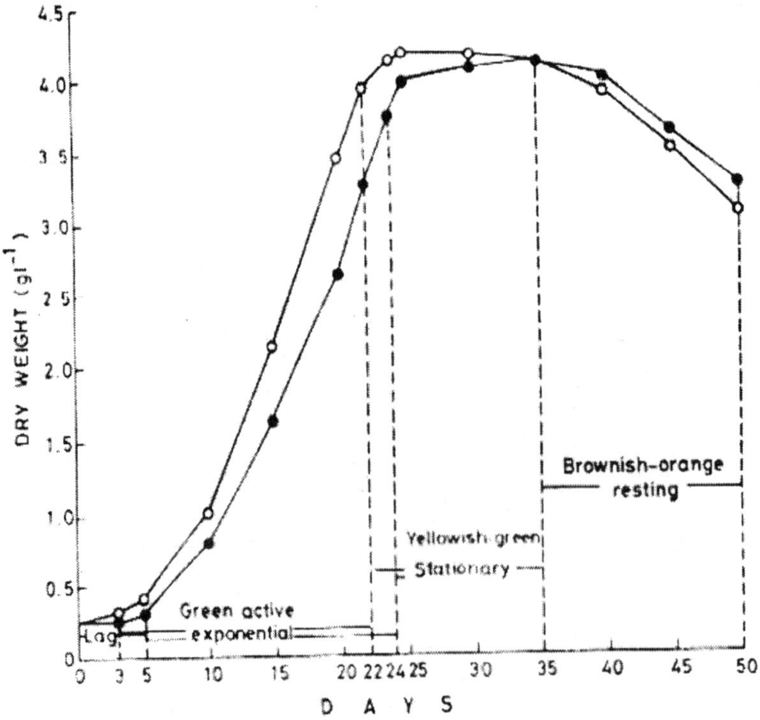

Figure 8.2: Physiological Growth Stages of *Botryococcus* spp. (O—O) *B. braunii*, (●—●) *B. protuberans*

this state, colonies were floating on the surface of the medium which attributed to the lipid accumulation. The cells lost their dry weight during the resting stage. This may be due to excretion and lysis of the cells (Singh, 1992). *Botryococcus* also exists in three carbon distributions ranges of hydrocarbons, C-17, C-27 and C-34 to C-40, depending on the algal growth phase. Approximately 1 per cent, 17 per cent and 85 per cent of dry weight is hydrocarbons in green resting, green active and brown resting growth phases, respectively (Tornabene, 1982; Casadevall *et al.*, 1985).

Chlorophyll-carotenoid Relationship

Regarding the relationship between total chlorophyll and carotenoid on dry weight basis, the carotenoids exceeds over chlorophylls after the stationary state (Table 8.1). On the 55th day of culture, chlorophylls and carotenoids occur in a proportion of 1:3.4 in *B.braunii* and 1:2.7 in *B. protuberans*. In the algae carotenoids seem to function primarily as photo protective agents (Krinsky, 1976) and as accessory pigments in photosystems. Carotenoids usually constitute 0.1 per cent of dry weight of microalgae, however, some algae (*Dunaliella salina*)produce greater amounts (Browitzka *et al.*, 1984) of carotenoids as beta-carotene under certain environmental conditions. Carotene is a hydrocarbon. There is carotenoid accumulation by green algae during senescence and resting cyst formation occur in the cytoplasm. Carotenoids are usually yellow isoprenoid polyene pigments derived from lycopene. They are synthesized *de novo* by all photosynthetic organisms and some other microorganisms. Beta-carotene, the most common carotenoid pigment, became one of the first lipid products to be extracted from microalgae. During the exponential growth state, most of the metabolic energy is used for synthesis of chlorophylls, proteins and intracellular carbohydrates. However, when the exponential state changes over to stationary and resting states, the energy is used predominantly for the synthesis of lipids. The excretion of hydrocarbons in *Botryococcus* species may be a way of disposing of surplus energy and thus avoiding damage to cells by photo-oxidation or by other adverse conditions.

Table 8.1: Chlorophyll and Carotenoid Relationship of *Botryococcus* Species During Cultures

Species	Contents* (% Dry Weight)	Cultures Time (Days)										
		0	5	10	15	20	25	30	35	40	45	55
B. braunii	Chlorophyll	0.03	0.07	0.32	1.12	1.46	1.68	1.57	1.42	1.10	0.96	0.70
	Carotenoid	0.01	0.03	0.14	0.6	0.87	1.00	1.28	1.5	1.72	2.10	2.34
	Chl: Carot.	1:0.3	1:0.4	1:0.4	1:0.5	1:0.6	1:0.6	1:08	1:1.1	1:1.2	1:2.2	1:3.4
B.protuberans	Chlorophyll	0.04	0.08	0.27	0.80	1.18	1.40	1.36	1.22	1.00	1.00	0.78
	Carotenoid	0.01	0.03	0.12	0.42	0.60	0.90	1.12	1.27	1.45	1.86	2.00
	Chl: Carot.	1:0.3	1:0.3	1:0.4	1:0.5	1:0.5	1:0.6	1:0.8	1:01	1:1.4	1:1.9	1:2.7

*: Means of three independent replicates under identical conditions.

5-10 per cent variations in the reading from mean.

Biomass and Chemical Profile

The biomass production and chemical profiles of the cells of *B. braunii* and *B. protuberans* grown in batch cultures under optimum conditions are given in Table 8.2. In both the species dry weight, protein, chlorophyll and carbohydrate contents declined from the stationary to the resting state but carotenoid and lipid content gradually increases. The highest biomass (4.2 gl^{-1} in *B. braunii* and

4.0 gl^{-1} in *B. protuberans*) was obtained at the end of the exponential state. In *B. braunii* about 23 per cent protein, 2 per cent chlorophyll, 1 per cent corotenoid, 16 per cent carbohydrate and 36 per cent lipid on dry weight basis were recorded. In *B. protuberans* the corresponding values were significantly less. Total lipid in both species gradually increases (43 per cent in *B. braunii* and 40 per cent in *B. protuberans*), and biomass (dry wt) decreases with increasing culture age. Whereas lipid production was slow and steady during the exponential state. During resting state, the cells lose their dry weight. This may be due to secreting behviour and lysis of cells of *Botryococcus*. Apparently most proteins and chlorophylls are synthesized during the early growth phase. The reduction in carbohydrate content at the stationary state may be due to the conversion of carbohydrate to lipids. In living cells, carbohydrates are converted into fats and *vice versa*, depending upon cellular conditions. Degradation of chlorophylls and increase in caroteniod often accompany lipid synthesis (Lien and Spencer, 1983).

Table 8.2: Biomass Production, Total Protein, Chlorophylls, Carotenoids, Carbohydrates and Lipid Contents of *Botryococcus* Species at Different Physiological States

Species	*Parameters*	*Exponential State (24th day)*	*Stationary State (35th day)*	*Resting State (50th day)*
B. braunii	Dry weight (gl^{-1})	4.27±0.31*	4.10±0.18	3.41±0.13
	Protein	22.80±1.32	20.53±1.17	18.26±1.24
	Chlorophyll	1.62±0.07	1.10±0.08	0.89±0.03
	Carotenoid	0.98±0.02	1.23±0.09	2.40±0.12
	Carbohydrate	16.20±1.16	14.64±1.15	12.32±0.15
	Lipid	36.15±1.85	39.64±1.70	43.80±2.14
B. protuberans	Dry weight (gl^{-1})	4.00±0.27	4.10±0.16	3.71±0.14
	Protein	19.61±1.23	20.00±1.13	18.76±1.29
	Chlorophyll	1.40±0.08	1.00±0.03	0.86±0.05
	Carotenoid	0.90±0.06	1.14±0.11	2.00±0.13
	Carbohydrate	13.62±1.12	12.80±1.10	10.45±1.10
	Lipid	34.30±1.70	37.94±1.82	40.72±2.10

*: Means ± SD of three independent replicates.

All the values are per cent of dry weight.

Hydrocarbons Production

For the estimation of hydrocarbons first of all total lipids were extracted with chloroform-methanol-water (5:10:4 by volume) according to the methods of Bligh and Dyer (1959) modified by Kates (1972). Solvents were evaporated under a stream of N_2 at 42°C, dried, and the weight of lipid was determined gravimetrically. Total lipid extracts (10mg) were fractionated into hydrocarbons, nonpolar and polar lipid fractions on a column (1x20cm) packed with silicic acid (silica gel $60F_{254}$, E. Merck) with chromatography grade solvents. Hexane (25ml), Chloroform (15 ml) and methanol (15 ml) are used to isolate hydrocarbons, nonpolar lipids (except hydrocarbons) and polar lipids, respectively (Kumar and Singh, 1996). The isolated lipids in each eluate ware estimated gravimetrically after evaporaion of the solvents under N_2 at 42 °c. The eluates were further identified by thin-layer chromatography (TLC).

Under optimum culture conditions, total nonpolar and polar lipids of *Botryococcus* species decreased by about 60 per cent from the late exponential to resting states in *B. braunii*, while hydrocarbons increased upto 70 per cent in *B. braunii* and 59 per cent in *B. protuberans* of total lipids (Table 8.3). Radioactive carbon labelling studies also showed that contents of hydrocarbons and carotenoids increased in older cultures. In *B. braunii*, hydrocarbon ranged from 17 to 29 per cent of total nonpolar lipids with the maximum values at resting state, but triglycerols (13.5 per cent of total nonpolar lipids) appeared during the exponential state and decreased markedly at the resting state. Polar lipids were mainly made of monogalactosyl diglyceride, digalactosly diglyceride, phosphatidyl ethanolamine and phosphatidyl glycerol at each state and their highest values were noted during the exponential state. Sulfouinovosyl diglyceride, phosphatidyl inositol and phosphatidyl choline remained constant throughout the culture growth. *B. protuberans* showed similar lipid composition as those found in *B.braunii*. The contents of aldehydes and sterols (per cent of total nopolar lipid) were maximum at exponential growth state and decreased with increasing cultures age in both the species (Data not shown).

Table 8.3: Hydrocarbon Production by *Botryococcus* Species

Species	*Lipid Composition (% of total lipid)*	*Pnysiological States*		
		Exponential State (24th day)	*Stationary State (35th day)*	*Resting State (50th day)*
B. braunii	Hydrocarbons	18.2±0.6*	48.7±1.1	70.2±2.6
	Nonpolar lipids**	52.5±1.2	32.4±1.6	18.3±1.4
	Polar lipids	29.3±1.8	18.9±0.5	11.5±0.8
B. protuberans	Hydrocarbons	20.7±1.1	47.5±1.4	59.4±2.7
	Nonpolar lipids	49.2±2.5	31.3±1.2	23.1±1.2
	Polar lipids	30.1±1.2	21.2±1.1	17.5±0.7

*: Means ± SD of three independent replicates; **: Except hydrocarbons.

The decrease in polar and nopolar lipids and increase in hydrocarbons with increasing culture age in both species may be due to the conversion of major nonpolar lipid components such as fatty acids, aldehydes, triglycerols and sterols into hydrocarbons. The reduction in algal polar lipids during the late stationary growth phase when nutrients such as nitrogen become limiting, has been reported by Piorreck *et al.* (1984). Both the species of *Botryococcus* shows higher efficiency of hydrocarbon formation during the exponential and early stationary states. Lipid accumulations in algae are chiefly esterfied fatty acids and are frequently associated with substantial increases in carotenoids while other nonsaponifiable lipids do not generally show large variations. Lipid accumulations in microalge occur mainly during the stationary state and exhibit generally very low productivities. Fats accumulate as a result of still greater decrease in the other cell constituents (Shifrin and Chisholm, 1981). In sharp contrast, *B.braunii* and *b. protuberans* accumulate hydrocarbons mainly during the stationary and resting states. In these species the hydrocarbon productivities have been found to be higher during the exponential and early stationary states (Kumar and Singh, 1996).

It appears that *Botryococcus* behaviour differs somewhat from the one commonly reported for many microalge. Both species accumulate mostly hydrocarbons (75 per cent of the total nonpolar lipids) instead of triglycerides and this accumulation does not require blockage of cell division because it can proceed during the exponential growth state. Hydrocarbon production and accumulation are affected by the changes in physiological status of the cells. Conspicuous formation of lipids by microalgae is the result of accumulation of storage triglycerols. Many genera of Chlorophyceae, Bacillariophyceae and Xanthophyceae have oil storage as a diagnostic feature instead of or in addition to starch. With the exception of *Botryococcus* species all the algae primarily produce triglycerols, not hydrocarbons. The fundamental differences between triglycerols formation in higher plants and microalgae are that in case of algae the oil-accumulating cells are themselves fully photosynthetic. This means that the complete pathway from carbon dioxide fixation to triglycerol synthesis can be modulated by the cells.

Hydrocarbons Production Under Nitrogen Limitation and Anaerobiosis

The production and storage of lipids by microalgae are usually regulated by environmental factors and are generally species specific (Shifrin and Chisholm, 1981). The enhancement of lipid storage during nitrogen-deficient condition has been demonstrated in other green algae (Ben-Amotz *et al.*, 1985). Highest concentration of lipid has been reported in *Monollantus salina* followed by *Nannochloropsis salina* which was deprived of nitrogen for nine days (Shifrin and Chisholm, 1981). Marked exception to this is *Dunaliella salina* (Ben-Amotz *et al.*, 1985). Where there is an actual decrease in lipid content and increase in carbohydrate content under nitrogen-limiting conditions. Lipids and carbohydrates are both used as energy reserves in algae and their synthesis is regulated qualitatively and quantitatively by a number of environmental factors. Besides lipids, changes in protein, carbohydrate and fatty acids are also measured in phytoplanktons (Harrison *et al., 1990*), and growth became limited by nitrogen deficiency. Variations in nitrogen content of the medium also cause changes in composition of carotenoids. Increasing nitrogen level in the medium leads to an increase in the biomass, protein and chlorophyll contents in freshwater green and blue-green algae (Piorreck and Pohl, 1984). In *Dunaliella* species, N-deficiency results in the accumulation of C_{40} isoprenoid beta-carotene in the chloroplast (Ben-Amotz and Avron, 1983), and in *Isochrysis* species, in the accumulation of C_{37} alkenone (Ben-Amotz *et al., 1985*). In these algae it appears that the acyclic and isoprenod hydrocarbons are non-degradable end products that accumulate under conditions of slow growth and stress. In N-limited *Chlorella vulgaris* cells starch comprised upto 55 per cent of the dry weight (Behrens *et al., 1989*) in comparison to 20 per cent under N-sufficient condition.

Anaerobiosis appears to trigger lipid synthesis by activating an oxygen-sensitive pyruvate dehydrogenase, which is located in the mitochondria (Inui *et al.*, 1987). Inui *et al.* (1982) showed the degradation of lipid and synthesis of carbohydrate in a green microalga, *Euglena gracilis* when cells were transferred from anaerobic to aerobic condition. Coleman *et al.* (1988) reported that under anaerobic conditions, nitrogen-deficient cells degraded carbohydrate and formed lipids by wax ester fermentation resulting in an increase in lipid content. Ben-Amotz *et al.* (1985) observed low production of hydrocarbons in green algae including *Botryococcus braunii* under optimum growth conditions and high production of hydrocarbons under growth limiting conditions. Besides nutrition, anaerobiosis also affects fatty acid and lipid composition. The accumulation of hydrocarbons and other neutral lipids is stimulated under anaerobic condition because they are incompletely oxidised products. Tornabene (1982) showed that both hydrocarbons and oxygenated compounds are synthesized through fatty acid precursors. Available literature reveals that most previous work on *Botryococcus* species has focussed upon growth, lipid and hydrocarbon production and composition under normal growth conditions (Wolf *et al.*, 1985; Vazquez-Duhalft and Greppin, 1987; Kumar and Singh, 1996) and in hydrocarbon production in relation to physiological state (Casadevall *et al., 1985,* Singh, 2000)

Little, if any, information is available concerning the lipid production of *Botryococcus* species under N-deficiency and anaerobic conditions (Singh and Kumar, 1992).

In the studies, on lipid and hydrocarbon production by *Botryococcus* species under nitrogen limitation and anaerobiosis, the growth rate declined under nitrogen-deficient conditions and chlorophylls content decreased significantly. Nitrogen deficiency apparently decreases the rate of synthesis of alpha–aminolevulenic acid, the first committed precursor of chlorophyll (Ohmori *et al.*, 1984), which in turn lowers the extent of chlorophyll synthesis. The lipid content of both algae were increased by nitrogen-deficient growth (Singh and Kumar, 1992). Anaerobic cells produced more lipid than aerobic cells and the highest values of lipid (72 per cent in *B.braunii* and 62 per cent in *B. protuberans*) were with nitrogen–deficiency under anaerobic conditions. In order to understand the role of nitrogen availability in regulating the utilization of available carbon for lipid synthesis, attempts were made to alter lipid contents by incubating cells under anaerobic condition for 12 days in nitrogen-sufficient or nitrogen deficient growth media. The increase in lipid content in both the species of *Botryococcus* was greater in aerobic cultures grown in nitrogen-free medium than in those grown anaerobically in nitrogen supplemented medium (Table 8.4).

Table 8.4: Contents and Composition of Lipids in Two Species of *Botryococcus* Grown Under Optimum Nitrogen-sufficient, Nitrogen-deficient and Anaerobic Conditions

Lipid	*Optimum*		*Without Nitrogen*		*Anaerobiosis*	
	B. braunii	*B. protuberans*	*B. braunii*	*B. protuberans*	*B. braunii*	*B. protuberans*
Content*	38.6±1.2	35.2±1.2	61.41±2.1	52.2±1.6	47.8±0.3	44.6±2.3
Composition: **						
Hydrocarbons	56.4±1.8	47.2±1.6	64.6±1.1	56.0±0.5	72.0±1.2	62.0±1.8
Nonpolar lipids***	18.0±0.8	32.5±2.1	19.1±1.3	33.4±1.8	10.5±0.7	25.5±1.1
Polar lipids	25.6±1.6	20.3±1.1	16.3±1.1	10.6±1.5	15.5±1.2	11.3±0.4

*: Percentage of dry weight; **: Percentage of total lipid; ***: Except hydrocarbons.

Mean ± SD of three independent replicates.

The proportion of hydrocarbons in the total lipids in both the species of *Botryococcus* was greater in anaerobic cultures (about 18 per cent) than under optimum condition. In nitrogen-deficient aerobic cultures there was only about 8 per cent increase in ratio of hydrocarbons in the total lipids. Anaerobiosis decreased the production of the non-polar lipid (except hydrocarbons) and polar lipid fractions, whereas nitrogen deficiency slightly increased the proportion of non-polar lipids to polar lipids relative to total lipid in both the species.

Although the concentration of the major polar lipids changed when the cells were grown under nutritional stress, the relative proportions of individual components remained essentially constant (Ben-Amotz *et al.*, 1985). Tornabene *et al.* (1983) showed that, at low nitrogen concentrations, the green alga *Neochloris oleoabundans* had a high lipid content and more than 70 per cent of these lipids were non-polar lipids such as triglycerols, with traces of hydrocarbons. In the work on *Botryococcus* species by Singh and Kumar (1992) have found high amounts of hydrocarbons under both the conditions. It appears from the results that since hydrocarbons are chemically-reduced products, anaerobiosis may have promoted the biosynthesis of hydrocarbons more effectively than that of other non-polar and polar lipid fractions. This study appears to confirm previous reports that the *Botryococcus* species are indeed unique among the algae in being able to produce and accumulate large amounts of lipids, comprising mainly hydrocarbons, under specific stress conditions.

Effects of Ultraviolet-B Radiation on Hydrocarbons Production

Ultraviolet (<400 nm) is an integral part of our solar system and it is universally accepted that the ozone layer in the stratosphere absorbs UV-light and protect us from its harmful effects. On the basis of wavelength, UV-radiation is divided into three categories: UV-A (320-400nm), UV-B (280-320 nm) and UV-C (100-280 nm) by Newton *et al.* (1979). The consequences of green house gases are mainly due to increased UV-B radiation on the earth. Shorter the wavelength of UV-B greater are the detrimental effects (Raghubanshi and Singh, 1991). Under green house conditions, plants exposed to UV-B radiation showed about 50 per cent reduction in growth. The impact of increasing UV-B radiation on biological system has been investigated only recently. In higher plants, reduction of leaf area, fresh and dry weight and photosynthetic activity have been reported in number of UV-B sensitive plants species (Worrest, 1986). Microalgae form the basis of food chain and biotic existence in any aquatic ecosystem. If these phytoplanktons are affected by increased UV-B radiation, the whole ecosystem may change. In marine plants UV-B radiation affects many metabolic processes, pigmentation and community composition (Worrest, 1986). The effect of UV-B radiation on diatoms and cyanobacteria have been studied by Hader *et al.*, (1986) and Dohler *et al.* (1986).

The effects of UV-B radiation on growth and metabolism of *Botryococcus* species have been studied as an environmental factor to observe the metabolism of hydrocarbons (Singh, 2001). There was about 50 per cent killing of cells at 6 Wm^{-2} dose of UV-B radiation following 40 min treatment in *B. protuberans* and 43 min in *B. braunii.* Complete loss of cells occurred after 110 min exposure in both the species (Data not shown). An LD_{50} dose reduced the growth of alga by 50 per cent as compared to control and the photosynthetic efficiency in terms of carbon assimilation capacity was reduced by 70 per cent in both the species. The hydrocarbon production in both the species was enhanced after 40min treatment of UV-B radiation but reduced after 60 min. (Data not shown).

In aquatic plants UV-B radiation impairs important physiological functions such as photosynthesis and enzymatic reactions in protein and carbohydrate biosynthesis. Increased dose of UV-B radiation decreases biomass and phytoplankton primary production (USEPA, 1987). The result presented here demonstrate adverse effects of UV-B radiation on *Botryococcus* spp. which seems to be a UV-B sensitive microalga. Here the effect of UV-B does not appear to be specific for a particular metabolic process; rather it effect a number of physiological and biochemical processes. It is an established fact that algae produces some extra secondary metabolites to cope with adverse effect or stress conditions. Stimulation of lipids and hydrocarbons in *Botryococcus* may be the result of this phenomenon. The source of UV-B radiation was a Fotodyne, Inc. (USA) lamp giving its main out put at 312.67nm. The desired radiation dose was obtained by adjusting the distance between UV-B lamp and the algal sample. Irradiations were performed in 75mm sterile petri plates with lid removed during the treatment.

Immobilization of Cells

Immobilization of algal cells may create anaerobic conditions and enhance enzymic activity. It has been proposed to digest the algal biomass to produce organic compounds by fermentation with cell immobilization by entrapment in alginate gel (Friedl *et al.*, 1991). In comparison to batch cultures where free cells are used, immobilized cells offer certain specific advantages such as (*i*) physical stabilization of cells, (*ii*) accelerated reaction rates due to increased cell density per unit volume, (*iii*) no washout of cells, (*iv*) biomass retention and prevention of overgrowth, and (*v*) easy seperation of cells and excreted products. Immobilized algal cells have many potential applications including biocatalysts for biotransformation and biosynthesis, waste water treatment through bioaccumulation of heavy metals. Biotransformations by immobilized algae include the production of sulphated polysaccharides by a redalga, *Porphyridium* (Gudin *et al.*, 1984), glycolate and glycerol from green microalga *Chlorella* and *Dunaliella*, respectively (Kerby and Stewart, 1988) and hydrocarbons from *Botyococcus braunii* (Bailliez *et al.*, 1985). Growth and hydrocarbon production by *B. braunii* entrapped and adsorbed in polyurethane foam have been investigated by Bailliez *et al.* (1988).

The effects of whole-cell immobilization in sodium alginate beads on growth, photosynthetic activity and lipid and hydrocarbon production by *Botryococcus* species in relation to growth phase have been studied recently by Singh (2003). It was observed that after 30 days of incubation, a rapid bleaching took place in free cultures of *B.braunii* and *B.protuberans*, while immobilized algae were still deep green even at the end of stationary state. This may be because immobilized cells offer biomass retention in microorganisms. Immobilization stimulated and stabilized photosynthetic O_2 evolution, chlorophyll and carotenoid contents even at the end of stationary and resting states in both the species of *Botryococcus.* Carotenoid content in entrapped cells increased significantly in both the species after exponential state and this increase was maintained upto the resting state. Maximum lipids (46 per cent in *B. braunii* and 45 per cent in *B. protuberans*) and hydrocarbons (72.5 per cent in *b. braunii* and 70 per cent in *B.protuberans*) were recorded in immobilized cells of resting state followed by stationary state (Table 8.5). Degradation of the chlorophyll and an increase in certain carotenoid pigments often accompany lipid synthesis during stationary and resting states in *Botryococcus* species (Singh, 1992). The observed increase in lipid and hydrocarbon accumulation by Singh (2003) was in agreement with the finding of Bailliez *et al.* (1988). Algae immobilized in alginate beads can transform oleic acid a fatty acid, into hydrocarbons. Immobilization of cells may create anaerobic conditions and anaerobiosis appears to trigger lipid synthesis in *Botryococcus* species (Singh and Kumar, 1992). In both free and entrapped cells of *Botryococcus* species, lipids and hydrocarbon accumulation occurred mainly during the stationary and resting states, and the hydrocarbon productivities were higher during exponential and early stationary states.

Table 8.5: Influence of Immobilization of *Botryococcus* Cells on Lipids and Hydrocarbons Production at the End of Each Physiological State

Species	*Contents (% Dry Weight)*	*Exponential State (24th day)*		*Stationary State (35th day)*		*Resting State (50th day)*	
		Free	*Immobilized*	*Free*	*Immobilized*	*Free*	*Immobilized*
B. braunii	Lipids	37.62±1.80	37.80±2.00	39.50±2.20	42.48±1.84	43.92±2.00	45.80±2.40
	Hydrocarbons*	20.15±1.16	19.82±1.20	47.65±2.40	52.45±2.62	68.15±2.80	72.50±2.86
B.protuberans	Lipids	38.80±1.30	36.12±1.25	37.84±1.38	40.72±2.00	41.25±2.20	44.6±2.00
	Hydrocarbons	21.30±1.70	20.64±1.60	45.86±1.42	60.75±2.50	63.40±2.64	70.34±2.90

Results are the means ± SD of three independent replicates.

Cultures were harvested on indicated days.

* The value of hydrocarbon is per cent of total lipid.

Acknowledgement

Y S is thankful to Dr. H.D. Kumar Ex-Professor of Botany, BHU Varanasi for guidance and lab facilities; and UGC, New Delhi for the sanction of Minor Research Projects: F.6-63/SPT/2000 (MRP/

NRCB) and (F.5.1.3. (140)/2004 (MRP/NRCB). Principal and staff members of Dr. SPM Govt. Degree College, Bhadohi are also thankfully acknowledged for support and facilities.

References

Aaronson S and Dubinsky Z (1982) Mass production of microalgae. *Experientia* 38: 36-40.

Bachofen R (1982) The production of hydrocarbons by *Botryococcus braunii. Experientia* 38: 47-49

Bailliez C, Largeau C and Casadevall E (1985) Growth and hydrocarbon production of *Botryococcus braunii* immobilized in calcium alginate gel. *Appl Microbiol Biotechnol* 23: 99-105.

Bailliez C, Largeau C, Casadevall E, Yang LW and Berkaloff C (1988) Photosynthesis, growth and hydrocarbon production of *Botryococcus braunii* immobilized by entrapment and adsorption in polyurethane foams. *Appl Microbiol Biotechnol* 29: 141-147.

Barclay W, Wyman C, Lewin RA and Cheng L (1988) Development of microalgal systems for the production of liquid fuels. In: Stadler T, Mollion J, Verdus MC, Karamanon Y, Morvan H and Christaen D (eds), *Algal Biotechnology*. Elsevier, Amsterdam, 55-64.

Behrens PW, Bingham SE, Hoeksema SD, Cohoon DL and Cox JC (1989) Studies on the incorporation of CO_2 into starch by *Chlorella vulgaris. J Appl Phycol* 1: 123-130.

Ben-Amotz A and Avron M (1983) Accumulation of metabolites by halotolerant algae and its industrial potential. *Ann Rev Microbiol* 37: 95-119.

Ben-Amotz A, Tornabene TG and Thomas WH (1985) Chemical profile of selected species of microalgae with emphasis on lipids. J Phycol 21: 72-81.

Bligh EG and Dyer WJ (1959) A rapid method of total lipid extraction and purification. *Can J Biochem Physiol* 37: 911-917

Borowitzka LJ, Borowitzka MA and Moulton TP (1984) The mass culture of *Dunaliella salina* for fine chemicals: From Laboratory to pilot plant. *Hydrobiologia* 116: 115-121.

Borowitzka MA (1988) Fats, Oils and hydrocarbons. In: Barowitzka MA and Borowitzka LJ (eds), *Microalgal Biotechnology*. Cambridge University Press, Cambridge, 257-287.

Cane RF (1977) Coorongite, Balkaschite and related substances-an annotated bibliography. *Trans R. Soc Australlia* 101: 153-164.

Casadevall E, Dif D, Largeau C, Gudin D, Chaumont D and Desanti O (1985) Studies on batch and continuous cultures of Botryococcus braunii: A hydrocarbon production in relation to physiological state, cell ultrastructure and phosphate nutrition. *Biotechnol Bioeng* 27: 286-295.

ColemanLW, Rosen BH and Schwartzbach SD (1988) Environmentral control of carbohydrate and lipid synthesis in *Euglena. Plant cell Physiol* 29: 423-432.

Dohler G, Biermann I and Zink J (1986) Impact of UV-B radiation on photosynthetic assimilation of ^{14}C-bicarbonate and inorganic ^{15}N-compounds by cyanobacteria. *Z Naturforsch* 41C: 426-432.

Dubinsky Z, Berner T and Aaronson S (1978) Potential of large scale algal culture for biomass and lipid production in arid lands. *Biotech Bioeng Symp* 8: 51-68.

Friedl A, Qureshi N and Maddox IS (1991) Continuous acetone-butanol-ethanol fermentation using immobilized cells of *Clostridium actobutylicum* in packed bed reactor and integration with product removal by prevaporation. *Biotech Bioeng* 38: 518-527.

Gelpi E, Oro J, Schneider HJ and Bennett ED (1968) olefins of high molecular weight in two microscopic algae. *Science* 161: 700-702.

Gudin C, Bernard A, Chaumont D, Thepenier C and Hardy T (1984) Direct bioconversion of solar energy into organic chemicals. *In* Biotech 84: The World Bitoech Report 1984, Online Publications Ltd London, 1: 541-559.

Hader D-P, Watanabe M and Furuya M (1986) Inhibition of motility in the cyanobacterium *Phormidium uncinatum* by solar and monochromatic UV irradiation. *Plant Cell Physiol* 27: 887-894.

Harrison PJ, Thompson PA and Caderwood GS (1990) Effects of nutrients and light limitation on the biochemical composition of phytoplankton. *J Appl Phycol* 2: 45-56.

Inui H, Miyatake K, Nakano Y and Kitaoka S (1982) Wax ester fermentation in *Euglena gracilis. FEBS letter* 150: 89-93.

Inui H, Ono K, Miyatake K, Nakano Y and Kitaoka S (1987) Purification and characterization of pyruvate: $NADP^+$ oxidoreductase in *Euglena gracilis. J Biol Chem* 262: 9130-9135.

Kates M (1972) Isolation, analysis and identification of lipids In: Work TS and Work E (eds), *Techniques in Lipidology*. Elsevier, Amsterdam, 268-618.

Kerby NW and Stewart WDP (1988) The biotechnology of microalgae and cyanobacteria. In: Rogers LJ and Gallon JR (eds), *Biochemistry of Algae and Cyanobacteria*; Proceedings of the Phytochemical Society of Europe, Clarendon Press, Oxford, 28: 319-334.

Krinsky NI (1976) The survival of negative microbes. In: Gray TGR and Postgate RJ (eds), *Symposia of the Society for General Microbiology*. Cambridge University press, Cambridge, 26: 209-239.

Kumar HD and Singh Y (1996) Lipid profile and hydrocarbon production by Botyococcus. *Current Res Plant Sc* 2: 71-80.

Lien S and Spencer KG (1983) Microalgal production of oils and lipids. In: Klass DL and Elliot HH (eds), Energy from Biomass and Wastes. Institute of Gas Technology, Chicago, 1107-1121.

Metzger P and Casadevall E (1988) Chemical variability in the oil rich alga *Botryococcus braunii*. In: Rogers LJ and Gallon JR (eds), *Biochemistry of the Algae and Cyanobacteria*. Proceed Phytochem Soc. Europe, Clarendon Press, oxford, 99-100.

Metzger P, Allard B, Casadevall E, Berkaloff C and Coute A (1990) Structure and chemistry of a new chemical race of *Botryococcus braunii* that produces lycopadiene. *J Phycol* 26: 258-266.

Moldowan JM and Seifert WK (1980) First discovery of botryococcene in petroleum. *J Chem Soc: Chemical Communications* 912-914.

Newton JW, Tyler DD and Slodki ME (1979) Effect of UV-B radiation on blue–green algae. *Appl. Environ Microbiol* 37: 1134-1141.

Ohmori M, Wolf FR and Bassham JA (1984) *Botryococcus braunii* C/N metabolism as affected by ammonia addition. *Arch Microbiol* 140: 101-106

Piorreck M, Basch K and Pohl P (1984) Biomass production, total protein, chlorophylls, lipids and fatty acids of fresh water green and blue-green algae under different nitrogen regimes. *Phytochemistry* 23: 207-216.

Piorreck M and Pohl P (1984) Formation of biomass, total protein, chlorophylls lipids and fatty acids in green and blue-green algae during one growth phase. *Phytochemistry* 23: 217-223.

Rattray JBM (1984) Biotechnology and fats and oils industry: An overview. *J Am Oil Chem Soc* 61: 1701-1712.

Renn DW (1988) Algal polysaccharides and their applications in biotechnology, food and other items of commerce. In: Ramus J and Jones MC. (eds), *Polysaccharides from Microalgae: A New Agroindustry.* Proc, Intl Workshop, Duke Univ. Marine Lab, Beaufort, North Carolina, 42-50.

Raghubanshi AS and Singh JS (1991) Increasing atmospheric abundance of radioactively active trace gases: Causes and Consequences. *Tropical Ecol* 32(1): 1-23.

Schepf E and Koch W (1978) Uber den Feinbau der olalge *Botryococcus braunii* kutzing. *Bot Jahrb Syst* 99: 370-379.

SERI (1985) Aquatic species programme review: Procedding of March 1985 Principal Investigator Meeting. SERI/CP-23/2700, Golden, Colorado: Solar Energy Res Institute.

Shifrin NS and Chisholm SW (1981) Phytoplankton lipids: Environmental influence on production and possible commercial applications. In: Shelef G and Soeder CJ (eds), *Algae Biomass.* Elsevier/ North Holland Biochemical Press, Amsterdam, 627-645.

Singh Y (1992) Physiological and Biochemical studies on the alga, *Botryococcus. Ph. D Thesis* BHU, Varanasi, India.

Singh Y (2000) Studies on a hydrocarbon rich microalga, *Botryococcus* at different physiological states. *J Indian Bot Soc* 79: 115-118.

Singh Y (2001) Effect of increased solar UV-B radiation on growth and metabolism of green alga, *Botryococcus. Nature and Biosphere* 6 (1&2): 14-20.

Singh Y (2003) Photosynthetic activity and lipid and hydrocarbon production by alginate immobilized cells of *Botryococcus* in relatioon to growth phase. *J Microbiol Biotechnol* 13 (5): 687-691

Singh Y And Kumar HD (1992) Lipid and hydrocarbon production by *Botryococcus* species under nitrogen limitation and anaerobiosis. *World J Microbiol Biotechnol* 8: 121-124.

Singh Y and Kumar HD (1994) Growth of *Botryococcus* species in improved medium. *Phykos* 33 (1&2): 77-87.

Templier J, Largeau C and Casadevall E (1984) Mechanism of non-iosprenoid hydrocarbon biosynthesis in *Botryococcus braunii. Phytochemistry* 23: 1017-1028.

Tornabene TG (1982) Microorganisms as hydrocarbon producers. *Experientia* 38: 43-46.

Tornabene TG, Holzer G, Lien S and Burris N (1983) Lipid Composition of the nitrogen starved green alga *Neochloris oleoabundans. Enzyme Microbiol Technol* 5: 435-440.

USEPA (1987) An assessment of the effects of ultraviolet-B radiation on aquatic organisms. *In*: Assessing the Risks of Trace Gases that can Modify the Stratosphere, EPA 400/1-87/00IC, 12: 1-33.

Vazquez-Duhalt R and Greppin H (1987) Growth and production of cell constituents in batch cultures of *Botryococcus sudeticus. Phytochemistry* 26: 885-889.

Wake LV and Hillen LW (1980) Study of a bloom of the oil rich alga *Botryococcus braunii* in the Darwin River Reservoir. *Biotech Bioeng* 22: 1637-1656.

Wolf FR Nemethy EK Blanding JH and Bassham JA (1985) Studies on the biosynthesis of unusual acyclic isoprenoids in the alga *Botryococcus braunii. Phytochemistry* 24: 733-737.

© R.K. Gupta & V.D. Pandey (*Editors*)
Advances in Applied Phycology, 2007
Daya Publishing House, New Delhi

131–141

Chapter 9
Food, Feed and Nutraceutical Applications of Algae

V. Prasad[1*] and R.K. Gupta[2]

[1]*Department of Botany, Thakur Ram Multiple Campus, Tribhuvan University, Birgunj, Nepal*
[2]*Department of Botany, Govt. P.G. College, Rishikesh – 249 201, Uttaranchal, India*

ABSTRACT

Algae are globally popular for their food, feed and nutraceutical values. Worldwide, over 160 species are commercially important for food sources belonging to the Rhodophyta, Phaeophyta, Chlorophyta and Cyanophyta. Various marine and freshwater algae are highly poteinaceous and rich in vitamins, minerals as well as primary and secondary metabolites. Nutraceuticals (health enhancing food) are used for the treatment of various diseases and pathological conditions in human beings and animals.

Keywords: *Algae, Feed, Food, Metabolites, Nutraceuticals.*

Introduction

Global population explosion and industrial growth has developed some major problems in nature such as quality and quantity of food, feed, drinking water,human health, soil, air and water pollutions. Algae can be utilized to solve these problems to some extent. Algae can be applied to solve the food, feed, fertilizer, energy, sanitational, waste water treatment, nutraceutically, industrially important secondary metabolite, pharmaceutically important chemicals and antibiotic scarcity problems (Kumar, 1999).

* Corresponding Author: Tel: 00977-51-526956; Fax: 00977-51-520332; Email: mpal_vp@yahoo.com.

In the age of toxic food production by the use of genetically modified plants and high doses of chemical fertilizers as well as herbicides and insecticides, algae can provide natural food with essential amount of chemicals in appropriate ratio for proper human health. Therefore, *Spirulina* is regarded as novel food. Besides, many algae may enrich soil fertility by natural nitrogen fixation. Various algae rich in several primary and secondary metabolites may be variously used for pharmaceutical and nutraceutical purposes.

Some of the chemicals are therapeutically used to treat skin-cancer, lactation, diabetes, thyroid stimulation, cancer, immune system stimulation, prostaglandin stimulation, chlosterol synthesis and human enzyme reactivation (Venkataraman and Becker, 1989). Phycobiliprotein of cyanobacteria is termed as phycoflour probe as fluorescence tag in connection with the detection of particular biological molecules (Klausner, 1986).

High protein and important useful chemicals presence in some algae has increased their scope in the sericulture, moriculture, pisciculture and feed for different poultry, cattles, molluscs and crustaceans. High protein and vitamin presence in algae have great potential in the field of cosmetics for skin care (Jha *et al.*, 2003).

Global environmental carbondioxide concentration can be reduced by the biomass of algae. Hydrogen and biogas are easily produced by algae can be used as alternate source of energy. Toxic heavy metals from the polluted soil as well as water can be removed by the use of phycoremediation technology.

Besides, there are so many aspects in which algae can be applied to solve several problems of human beings, animals and environment. Genetic engineering researches in useful algae would also give new dimensions in future. Among all the scopes of algae, food feed and nutraceutical aspects are more directly related to human health and economy.

Food Value of Algae

Algal biotechnology as known now combines the traditional and new technologies where algal biomass can be a source of food, feed, biochemicals, lipids, polysaccharide, vitamin and pigments used for various pharmaceutical purposes. The production of food crops through conventional agriculture is insufficient to meet the ever increasing demand of the fast growing population. It is more so in undeveloped country. Hence, for the production of unconventional food products the single cell protein is gaining momentum. Several species of cyanobacteria like *Nostoc, Phormidium, Chroococcus* were used as food in the past and is still being used in several countries like China, Figi, Ecnader, Monogolea etc.

But, *Spirulina* has received tremendous attention due to its higher protein efficiency ratio (PER) than those of cereals, vegetable and soya protein (Venkataraman LV, 1993). *Spirulina* has been promoted as natural health and slimming food in the US and Europe primarily to get profitable returns. *Spirulina* is being marketed in different forms, recipes, and products and a large number of formulations are available for the preparation of foods, using *Spirulina* powder, including fermented foods, comparable to cheese joghurt, tofu etc (Jha *et al.*, 2003). Decolourized, odorless and tastless *Spirulina* concentrates are being considered for the preparation of novel food (Venkataraman and Becker, 1989).

In the Far East and the Pacific Islands, people eat seaweeds such as *Porphyra, Laminaria, Undaria, Gracilaria, Alaria* and *Asparagopsis.* In Hawaii, about 50 sp. of algae are eaten worldwide (Kumar, 1999). Marine as well as freshwater algae are used as food in different countries. Marine algae are treated as sea weed most of which are highly proteinaceous and rich in vitamins. Therefore, algal studies for food purpose can be categorized into marine algal food and freshwater algal food.

Marine Algal Food

Table 9.1: Chemical Composition of Some Edible Sea Weeds (g)

Chemical Composition	*Sea Weed (100g)*							
	P. tenura	*U. sp.*	*L. sp.*	*E. sp.*	*U. pinnatifida*	*G. sp.*	*C. crispus*	*E. bicyclis*
Protein	31	24	7	21	13	8	11	7.5
Fat	0.6	0.6	2.0	0.3	2.0	0.05	3.0	0.1
Carb.	39	42	49	61	48	58	55	51
Ca (μg)	470	730	800	600	1300	–	–	1170
I_2 (ng.)	0.5	–	190–470	–	15–35	–	1–1.2	56–59
Vitamins								
A (IU)	44500	960	430	500	140	–	–	50
B_1 (mg)	0.25	0.06	0.08	0.04	0.11	–	–	0.02
B_2 (mg)	1.24	0.03	0.32	0.52	0.14	–	–	0.02
B_6 (μg)	1.04	–	0.27	–	–	–	–	–
B_{12} (μg)	13–29	6.3	0.3	1.3	–	–	–	–
C (mg)	20	10	11	10	15	–	–	11
Niacin (mg)	10	8.0	1.8	1.0	10	–	–	1.8
Folic acid (μg)	8.8	11.8	–	42.9	–	–	–	–

Porphyra tenura, *Ulva* sp., *Laminaria* sp., *Enteromorpha* sp., *Undaria pinnatifida*, *Gracilaria* sp., *Chondrus crispus*, *Eisenia bicycli*.

Source: Kumar, 1999.

Table 9.2: Amino Acid Composition of Protein in Some Sea Weeds (g of amino acid–N/100 g of protein-N)

Amino acid	*Sea Weed (100g)*							
	P. tenura	*U. sp.*	*L. sp.*	*E. sp.*	*U. pinnatifida*	*G. sp.*	*C. crispus*	*E. bicyclis*
Try.	1.3	0.3	1.4	0.8	1.3	8.0	12.6	0.8
Lys.	4.5	4.5	0.8	4.3	7.8	–	3.6	4.3
Hist.	1.4	4.0	–	2.7	4.0	–	3.3	1.8
Arg.	16.4	14.9	1.9	7.5	18.9	1.6	–	15.8
Asp.	7.0	6.5	13.7	5.6	5.0	27.5	87.1	5.7
Thr.	4.0	3.1	3.8	2.4	2.3	12.0	12.0	3.0
Ser.	2.9	3.0	33.7	2.8	2.3	51.6	12.9	2.8
Glu.	7.2	6.9	55.0	5.1	7.6	28.5	49.5	6.3
Pro.	6.4	4.0	51.0	2.8	4.5	24.2	50.8	5.1
Gly.	7.2	5.2	5.2	4.4	6.5	3.7	29.1	5.0
Aln.	7.4	6.1	23.8	4.8	7.0	27.8	71.4	5.5
Cys.	0.3	1.2	13.1	0.5	0.7	3.7	13.3	0.7
Val.	6.4	4.9	3.9	4.1	5.9	14.0	7.6	4.9
Meth.	1.7	1.6	–	2.2	2.4	–	–	1.7
Isol.	4.0	3.5	5.8	2.9	4.4	9.7	6.4	4.4
Leu.	8.7	6.9	6.8	5.1	7.3	14.3	5.3	6.3
Tyr.	2.4	1.4	2.1	1.6	2.1	7.5	–	0.9
Pher.	3.9	3.9	3.8	3.7	4.0	11.3	–	3.7

Porphyra tenura, *Ulva* sp., *Laminaria* sp., *Enteromorpha* sp., *Undaria pinnatifida*, *Gracilaria* sp., *Chondrus crispus*, *Eisenia bicycli*.

Source: Kumar, 1999.

Freshwater Algal Food

Table 9.3: Biochemical Composition of Different Species of Outdoor Grown Algae (per cent dry weight)

Biochemical Composition	*Chlorela vulgaris*	*Spirulina platensis*	*Spirulina maxima*	*Scenedesmus acultus*
Protein	40	52	50	47
Carbohydrate	18	19	16	17
Lipid	20	12	15	14
Nucleic acid	–	4.3	4.3	5.6
Amino acids (g/16gN)				
Arg.	7.2	7.4	7.5	6.3
Ile.	4.4	6.9	6.9	4.9
Ser.	5.0	4.3	4.2	4.8
Leu.	10.4	10.8	10.8	10.6
Pro.	5.0	4.1	4.1	5.3
Lys.	6.8	5.4	5.5	6.9
Hys.	2.0	2.0	2.1	2.3
Phe.	6.1	5.7	5.6	6.4
Tyr.	4.1	5.0	4.9	4.6
Cys.	1.0	0.7	0.6	0.8
Met.	2.4	2.7	2.6	2.7
Thr.	5.1	5.3	5.5	5.8
Trp.	1.9	1.5	1.5	–
Val.	6.6	7.9	7.9	6.9
Ala.	9.4	9.3	9.0	9.7
Glr.	12.8	16.8	16.3	12.7
Gly.	6.8	6.3	6.3	7.0
Asp.	10.3	11.7	11.7	11.1

Source: Materassi and Vincenzini, 1989.

Among the thousands of microalgae that are known in nature only very few have been shown to be amenable for food technology. The suitable forms include *Chlorella, Scenedesmus* and *Spirulina.* During the last twenty years, commercial production of *Spirulina* has slowly increased in Taiwan, Thailand, Israel, Peru, Japan, U.S. and Mexico. In south India, MCRC at Madras in Tamilnadu, has a large scale *Spirulina* Production unit with a production capacity of 7-10t/y *Spirulina*. Later on, CFTRI, Mysore and Auroville centre, Pondicherry were developed as algal food production centers (Venkataraman and Becker 1989, Venkataraman, 1983).

Nutraceutical Value of Algae

Any product isolated from food that is generally sold in medicinal form not usually associated with food and shown to be physiologically beneficial or that protect against chonic disease is called nutraceutical. The health enhanced food stuffs are becoming popular among consumers. The global market of such food is growing rapidly (Kumar, 2004).

Many vegetarian people suffer from chronic malnutrition, often accompanied by deficiency of Vitamin B_{12}. Many urban and rural people also suffer from iodine deficiency disorders such as stunted growth, deaf nutism, goitre and congenital mental and physical disabilities. The cobalt and iodine tolerant strains of *Spirulina platensis* have potential application in relieving the deficiency of Vit.B_{12} and iodine in vegetarians (Kumar and Singh, 1991).

Recombinant DNA technology application has produced numerous useful plants, popularly known as GM plants. The novel food status or food additive status of food items derived from GM crops may provide access to a broad market. Pharmaceuticals in GM plants are exemplified by recombinant proteins such as antibodies, hormones and blood clotting factors whereas in functional food and nutraceuticals they are primary or secondary metabolites (Kleter *et al.*, 2001).

Use of Algae for the Treatment of Diseases

Various algae have long been used for treatment of a wide range of diseases and pathological conditions. Some of them are shown in Table 9.4.

Table 9.4: Use of Algae for the Treatment of Diseases

Algae	*Treatment*	*Reference*
(*a*) **Marine Algae**		
Laminaria sp.	Neuroses, palsy, hypertension, edema and gynaecological disorders	Kumar, 1999
Eisenia bicyclis	Gynaecological disorders	Kumar, 1999
Undaria pinnatifolia	Cleans blood	Kumar, 1999
Nemacystus decipiens	Swellings	Kumar, 1999
Porphyra sp	Beriberi	Kumar, 1999
Digenea simplex	Anthelmintic	Kumar, 1999
Chondrus crispus	Anthelmintic, respiratory ailments	Kumar, 1999
Alsidium helminthocorton	Anthelmintic	Kumar, 1999
Enteromorpha sp., *Laminaria* sp. *Saragassum* sp.	Haemorrhoids, stomach disorders and cancer	Kumar, 1999
(*b*) **Freshwater Algae**		
Phormidium sp., *Corium* sp., *Lingbya* sp., *Synechococcus elongatus*, *Trichodesonium erython*	Antibacterial, antifungal and antiviral (Hepatitis B virus and *Herpes simplex* virus)	Gopalkrishnan *et al.*, 1996
Spirulina sp.	Hyperglycemia (diabetes mellitus) and Vit A deficiency	Mani *et al.*, 1996 Malandkar and Subbulakshmi, 1996
S. platensis	Anaemia	Kapoor and Mehta,1991
S. fusiformis	Tropical pancreatitis	Shenoy et al,1991
S. fusiformis	Rheumatoid arthritis	Malathi, 1991
Spirulina sp.	Malnutrition, anaemia,cataracts, constipation, stress, ulcer, allergy,oral cancer and weight control	Jassby, 1988 Richmond, 1988.
Spirulina sp., *Chlorella* sp., *Euglena* sp., *Scenedesmus* sp., *Chaetoceros* sp., *Peridium* sp., *Lyngbya spiralis*	Malnutrition antiparasitic activity against *Trichomonas* sp.	Borowitzka, 1988
Spirulina cell extracts or culture filtrate	Foliar spray enhances growth and yield of mulberry Foliar spray or presoaked seeds enhances growth and yield of crop plants.	Bongale and Rao, 1991

Algal Primary and Secondary Metabolites in Treatment of Diseases

Algae are the source of carotenoid pigments, vitamins, protein, lipids, oils, sugar and polysaccharides. According to Borowitzka (1988), among a large number of algae, the following algae are famous for: b-carotene and glycerol (*Dunaliella salina*), protein (*Scenedesmus* sp.), health food (*Chlorella* sp), human and animal nutrition and y–linolenic acid (*Spirulina* sp.), aquaculture (*phaeodactylum sp, Nannochloropsis* sp, *Dunaliella* sp, *monochrypsis* sp. and *Chaetoceros* sp.) and polysaccharide (*Porphyridium sp.*). Some antibiotic and essential chemical produced by different algae and their use against some diseases is shown in the Table 9.5.

Table 9.5: Primary and Secondary Metabolites in Treatment of Diseases

Algal Organisms	*Chemical/Antibiotics*	*Treatment/References*
Dunaliella sp., Botryococcus sp., Spirulina sp., Chlorella sp. and Scenedesmus sp.	Carotenoids (Vit A) β-carotene	Anticancer agent (Klausner, 1986)
Spirulina sp.	Phycobiliproteins	Fluorescencebags (immuno-diagnostics) Jha *et al.*, 2003.
Laurencia thyrsifera, Spirulina sp.	Thyrsiferol acetate Superoxide dismutase (SOD)	Wrinkling of skin (Venkataraman, 1993)
Spirulina sp., Chlorella sp., Euglena sp., Scenedesmus sp., Chaetoceros sp., Peridium sp.	Vit $B_{12,}B_{6,}$ B_1, Riboflavins, Nicotinic acid, Pantothenate	Malnutrition (Borowitzka, 1988)
Microcystis pulverana, Chlamydomonas reinhardtii, Chlorella vulgaris, Coccomyxa sp., Euglena sp., Scenedesmus sp., Chaetoceros simplex, Peridium cinctum	Biotin	Malnutrition (Borowitzka, 1988)
Nostoc sp., Anabaena sp., Chroococcus sp., Porphyridium sp., Laminaria sp., Macrocystis sp., Chlorella sp., Spirulina sp., Scenedesmus sp.	Polysaccharide	Emulsifiers and thickening agent (Borowitzka, 1988)
Chlorella sp., Spirulina sp., Scenedesmus sp.	β-alanine betaine	Antitumor agent (Borowitzka, 1988)
Monstroma nitinum	Ulvaline	Blood chlosterol (Glombitz and Koch, 1989)
Laminaria sp., Ecklonia cava, Eisenia bicyclis	Fucoidan	Leukaemia (Glombitz and Koch, 1989)
Fucus vesiculosus, Ulva pertusa, Chondrus crispus	Laminaran	Blood articoagulating activity (Glombitz and Koch, 1989)
Laminaria angustata	Laminine	Blood pressure, heart and muscle motion (Glombitz and Koch, 1989)
Saragassum sp.	Monoiodotyrosine Diiodotyrosine, Thyroxine	Blood chlosterol (Glombitz and Koch, 1989)
Saragassum muticum	Fucosterol/Sagasterol	Antihyperlipidaemic activity (Glombitz and Koch, 198 9)

Contd...

Table 9.5–Contd...

Algal Organisms	*Chemical/Antibiotics*	*Treatment/References*
Codium pugniformis	Polysaccharide	Anticarcinoma activity (Glombitz and Koch, 1989)
Bifurcaria galapagensis, Cystoseira mediterranea, Dictyopteris undulata	Bifurcarenone, Mediterraneol hydroquinone	Antibacterial and antifungal activity (Glombitz and Koch, 1989)
Laurencia hybrida	Formylundecatrienoic acid Hydroxycicesapentaenoic acid	Antibacterial and antifungal activity (Glombitz and Koch, 1989)
Alsidium sp.	Domoic acid	Anthelmintic (insecticidal activity) (Glombitz and Koch, 1989)
Digenea sp.	Kainic acid	Epilepsy (Glombitz and Koch, 1989)
Trididemnum sp.	Tunicate	Acticancer (Glombitz and Koch, 1989)
Chondrus sp., Eucheuma sp., Hypnea sp., Iridaea sp., etc.	Carrageenan	Cough (antiinflammatory activity) (Glombitz and Koch, 1989)
Gelidium sp., Pterocladia sp., Gracilaria sp.	Agar	Blood anticoagulant (Glombitz and Koch, 1989)
Lamiunaria sp., Macrocystis sp., Ascophyllum sp., Saragassum sp.	Alginate	Dental molds. Hemstatic agent in surgery, Radiography contrast agent. (Glombitz and Koch, 1989)
Saragassum linifolium, Ecklonia sp., Laminaria sp., Hizikia sp., Lessonia sp., Myagropsis sp.	Sulhatized polysaccharide (SPS)	Blood anticoagulant activity. (Glombitz and Koch, 1989)
Cyanobacterials and *Chlorophytes*	Chlorophyll, carotenoid and phycocyanin pigments	Colorants of various food stuffs (Glombitz and Koch, 1989)
Porphyra sp., *Lithothamnion sp.*	Y–aminobutyric acid (GABA) 1-aminocyclo propane-1-carboxylic acid (produce natural ethylene)	Metamorphosis of Larvae (Glombitz and Koch, 1989) Plant growth stimulating effect (Borowitzka, 1988)
Chlorella sp., *Scenedesmus sp.*	Growth regulating substances AA, GA, KN, and others	Growth of plant tissue and cell culture (Borowitzka, 1988)
Chlorella vulgaris	Chlorellin	Antibiotic (Borowitzka, 1988)
Phaeocystis pouchetii, Polysiphonia sp., *Rhodomela sp.*	Acrylic acid	(Borowitzka, 1988)
Spirulina sp., *Phormidium* sp., *Phaeodactylon sp.*	β-lactamase	Hall and Rao, 1996

Source: Jha *et al.*, 2003; Materassi and Vincenzini, 1989; Chapman and Gellenbeck, 1989.

Folk Medicines

In Russia and Europe, various algae have long been used as folk medicine to cure and treat many pathological conditions (Kumar, 1999), such as: *Laminaria japonica* (hypertension, edema), *Eisenia bicyclis* (gynaecological disease), *Undaria pinnatifida* (clean blood after child birth), *Nemacystus decipiens* (swellings),*Porphyra sp* (beriberi), *Digenea simplex* (anthelmintic), *Chondrus crispus* (respiratory ailments) and *Enteromorpha sp* (haemorrhoids and stomach diseases).

Feed Value of Algae

Microalgae are used as live feed for commercially important molluscs fishes and crustaceans. The most frequently used species are diatoms, *Dunaliella, Chlorella, Scenedesmus, Chlamydomonas* and *Spirulina.* Natural algal bloom production of microalgae used for feeding the aquaculture is given special attention. A new tendency has developed to use harvested algae as feed in pisciculture (Pauw and Persoone, 1988). *Spirulina* and *Nostoc* have been used as feed, supplements for poultry, swine, fish, silkworm caterpillars, shrimps, daphnia, artemia and crustaceans (Ciferri and Tiboni, 1985; Jha *et al.*, 2003; Materassi and Vincenzini 1989). *Spirulina* is used for lactation (Kapoor and Mehta, 1991) in cattles, and growth of silvercarp and silkworm (Bongale and Rao, 1999; Thilagavathy and Kumuthakalavalli, 1991).

The nutritional status of algae used as feed is shown in Table 9.6.

Table 9.6: Gross Chemical Composition of Some Algae (per cent of dry matter)

Chemical Composition	*Algal Species*										
	S.sp.	*C.sp.*	*Sc.sp.*	*D.sp.*	*Sy.sp.*	*E.sp.*	*Pr.sp*	*Ph.sp*	*U.sp.*	*Ur.sp.*	*St.sp.*
Protein	46-71	51-58	47-56	49-57	63.0	39-61	28-45	41.0	45.0	58.0	51.0
Carbohydrate	5-20	12-26	10-17	4-32	15.0	14-18	25-33	-	-	-	-
Lipid	2-9	40-22	2-14	6-8	11.0	14-18	22-38	3.8	1.1	1.7	1.2
Nucleic acid	2.5-4.5	-	3-6	-	5.0	-	1-2	-	-	-	-
Vitamin contents (μg/g dry wt.$^{-1}$) B_6	3.7	11.23	-	-	-	10.3	-	2.0-12.36	-	-	-
B_{12}	1.2-2.5	0.02-0.09	1.4	-	-	-	-	0.41-6.65	-	-	-
C	-	-	760	-	-	-	-	-	-	-	-
Biotin	-	0.14–2.5	0.8	-	-	0.15	-	0.58	-	-	-
Riboflavin	25–40	27–80	150	-	-	39	-	20-60	-	-	-
Nicotinic	105-118	120-240	26.2	-	-	30	-	30.6-73.5	-	-	-

Spirulina sp., *Chlorella* sp., *Scenedesmus* sp., *Dunaliella* sp., *Synechococcus* sp., *Euglena* sp., *Prymnesium* sp., *Phormidium* sp., *Ulothrix* sp., *Uronema* sp., *Stigeoclonium* sp.

Source: Becker, 1988; Borowitzka, 1988.

Nature of feeding and algal organisms are shown in Table 9.7.

Conclusion

Algae and cyanobacteria are very useful organisms of this beautiful planet,the earth. They can be applied to solve major poblems and to alleviate the economic status of human beings. Among several applications,algae are globally popular for food, feed and nutraceutical values. *Spirulina* is considered as novel food, feed and nutraceutically important cyanobacterium and thus it is cultivated in biomass in various developed and developing countries. Various marine and freshwater algae such as *Porphyra, Vlva, Enteromorpha, Undaria, Chlorella, Spirulina, Scenedesmus* are highly proteinaceous and rich in vitamins, amino acids and minerals and thus used as human food in the Far East and Pacific islands as well as various developed countries. Worldwide, over 160 sp. are commercially important for food sources mostly belonging to the Rhodophyta, Phaeophyta, Chlorophyta and Cyanophyta (Kumar,1999).

Table 9.7: A List of Algal Organisms Used as Feed for Different Animals

Feeder	*Algal organisms*	*Reference*
Daphnia	*Anacystis, Synechococum, Anabaena, Chlorella, Ankistrodesmus, Chlamydomonas*	Kumar, 1999
Tonypus	Cyanobacteria, *Synedra, Navicula*	Kumar, 1999
Ostracod *Cypris sp*	*Nostoc linkia*	Pandey and Kumar, 1989
Frog todpole	*Phormidium valderianum*	Uma *et al.*, 1996
Chicks and Poultry	*Gracilaria, Spirulina, Oscillatoria, Enteromorpha Chlorella, Scenedesmus.*	Jha *et al.*, 2003; Nayak and Padhi, 2003; Becker,1988.
Fish	*Enteromorpha, Spirulina*	Nayak and Paddhi, 2003
Swine, molluscs silkworm	*Spirulina, Nostoc, Chlorella, Scenedesmus*	Jha *et al.*, 2003; Becker, 1988.
Molluscs, shrimps and fishes	*Chlorella,Phaeodactylum, Skeletonema, Dunaliella*	Materassi and Venecenzini, 1989
Gold fish	*Tiffaniella, Pterosiphonick, Murrayellopsis, Nostoc calcicola*	Jha *et al.*, 2003 Kumar, 1999
Coastal animals	*Ascophyllum nodosum*	Kumar, 1999
Livestock	*Alaria, Laminaria, Palmaria, Fucus, Ascophyllum*	Kumar, 1999
Cow	*Alaria esculenta*	Kumar, 1999
Caltles, sheep	*Scenedesmus dimorphus, S. bijugatus, Oscillatoria sp., Chlorella pyrenoidosa*	Shrivastav and Dadheech, 1989

Global market of nutraceuticals (health enhanced food stuffs) is growing rapidly. The novel food status and secondary metabolites present in various algae such as *Laminaria, Porphyra, Enteromorpha, Saragassum, Gelidium, Chondrus,Digenia, Spirulina, Chlorella,Dunaliella, Scenedesmus,Phormidium,Nostoc* and various others are nutraceutically important and used for the treatment of various diseases and pathological conditions. Hence, various algae have long been used as folk medicine to cure and treat many pathological conditions.Due to highly nutritive value of *Anabaena,Chlorella, Ascophyllum, Alaria, Laminaria* and various other algae are used as feed for cattles, chicks,poultry, daphnia,molluscs, swines,shrimps and silkworms.

Acknowledgements

Author, R.K. Gupta is thankful to UGC, New Delhi for financial support.

References

Becker EW (1988) Microalgae for human and animal consumption.In. Borowitzka LJ (eds) *Microalgal Biotechnology*. Cambridge University Press, NewYork, 222-256.

Bongale UD and Rao BVN(1991) Response of mulberry to *Spirulina* cell extracts and culture filtrates as foliar sprays. *Sprirulina ETTA Nat. Symp.* AMM MCRC, Madras, India.

Borowitzka MA (1988) Vitamins and fine chemicals from microalgae. In: Borowitzka MA and Borowitzka LJ (eds) *Microalgal Biotechnology*. Cambridge University Press, New York, 153-196.

Ciferri O and Tiboni O (1985) The biochemistry and industrial potential of *Spirulina. Ann. Rev. Microbiology*. 39:503-526

Chapman DJ and Gellenbeck KW (1989) An historical perspective of algal biotechnology. In. Cresswell RC, Rees TAV and Shah N (eds) *Algal and Cyanobacterial Biotechnology*, Longman Scientific Technical, New York, 1-23.

Glombitza KW and Koch M (1989) Secondary metabolite of pharmaceutical potential. In. Cresswell RC, Rees TAV and Shah N (eds) *Algal and Cyanobacterial Biotechnology*. Longman Scientific and Technical, New York, 161-219.

Gopalakrishnan V, Pramod NP, Sundraraman M, Subramanian G and Thyagarajan SP. (1996) Antimicrobial properties of marine cyanobacteria, *Int. Symp. on Cyanobacterial Biotechnology*, Tiruchirapalli, India. 76.

Hall DO and Rao KK (1996) Cyanobacterial Biotechnology: present status and future opportunities. *Int. Symp. on Cyanobacterial Biotechnology*, Tiruchirapalli, India. 17-18.

Jassby A (1988) Spirulina a model for microalgae as human food. In. Lambi CA and Waaland JR (eds) *Algae and Human affairs*, CUP, Cambridge, 149-179.

Jha MN, Sharma SG, Prasad AN and Tripathi KK (2003) Cyanobacterial, Biotechnology. In. Verma BN, Kargupta AN and Yadav RN (eds) *Phycology Nature and Nurture*, Kalyani Publishers, New Delhi. 225-237.

Kapoor R and Mehta U (1991) Bioavailability of betacarotene from *Spirulina platensis, Spirulina* ETTA Nat. Symp. AMM MCRC, Madras, India.

Klausner A (1986) Algal culture-Food for thought. *Biotechnology* 4(11): 947.

Kleter GA, Van der Krieken WM, Kok EJ, Bosch D, Jordi W and Gilissesn LJWR (2001) Regulation and exploitation of genetically modified crops. *Nature Biotechnology*. 19: 1105-1110.

Kumar HD (2004) *Crop Modification: Biotech and Non-Biotech Approaches*. Daya publishing House, Delhi, 66.

Kumar HD (1999) *Introductory Phycology*. East West Press Pvt. Ltd, New Delhi, 545-598.

Kumar HD and Singh Y (1991) Iodized and cobalt enriched strains of *Spirulina platensis. Spirulina* ETTA Nat. Symp. AMM MCRC, Madras, India.

Malandkar M and Subbulakshmi G (1996) Effect of Spirulina supplementation on vitamin a status during pregnancy. *Int. Symp. on Cyanobacterial Biotechnology*. Tiruchirapalli, India. 67.

Malathi M (1991) Efficacy of *Spirulina fusiformis* in a case of rheumatoid arthritis *Spirulina* ETTA Nat. Symp. AMM MCRC, Madras, India.

Mani UV, Sangeeta S. Iyer UM and Subramanian SR (1996) Effect of Spirulina supplementation in control of diabetes mellitus. *Int. Symp. on Cyanobacterial Biotechnlogy*, Tiruchirapalli, India. 64.

Materassi R and Vincenzini M (1989) A general survey of feed and food utilization of autotrophic microorganisms. *Proc.3rd Int. Conf. on LEAF*–PRO 89, Pisa–Perugia–Vitorbo, Italy. 366-373.

Nayak S and Padhi S (2003) Biochemical screening of algae for their nutritional value in Pisciculture. In. Verma BN, Kargupta AN and Yadav RN (eds) *Phyoclogy Nature and Nurture*. Kalyani Publisher, New Delhi. 239-242.

Pandey AB and Kumar HD (1989) Feeding of Blue-green algae isolated from flooded soils by the ostracod *cypris sp.* Proc 3rd Int. Conf. on LEAF-PRO, 89, Pisa-Perugia-Viterbo, Italy. 506.

Pauw ND and Persoone G (1988) Microalgae for aquaculture. In. Borowitzka MA and Borowitzka LJ (eds) *Microalgal Biotechnology*. Cambridge University Press, Cambridge, 197-221.

Richmond A (1988) Spirulina. In Borowitzka MA and Borowitzka LJ (eds) *Microalgal Biotechnology*. Cambridge University Press, Cambridge, 85-121.

Shenoy KT, Bai J, Sarah V and Leena KB (1991) *Spirulina fusiformis* (Multinal) for pain relief in tropical pancreatitis: Placebo controlled trial. *Spirulina* ETTA Nat. Symp. AMM MCRC, Madras, India.

Shrivastava P and Dadheech PK (1989) Chlorococcalean algae for utilization as a source of proteins. Pro. 3rd Int. Conf. on LEAF–PRO 89, Pisa–Perugia–Viterbo, Italy. 502.

Thilagavathy D and Kumuthakalavalli (1991) Evaluation of the suitability of *Spirulina* algae in the feed for silvercarp fingerlings, *Spirulina* ETTA Nat. Symp. AMM MCRC, Madras. India.

Uma L, palanisamy M, Murugian P. Akbarsha M, Subramanian G and Subramanian G (1996) Novel Cyanobacterial feed for frog tadpoles. *Int. Symp. On Cyanobacterial Biotechnology*. Tiruchirapalli, India. 66.

Venkataraman LV (1983) *Spriulina* state of art and emerging prospects. *Phykos*, 28 (1&2): 321.

Venkataraman LV (1993) *Spirulina* in India. *Proc. Nat. Sem Cyanobacterial Research-Indian Scene.* NFMC,BARD, Tiruchirapalli. 92-116.

Venkataraman LV and Becker EW (1989) Challenges in commercialization of microalge. *Proc. 3rd Int. Conf. LEAF*–PRO 89, Pisa-Perugia-Viterbo, Italy. 404–411

© R.K. Gupta & V.D. Pandey (*Editors*)
Advances in Applied Phycology, 2007
Daya Publishing House, New Delhi

Chapter 10

Nitrogen Fixing Cyanobacteria and their Potential Applications

K.K. Choudhary, S.S. Singh and A.K. Mishra*

Department of Botany, Banaras Hindu University, Varanasi – 221 005

ABSTRACT

Cyanobacteria are unique organism confining the cellular structure of prokaryotes and metabolic activities of eukaryotes with varieties of species of diverse habitats. Cyanobacteria produce large number of metabolites in response to their environment to cope with the adverse condition. These small organisms are on high demand among scientific world due to their high potential for the production of different metabolites. These include conversion of atmospheric nitrogen into ammonia *i.e.* bio-fertilizer, pigments production, antibiotic production, bio-flocculant, vitamins, anti-viral drugs, enzymes, growth hormones, bioremediation, polysaccharides, etc. In this article, some of the bio-potential of nitrogen fixing cyanobacteria in respect to bio-fertilizer, polysaccharides and pigment in particular reference to carotenoids and phycobiliproteins have been discussed.

Introduction

Cyanobacteria are the only group, which shares the metabolic activity of eucaryotes and cellular structure of prokaryotes. Many cyanobacteria metabolize molecular nitrogen from the air and convert it to ammonia with the help of enzyme nitrogenase deriving the required energy from sun (Adams and Duggan, 1999). In addition to augmenting, the cyanobacteria increase the availability of fertilizer N to the crop. Substantial quantities of amino acids like aspartic and glutamic acid and alanine, vitamins like B_{12} and auxin like substances are liberated by nitrogen fixing blue green algae (Gupta and Shukla, 1969). These extrametabolites benefit the crop growth enabling the crop plants to utilize more of the available nutrients. Many cyanobacteria have been also shown to mobilize the insoluble phosphate in

* Corresponding Author: E-mail: akmishraau@rediffmail.com.

the soil thereby increasing its availability to the crop plants (Roger *et al.*, 1987). The present day input intensive agriculture has become a molar source of pollutants originating through indiscriminate use of agrochemicals. Selective uptake and bioaccumulation of heavy metals and biodegradation of toxicants by cyanobacteria attain importance in remedying the polluted habitats like ponds and estuaries. Many cyanobacteria like *Microcysts, Oscillatoria* and *Lyngbya* act as bio-indicators of degree and type of pollution. Harvesting of nitrogen, phosphate and potassium from the wastewater through the production of algal biomass is an important component of integrated recycling of plant nutrients. Biological reclamation of sewage using algal photosynthesis is highly efficient and cost effective and generated products with varied end uses. A relatively new field of possible exploitation of cyanobacteria has arisen in last decade by the growing industrial interests towards polysaccharides and pigments of microbial origins. Cyanobacteria envelope outermost layer by developing mucilaginous sheath of polysaccharidic nature to protect them from desiccation (Hill *et al.*, 1995). These polysaccharides are found to be commercial interest in respect to human welfare. These polysaccharides are highly charged particles that make them compatible in removal of toxic metals from the polluted water by forming stable gel by binding with metallic ions. The viscous nature of polysaccharides enables them to make the soil more fertile by increasing the water holding capacity of soil. They also have capability of forming gels with good tensile strength and serve as an emulsifying agent. The immunological potential of phosphate containing polysaccharides is under investigation. The pigments, particularly carotenoids and phycobiliproteins, are of commercial interest because of their safer and non-toxic application as colouring agent of food, beverages, dyes, drugs and cosmetics products as a substrate to the synthetic pigments. Carotenoids also serve as a second line of defense against modification of low density proteins (Ausich, 1977). Recent studies have revealed the significance of phycocyanin in fluorescence immunoassays and fluorescence microscopy for diagnostic and biomedical research.

In this article, attempts have been made to discuss some of the important aspects of research in cyanobacteria for different purposes of human welfare particularly with reference to their use as biofertilizers, polysaccharides and pigments.

Cyanobacteria as a Biofertilizers

The ever-increasing population of human kind is posing a problem on limited agricultural fields to meet the demand of food and fodder for the humans with maintaining the ecosystem sustainable. To meet this challenge, new high yielding varieties with improved agricultural technologies have been developed which requires more chemical fertilizers for maximum production. The excess use of chemical fertilizers is resulting into the accumulation of chemicals in insoluble forms in the soil that results into the multi-nutrient deficiencies and soil infertility and also, accumulated insoluble forms of fertilizers are passing into the grains which are hazardous for the humans as well as animals.

Cyanobacteria are prokaryotic organisms which possess a number of unique biological characteristics. They are the only prokaryotes (except *Prochloron*) which carry out phototrophic mode of growth which utilizes water, sunlight and carbon dioxide from the nature and capable of nitrogen fixation because of possession of nitrogenase complex, an enzyme restricted to a few selected procaryotes. Therefore, diazotrophic cyanobacteria have a great potential as biofertilizers and their use will certainly decrease the demand for chemical fertilizer that is hazardous for plants as well as animals. The agronomic potential of cyanobacteria, either free–living or in symbiotic association with water fern *Azolla*, has long been recognised (Singh, 1961; Venkatraman, 1972; Whitton and Roger, 1989; Rai, 1990). Distribution of these organisms in diverse habitats with highly variable morphological features *viz.* unicellular, colonial, unbranched filamentous, heterocystous and branched heterocystous forms are well documented (Desikachary, 1959). Heterocystous forms consists of heterocyst–a specialized cells that contain the nitrogen fixing mechanism (Stewart, 1980; Adams and Duggan, 1999)

The different rice fields of India and abroad have been surveyed for the presence of cyanobacteria and a number of papers have been published from different regions of globe and from India (Venkataraman, 1972; Singh *et al.*, 2001). In majority of the cases studied so far regarding the presence of cyanobacteria in the rice fields, the heterocystous forms *i.e.* nitrogen fixers are in abundance. Most commonly appearing cyanobacteria are *Anabaena* sp. and *Nostoc* sp. In addition to this, *Scytonema* sp., *Calothrix* sp., *Nodularia*, *Gloeotrichia* sp., *Aulosira*, are common in rice field.

Heterocystous forms of cyanobacteria have been studied extensively for the development and function of 'heterocyst' that contain an enzyme complex *i.e.* nitrogenase. This enzyme consists of dinitrogenase reductase (homodimeric Fe protein) and dinitrogenase (a heterotetrameric Mo Fe protein). The enzyme nitrogenase is extremely oxygen sensitive and reduces atmospheric dinitrogen to ammonia only under anaerobic condition. The dinitrogen reduced to NH_3 (ammonia) is made available to rice plant through exudation autolysis and microbial decomposition (Roger *et al.*, 1987). The study reveals that the inoculation of cyanobacterial as a nitrogen fertilizer contributes about 25-30 kg of nitrogen per hectare per season and increases productivity of rice upto 10-15 per cent. Encouraged from such results, cyanobacterial biofertilizer programme (Patterson, 1996) was developed which primarily aims at preparation and distribution of biofertilizers to farmers. For this purpose, the cyanobacterial flora of different region for their nitrogen fixation is studied and the strains showing better nitrogen fixation efficiency were selected and used as a biofertilizer mother inoculum.

The algalization of field with cyanobacterial biofertilizers resulted into the enhanced grain yield and strain yields (Nierzwicki-Baur, 1990). Singh (1961) reported that alkaline usar soils in Uttar Pradesh could be reclaimed by using cyanobacteria to neutralize the pH of the soil. The cyanobacteria are also known to increase soil fertility by enhancing the available N and P (Roger *et al.*, 1987). Roger and Kulasooriya (1980) reported the relative increase to grain yield over control by 28 per cent in pot experiments while15 per cent grain yield in field experiments in India has been reported by Venkataraman (1993). Further studies on application of cyanobacterial biofertilizer in field reveals that in addition to increasing the grain and straw yield, the increase of N content of grain and straw has been also reported. The other benefits are increase in plant height, number of spikelet per panicle, and amount of dry matter (Singh, 1961; Rao *et al.*, 1977).

Encouraged from the results obtained so far from the inoculation of cyanobacteria as biofertilizer, scientists attempted to categorize the factors that are influencing the efficiency of cyanobacterial strains as a biofertilizer in the field and to develop genetically improved resistant strains. One, strategy, in the improvement of nitrogen fixation efficiency of heterocystous cyanobacteria is to increase the heterocyst frequency so that the expression of nitrogenase is increased. Strings of heterocyst have been reported in *Nostoc linckia* which fixes more nitrogen compare to the wild type (Singh and Tiwari, 1969). Buikema and Haselkorn (1991) reported the formation of multiple heterocysts in *Anabaena* 7120 carrying extra copies of *hetR* gene which is essential of heterocyst differentiation. Such strains with elevated level of nitrogen fixation ability can be used as an efficient biofertilizer. Further studies revealed that herbicides used to control the weeds are damaging the photosynthetic apparatus of the cyanobacteria and thus reduces the N_2-fixing activity of surviving cyanobacterial cells. Selection of genetically stable strains, which can efficiently fix nitrogen in the presence of field dose concentrations of herbicide commonly used in rice fields. Such strains combined with excretion of ammonia have been reported to serve as ecofriendly biofertilizer (Tiwari *et al.*, 1991). A number of herbicide resistant mutants of heterocystous and non-heterocystous cyanobacteria have been reported by several scientists (Astier *et al.*,1980; Tiwari *et al.*, 1981). Interest is also being increased with respect to ammonia liberation through immobilization of nitrogen fixing cells, which liberate ammonia in the presence of glutamate analogue (MSX) which is specific inhibitor of glutamine synthetase (GS), the primary enzyme of ammonia assimilation (Kerby *et al.*, 1986; Spiller *et al.*, 1987, Thomas, 1990). Using transposon mutagenesis, MSX resistant glutamine synthetase impaired mutants have been isolated and characterized for their ability to release N_2–fixed ammonia extracellularly in the growth medium (Mahasneh *et al.*, 1994).

The future of Water fern *Azolla* as nitrogen nutrition is also very promising because it fixes atmospheric nitrogen with the help of its cyanobiont, *Anabaena azollae*, and serves as a potential biofertilizer for rice crop (Nierzwicki-Baur, 1990; Watanabe and Liu, 1992; Singh and Singh, 1997). Infact, *Azolla-Anabaena* symbiotic association fixes more nitrogen compare to the free living cyanobacteria and it could fix as much as 3.6 kg N/ha/day. There are several reports concerning the beneficial effects of *Azolla* on the growth and yield of rice (Watanabe, 1987; Singh and Singh, 1990 and 1992). Its application also improve the soil fertility and has a residual effects on the yield of other crops (Mahapatra and Sharma, 1989; Kannaiyan, 1990) and also suppresses weed growth (Satapathy and Singh, 1985) as well as methane efflux from the flooded rice fields (Bharati *et al.*, 2000). Although *Azolla* biofertilizer technology is very simple, cost effective and also ecofriendly, it has not yet spread in India mainly due to the problems associated with production and transport of huge amount of fresh inoculum.

Cyanobacteria as a Source of Polysaccharides

The exploitation of cyanobacteria as polysaccharides is emerging as new area of biotechnological application for the human welfare. The polysaccharides of microbial origin show advantages over the polysaccharides extracted from plants or macroalgae. The cyanobacteria have been considered for polysaccharide production due to their capability to excrete mucilaginous material (Drews *et al.*, 1982 and Bertocchi *et al.*, 1990) with potential useful properties. In fact, most of the cyanobacterial strains possess additional surface structures, mainly of a polysaccharide nature, that comprise of sheaths, capsules and slimes. During cell growth in batch cultures aliquots of the polysaccharides material of both capsules and slimes may be released as water soluble material into the surrounding medium, causing a progressive increase in its viscosity. These water soluble released polysaccharides (RPSs) can be easily recovered from liquid cultures, are attracting much attention because of their application for a variety of industrial purposes and make cyanobacteria an ideal organism for present day biotechnological sources of new polymers.

As it is evident from accumulated reports regarding the existence of cyanobacteria in diverse habitats, the production of polysaccharides is directly related with the environmentally stresses that affects the growth and ultimately survival of the cyanobacteria. The polysaccharides investments serve as a boundary between cyanobacterium and surrounding (Costerton *et al.*, 1981; Whitefield, 1988). They further showed their protective role against desiccation, antibacterial agents or predations by protozoans. Hill *et al.* (1994) proposed that *Nostoc commune* secretes a polysaccharide which acts as a repository of water and butter between cells and atmosphere. Hill *et al.* (1997) further showed that addition of polysaccharide to artificial membrane vesicles prevents membrane fusion which is the main damage process in the desiccation. In this way *Nostoc commune* stabilizes itself *in vivo* under adverse condition. Fattom and Shilo (1984) suggested that in benthic cyanobacteria, the attachment of cells to the sediment is facilitated by cell hydrophobicity which is determined by extracellular polymeric substances. Martin *et al.* (1974) suggested that the slimy layer surrounding trichomes facilitate the homogenous dispersion of trichomes into the liquid medium to improve light utilization and nutrient uptake. Similarly, mucilaginous sheath surrounding heterocysts of *Nostoc cordubensis* protects nitrogenase enzyme complex from the oxygen inactivation (Prosperi, 1994).

Reddy *et al.* (1996) while studying the role of mucilaginous envelope in *Cyanothece* BH68 suggested that in addition to acting as a physical barrier to the atmospheric oxygen, also serve as a chelates for iron and calcium which are both essential for nitrogen fixation. In several other cases, it has been suggested that cyanobacterial polysaccharides, which are mostly characterized by their anionic nature, play an important role in the sequestering or immobilization of metal ions, which are respectively essential or harmful to bacterial life (De Philippis *et al.*, 1991; Bender *et al.*, 1994). Robins *et al.* (1986) demonstrated the role of exopolysaccharides released by the cyanobiont *Anabaena azollae* in attachment of this cyanobacterium to the surfaces and the cavities of the fronds of the host plant *Azolla filiculoides*. Gantar *et al.* (1995) showed that the firm associations of the filaments of a *Nostoc* strain to the roots of wheat are due to the binding of polysaccharide matrix that surrounds the trichomes. These studies suggest that the different cyanobacterial strains release different polysaccharidic material, in response to environmental stresses to protect themselves from condition for their survival.

The cyanobacteria studied so far for their potential for polysaccharide production. 10 different monosaccharides have been found. These are (a) hexoses: glucose, galactose and manure; (b) pentoses: ribose xylose and arabinose; (c) deoxyhexoses: fercose and shamnose and (d) acidic hexoses: glucuronic and galactuionic acid. Some others (Bender *et al.*, 1994; Fischer *et al.*, 1997; Gloaguen *et al.*, 1995) have shown the presence of methyl sugars and amino sugars in released polysaccharides by cyanobacteria. Out of the above found monosaccharides. Most frequently observed monosaccharide is the glucose which is found in more than 90 per cent of the polymers followed by galactose, manure and shamnose (80-85 per cent of the polymers). Gloaguen et. al. (1996) showed the presence of shamnose, glucose, xylose, manure, galactose and fucose as the major components in the capsule of *Mastigocladus laminosus* and *Phormidium* sp. of a thermal spring.

The cyanobacterial production of polysaccharides have an advantages over the polysaccharides of other microbial origin in having: (i) the cyanobacterial polysaccharides of anionic nature containing two different uronic acids which is rarely found in the polymers released by strains belonging to other microbial groups, (ii) cyanobacterial RPSs show presence of one or two pentoses sugars that are usually absent in other polysaccharides of prokaryotic origin (iii) Most RPSs synthesized by cyanobacteria are quite composed of six or more monosaccharides whereas the polysaccharides of other microbal origin are usually composed of less then four (Sutherland, 1994). The most important feature of the cyanobacterial production of the polysaccharides over the other microbial origin of polysaccharide is that many other micro-organisms accumulating significant amounts of exopolysaccharides in natural habitats lose this property when cultivated under laboratory conditions, where the environmental pressure has been removed (Costerton *et al.*, 1981) whereas cyanobacteria retain this property and thus need to be tested with regard to their stability of RPS production.

The potential of polysaccharides to increase the viscosity of aqueous solution, forming gels with good tensile strength, stabilizing emulsion etc. make them to use as industrial scale for different purposes. Another important feature that contribute to the chemical and physico-chemical properties of the polysaccharides is the presence of a polypeptide moiety or other non-saccharide components such as organic (*e.g.* acetyl, pyruvyl, succinyl groups) or inorganic (*e.g.* sulphate or phosphate groups) substituents (Sutherland,1994). The release of phosphate containing polysaccharides by cyanobacteria is attracting much attention because of their possible immunological significance (Sutherland,1994). The cyanobacteria also release sulfated polysaccharides which have inhibitory properties against various types of viruses (Witvrouw *et al.*, 1997) and tumours (Riou *et al.*, 1996). Further it was showed that the presence of high concentration of charged components like uronic acids, sulphate or phosphate groups, pyruvato ketals) usually form stable gels in the presence of metallic ions and are the most promising for the removal of toxic metals from polluted water (Bender *et al.*, 1994; Gloaguen, 1996;).

The production of polysaccharides from other microbes and from plants are in progress at industrial scale so it will be very difficult, in spite of several advantages over the released polysaccharides by microbes and plants, for cyanobacterial originated polysaccharides to get an actual industrial perspectives for uses with a large market due to a wide gap between current knowledge and application as a biotechnology programme. But the use of photoautotrophic micro-organisms as a source of natural biopolymers will very likely attract more and more attention.

Cyanobacteria as a Source of Pigments

The extraction of pigments from cyanobacteria is another most important area of present day research. Cyanobacteria synthesize a number of pigments *viz.* chlorophyll (chlorophyll a), carotenoids and phycobiliproteins and scytonemin. Among these are carotenoids and phycobiliproteins, due to their biotechnological application, attracting the attention of global scientific community to evolve an effective and cost effective method to exploit cyanobacteria for pigment production at a large scale. In some species of cyanobacteria, the greater quantities of pigment has been observed and experimentally reported. The chlorophyll made by cyanobacteria is the same as that made by the majority of land plants. Chlorophyll has not made much more impact as a biotechnological application except their role in photosynthesis. The another pigment, scytonemin, considered to be UV blocking pigment has also not received much more attention due to very little information. This is generally found in cyanobacteria growing under stress condition particularly at Polar Regions.

The need of exploitation of cyanobacteria for pigment production is because of the carcinogenic effect of chemically produced pigments. Also, the intermediate products produced during the synthesis of pigments by artificial method are harmful. The carotenoids produced by cyanobacteria as well as by other biological organisms, have a wide commercial application and are used for natural food colouring (Emodi, 1978), as feed adhesives, to enhance flesh colour of fish and egg yolk colour (Schiedt *et al.*, 1985). They are also known to improve the health and fertility of lot-fed cattle (Jakson *et al.*, 1981), and also served as photoprotective agent (Krinsky, 1976), growth promoters and aid in cancer prevention (Peto *et al.*,1987). Another most desired pigment is phycobiliproteins which are fine chemicals and are used as natural colours and as fluorescent dye (phycoflour probe) in immunoassays and are important components of diagnostic kids.

Carotenoids are largest class of naturally occurring pigment that play various biological roles. More than 600 carotenoids have been identified till today. They provide various colours like yellow, orange and red, in original. Carotenoids are synthesized by all photosynthetic organisms as well as non-photosynthetic bacteria and fungi. Most carotenoids are composed of a C_{40} hydrocarbon backbone, constructed from eight C_5 isoprenoid units and contain a series of conjugated double bond, carotenes don't contain oxygen atoms and are either linear or cyclized molecules containing one or two end rings. Xanthophylls are oxygenated derivatives of carotenes. Various glycosylated carotenoids are ester carotenoids have been identified. The C_{40} backbone can be further extended to give C_{45} or C_{50} carotenoids or shortened to get apocarotenoids.

The carotenoids in photosynthetic organisms serve as light harvesting pigments in the range of 450-570 nm and in prevention of photo-oxidative damage. They are integral constituents of the protein-pigment complexes of the light harvesting antennae in photosynthetic organisms and are also an important component of the photosynthetic reaction centre. Newman and Sherman (1978) have described that carotenoids are associated with cyanobacterial photosystem PS II. There are two b-carotene molecules in the reaction centre core of PS II (Ohno *et al.*, 1986; Newell *et al.*, 1993) that probably protect chlorophyll P680 from photodamage in isolated PSII reaction centres (De Las Rivas *et al.*, 1995) and this may be related to the degradation against of the D1 subunit of PS II (Sandmann *et al.*, 1993). An additional function of carotenoids is to protect the photosynthetic apparatus caused due to the excited triplet state of chlorophyll by absorbing triplet state energy from chlorophyll (CAR: ^{3}CHL ® ^{3}CAR: CHL) due to presence of nine or more carbon-carbon double bonds, and thus prevent the formation of harmful singlet-state oxygen radicals (O_2 + ^{3}CHL ® $^3O_2$ + CHL). Buckley and Houghton (1976) showed the accumulation of b-carotene in a UV resistant mutant of *Gloecocapsa alpicola* probably protecting species damage by UV irradiation. The organisms produces carotenoids to bulbils their demand for their own existence in response to exposure to different environmental condition and also for their biochemical activities. Due their non-toxic nature colourful nature they are of commercial importance for present day demand for humans and animal consumption.

Cyanobacteria synthesize the carotenoids as do the higher plants but they produce some unique types of oxygenate carotenoids *i.e.* Xanthophytes such as Ketocarotenoids, 2-hydroxyderivatives and glycosides. The most common carotenoids synthesized by cyanobacteria are b-carotene, Zeaxanthus. Ketocarotenoid echineene, as the carotenoid-glycoside myxoxanthaphyll. Carotenoids are produced from general isoprenoid biosynthetic pathway (See review article Yokohama *et al.*, 1982; Britton, 1988). Carotenoids are extracellular components and cannot be excreted into the medium in a fermentation process. There are several key consequences of this physiological situation. First, the key productivity parameters for biological carotenoid production are the cost of biomass production, the concentration of the carotenoid of interest inside the cell, and the metabolic activity of the cells producing the carotenoids. Further, the cost of processing and purification are very high.

Carotenoids are currently being produced for animal and human consumption. Carotenoids, in addition to colouring material, are also used as constituents in vitamins, and dietary supplements. The studies have shown that a diet rich in carotenoids can lead to a reduced risk of heart disease, cancer, eye disease and other diseases. With this increased awareness by the vitamin manufactures as well as the consuming public, there is increasing interest in the biological production of carotenoids by many companies.

There are several commercial operational currently used to produce carotenoids for human as well as on animal consumption. The production of b-carotene by unicellular algae *Dunaliella* sp. is a well developed technology. Considerable research has been performed to develop other biological systems including cyanobacteria for the carotenoids. But due to cost sensitive production of carotenoids from cyanobacteria, these organisms are not making their place in industrial market. The production cost for these and for the other systems that have been developed is greater than that of the corresponding chemical synthesis. The researches on carotenoid production by biological systems, two strategies have been formulated. Firstly, to increase the efficiency of biomass production as to increase the efficiency of carotenoid synthesis. Further it has been suggested that carotenoid synthesis is governed in the cell by the level and the activity of the carotenoid biosynthetic enzymes which can be altered by the reconsignment DNA technology by isolating and characterizing the genes responsible for the synthesis of the compound as well as promotes that will be active when the genes need to be expressed maximally and their incorporation and expression into the desired hosts (Ausich, 1977). A very little approach has been made to transform the cyanobacterial system for carotenoid production to make them ideal candidate for industrial production.

Cyanobacteria need a consistent and effective approach to optimize the growth condition for biomass production to get maximum carotenoids and also efforts should be taken to minimize the cost of production and purification in comparison to chemically synthesized carotenoids.

The phycobiliproteins (PBPs) present in the cyanobacteria are natural pigments and their use in beverages, foods and cosmetics as biocolorants is spreading widely since they have been proved to be safer and non-toxic compared to the conventional natural colorducts. The aesthetic beauty of cyanobacteria is due to the presence of intense fluorescent blue and red, water soluble chromoproteins *i.e.* phycobiliproteins, which represent 50 per cent of the total cellular proteins. The components of their chromoproteins include phycocyanin (blue pigment), phycoerythrin (red pigment) and allophycocyanin. Recently, increasing attention has been paid to exploit the cyanobacterial potential for obtaining industrially valuable components.

Phycobiliproteins are high molecular mass complexes ranging from approximately 1,00,000 to 2,08,000 daltons and consist of two dissimilar subunits a(Mr10000-19000) and b (Mr 14000-21000), in a 1:1 molar ratio (Bennet and Bogorad, 1971; Gysi and Zuber, 1974). Teale and Dale (1970) reported that the phycoerythrin and phycocyanin contain two types of chromophores, a sensitizing ('s') and a fluorescing ('f'). According to Glazer *et al.*, 1973, the b-and subunits of phycocyanin carried two w-type and one f-type chromophores, respectively, with the energy absorbed by the b subunit being transferred, via the subunit, to the next acceptor. Such a process could indeed provide an efficient directional flow of energy to the chlorophyll-a located in the thylakoid membrane. These phycobiliproteins were together termed as 'phycobilisome' (Gan't and Conti,1966a, 1966b).

The phycobiliproteins play an important role in light harvesting and transfer them to chlorophyll. Arnold and Oppenheimer (1950) established that the migration of light energy proceeds from phycocyanin to chl-a by resonance energy transfer subsequent studies by other investigators suggested that phycoerythrin transfer light energy to chlorophyll-a via phycocyanin (French and Young, 1952). In the 1970s, it became clear that light energy collected by phycobiliproteins within the phycobilisomes was transferred mainly to photosystem II, with allophycocyanin acting as a link between phycocyanin and the membrane bound chlorophyll-a (Halldal, 1970; Lemasson *et al.*, 1973).

With the establishment of role of phycobiliproteins as a colouring materials for daily use materials and the presence of these proteins in the cyanobacteria, researchers attempted to develop a cost effective methods to exploit these organisms for biosafe colorants. Presently, chemicals used as colorants in foodstuffs cause health hazards to humans. As phycobiliproteins are found to be commercially important as natural colorants and which account for 20-28 per cent of the dry weight of cyanobacteria (Patterson, 1996), they would serve as potential alternate biosafe colorants. Though numerous reports have been accumulated so far but a little of cyanobacterial sp. have been projected for industrial production. Out of them, only *Spirulina* sp. is used industrially for production of phycobiliprotein and other biosafe compounds. The cost of processing and purification from cyanobacteria, like carotenoids, are so high in comparison to chemically produced food colorants, that they have not received much more attention at industrial scale. The nitrogen fixing heterocystous, filamentous cyanobacteria are particularly attractive for the photoproduction of phycobiliproteins and other chemicals because they don't require nitrogen source as they utilize atmospheric molecular nitrogen. The lack of combined nitrogen in the culture media, aside from its economical implication, restricts the problem of contamination by other organisms. Moreover, the filamentous nature of these organisms facilitates separation of biomass from the medium. Despite these obvious advantages for mass production, available information regarding outdoor production of filamentous N_2-fixing cyanobacteria is very poor

Future Prospects

The present information on cyanobacterial potential for various secondary products has not led to a resounding commercial success. This is only because of cost-intensive assessment and processing of products in comparison to other biological sources like *E. coli* and Yeast. The present informations have simply developed the expertize necessary to take advantage of the new opportunities that are projected for the future. Therefore, one should must take into account the unique characteristics of cyanobacteria to make cyanobacterial biotechnology became a significant presence in the commercial sector. The challenge for researchers working in this field is to concentrate on what cyanobacteria have that those other systems don't possess. The most obvious area to explore, as has been discussed above, is to look for production of secondary products use in the pharmaceutical industry by natural product screening and developing a cost-effective technique so that they can compete with the other biological systems which have received much attention in the industrial market. The future of cyanobacterial biotechnology is promising and with emergence and acceptance of "greener technology" with the change in environment, the photo-trophic microorganisms will receive more and more attention. It is time to take the basic research and fuse it with potential applications. This is a challenge that cyanobacterial biotechnology must meet in future.

References

Adams DG and Duggan PS (1999) Heterocyst and akinete differentiation in cyanobacteria. *New Phytol* 144: 3-33.

Arnold W and Oppenheimer JR (1950) International conversion in the photosynthetic mechanism of blue-green algae. *J Gen Physiol* 33: 423-435.

Astier C, Vernotte C, Der-Vertanium M and Joset-Espardellier F (1980) Isolation and characterization of two DCMU-resistant mutants of the blue-green alga *Aphanocapsa* 6714. *Pl Cell Physiol* 20: 1501-1510.

Ausich RL (1977) Commercial application of carotenoid production. *Pure and Appl. Chem* 69: 2169-2173.

Bender J, Rodriguez-Eaton S, Ekanemesang UM and Phillips P(1994) Characterization of metal binding bioflocculants produced by the cyanobacterial component of mixed microbial mats. *Appl Environ Microbiol* 60: 2311-2315.

Bennet A and Bogorad L (1971) Properties of subunits and aggregates of blue-green algal biliproteins. *Biochem.,* 10: 3625-3634.

Bertochhi C, Navarini L, Cesaro A and Anstasio M (1990) Polysaccharides from cyanobacteria. *Carbohydr Polym* 12: 127-153.

Bharati K, Mohanty SR, Singh DP, Rao VR and Adhya TK (2000) Influence of incorporation or dual cropping of *Azolla* on methane emission from a flooded alluvial soil planted to rice in Eastern India. *Agric Ecosyst Environ* 79: 73-83.

Britton G (1988) Biosynthesis of carotenoids. In: Goodwin TW(ed), *Plant Pigments.* Academic Press, London, 133–180.

Buckley CE and Houghton JA (1976) A study of the effects of near UV radiation on the pigmentation of the blue-green alga *Gloeocapsa alpicola. Arch Microbiol* 107: 93–97.

Buikema WJ and Haselkorn R (1991) Characterization of a gene controlling heterocyst differentiation in the cyanobacterium *Anabaena* 7120. *Genes Dev* 5: 321-330

Costerton JW, Irvin RT and Cheng KJ (1981) The bacterial glycocalyx in nature and disease. *Ann Rev Microbiol* 35: 299-324.

De Las Rivas J, Klein J and Barber J (1995) pH sensitivity of the redox state of cytochrome b559 may regulate its function as a protectant against donor and acceptor side photoinhibition. *Photosynth.* Res 46:193-202.

De Philippis R, Sili C, Tassinato G, Vincenzini M and Materassi R(1991) Effects of growth condition on exopolysaccharide production by *Cyanospira capsulata. Biores Technol* 38: 101-104.

Desikachary TV(1959) *Cyanophyta*. Indian Council of Agric. Res., New Delhi.

Drews G and Weckesser J (1982) Function, structure and composition of cell walls and external layers. *In*: Carr NG and Whitton BA (eds.),*The Biology of Cyanobacteria*. Blackwell Scientific, Oxford, 333-357.

Emodi A (1978) Carotenoids: properties and applications. *Food Technol* 32: 38-42.

Fattom A and Shilo M (1984) Hydrophobicity as an adhesion mechanism of benthic cyanobacteria. *Appl Environ. Microbiol* 53: 2226-2230.

Fischer D, Schlosser UG and Pohl P(1997) Exopolysaccharide production by Cyanobacteria grown in closed photobioreactors and immobilized using white cotton to welling. *J Appl Phycol* 9: 205-213.

French CS and Young VK(1952) The fluorescence spectra of red algae and the transfer of energy from phycoerythrin to phycocyanin and chlorophyll. *J Gen Physiol* 35:873-890.

Gantar M, Rowell P, Kerby NW and Sutherland IW (1995) Role of extracellular polysaccharide in the colonization of wheat (*Triticum vulgare* L.) roots by N_2-fixing cyanobacteria. *Biol Fertil Soils* 19: 41-48.

Gan't E and Conti SF (1966a) Granules associated with the chloroplast lamellae of the *Porphyridium cruentum*. *J Cell Biol* 29:423-434.

Gan't E and Conti SF (1966b) Phycobiliprotein localization in algae. *Brookhaven Symp Biol* 19:393-405.

Glazer AN and Fang S (1973) Spectroscopic properties of C-phycocyanin and it's a and b subunits. *J Biol Chem* 16: 5679-5685.

Gloaguen V, Morvan H and Hoffmann L (1995) Released and capsular polysaccharides of Oscillatoriaceae (Cyanophyceae, Cyanobacteria). *Algol Studies* 78: 53-69.

Gloaguen V, Vebret L, Morvan H and Hoffmann L (1996) Chemical composition of the capsular polymers of the blue green algal mats of the thermal spring of Neris-les-Bain (France). *Algol Studies* 83: 245-355.

Gysi J and Juber H (1974) Isolation and characterization of allophycocyanin II from the thermophilic blue-green alga *Mastigocladus laminosus* Cohn. *FEBS Lett* 48: 209-213.

Gupta AB and Shukla AC (1969) Effect of algal extracts of *Phormidium* sp. on growth and development of rice seedlings. *Hydrobiologia* 34: 77-84.

Halldal P (1970). The photosynthetic apparatus of microalgae and its adaptation to environmental factors. *In*: Halldal P (ed), *Photobiology of Microorganisms*. Wiley, London,17-55.

Hill DR, Keenan TW, Helm RF, Potts M, Crowe LM and Crowe JH (1995) Extracellular polysaccharide of *Nostoc commune* (cyanobacteria) inhabits fusion of membranc vesicles during desiccation. *J Appl Phycol* 9: 237-248.

Hill DR, Peat A and Potts M (1994) Biochemistry and structure of the glycan secreted by desiccation-tolerant *Nostoc commune* (cyanobacteria). *Protoplasma* 182: 126-148.

Jakson PS, Furr BJA and Johnson CT (1981) Endocrine and ovarian changes in dairy cattle fed a low B-carotene diet during an oestrus synchronization regime. *Res Vet Sci* 31: 377-383.

Kannaiyan S (1990) *Biotechnology of Biofertilizer for Rice Crop*. Tamil Nadu Agric Univ, Tamil Nadu, 1-225.

Kannaiyan S (1992) *Azolla Biofertilizer Technique for Rice*. Tamil Nadu Agric Univ, Mag, 1-56.

Kerby N W, Rowell P and Stewart WDP (1986) The uptake and metabolism of methylamine by nitrogen fixing cyanobacteria. *Arch Microbiol* 143: 353-358.

Krinsky NI (1976) The survival of vegetative microbes. In: Gray TGR and Postgate JR (eds), *Symposia of the Society for General Microbiology*. 26: 209-239.

Lemasson C, Tandeau de Marsac N and Cohen-Bazire G (1973) Role of allophycocyanin as a light-harvesting pigment in cyanobacteria. *Proc Natl Acad Sci, USA*. 70: 3130-3133.

Mahapatra BS and Sharma GL (1989) Integrated management of *Sesbania, Azolla* and urea nitrogen in lowland rice under rice wheat cropping system. *J Agric Sci* (Cambridge) 113: 203-206.

Mahasneh IA, Mishra AK and Tiwari DN (1994). Transposon-induced mutants of *Anabaena* sp. PCC7120 capable of ammonia liberation, *Biotechnol Lett* 16: 765-770.

Martin TJ and Wyatt JT (1974) Extracellular investments in blue green algae with particular emphasis on genus *Nostoc*. *J Phycol* 10: 204-210.

Newell RW, Van Averonge H, Barber J and Macino G (1993) Spectroscopic characterization of the reaction centre of PSII using polarized light: Evidence for beta-carotene excitons in PSII reaction centres. *Biochim Biophys Acta* 1051: 232-238.

Newman PJ and Sherman LA (1978) Isolation and characterization of Photosystem I and II membrane particles from blue-green alga *Synechococcus cedrorum*. *Biochim Biophys Acta* 503: 343-361.

Nierzwicki-Bauer SA (1990) *Azolla-Anabaena* symbioses: Use in agriculture. In: Haeselkorn R (ed), *Handbook of Symbiotic Cyanobacteria*. CRC Press, Florida, 119-136.

Ohno T, Satoh K and Katoh S (1986) Chemical composition of purified oxygen-evolving complexes from the thermophilic cyanobacterium *Synechococcus* sp. *Biochim Biophys Acta* 852: 1-8.

Patterson GML (1996) Bitechnological applications of cyanobacteria. *J Sci Ind Res* 55: 669-684.

Peto R, Doll R, Buckley J. and Sporn MB (1981) Can dietary beta-carotene materially reduce human cancer rates? *Nature* 290: 201-208.

Prosperi CH (1994) A cyanophyte capable of fixing nitrogen under high levels of oxygen. *J Phycol* 30: 222-224.

Rai A N (1990) *Handbook of Symbiotic Cyanobacteria*. Boca Raton, FL, USA: CRC Press.

Rao JL, Venkatachari A, Rao SWVB and Reddy RK (1977) Individual and combined effects of bacterial and algal inoculation on the yield of rice. *Curr Sci* 46: 50-51.

Reddy KJ, Soper BW, Tang J and Bradley RL (1996) Phenotypic variation in exopolysaccharide production in the marine, aerobic nitrogen-fixing unicellular cyanobacterium *Cyanothece* sp. World *J Microbio Biotechnol* 12: 311-318.

Riou D, Colliec-jouault S, Pinezon du Sel D, Bosch S, Siavoshian S, Le Bert V, Tamasoni C, Sinquin C, Durand P and Roussakis C (1996) Antitumour and antiproliferative effects of a fucan extracted from *Ascophyllum nodosum* against a non-small-cell bronchopulmonary carcinoma line. *Anticancer Res* 16: 1213-1218.

Robins RJ, Hall DO, Shi DJ, Turner RJ and Rhodes MJC (1986) Mucilage acts to adhere cyanobacteria and cultured plant cells to biological and inert surfaces. *FEMS Microbiol Lett* 34: 155-160.

Roger PA and Kulasooriya SA (1980) *Blue-green algae and rice.* Int Rice Res Inst., Los Banos,Philippines.

Roger PA, Santiago-Ardales S, Reddy PM and Watanabe I (1987) The abundance of heterocystous blue-green algae in rice soils and inocula used for application in rice fields. *Biol Fertility Soils* 5: 98-105

Sandman G, Kuhn M. and Boger P (1993) Carotenoids in photosynthesis-protection of D1 degradation in the light. *Photosynth Res* 35:185-190.

Satapathy KB and Singh PK (1985) control of weeds by *Azolla pinnata* in rice fields. *J Aquat Plant Manag* 23: 40-42.

Schiedt K, Leuenberger FJ, Vecchi M and Glinz E (1985) Absorption retention and metabolic transformation of carotenoids in rainbow trout, salmon and chicken. *Pure and Appl Chem* 57: 685-692.

Singh AL and Singh PK (1990) Phosphorus fertilization on growth and nitrogen fixation of *Azolla* and BGA in rice fields. *Ind J Plant Physiol* 33: 21-26.

Singh PK and Singh DP (1992) *Azolla*–a potential way towards organic farming in rice cultivation in Indial conditions. In: Rai MM and Verma LN (eds), *Proc Natl Seminar Organic Farming.* Ravi Printing Press, Jabalpur, 233-240.

Singh RN and Tiwari DN (1969) Induction of ultraviolet irradiation of mutatin in the blue-green alga *Nostoc linckia* (Roth.). *Nature* (Lond.) 221: 62-64.

Singh PK and Singh DP (1997) *Azolla* and Rice cultivation. *In*: Somani LL, Bhandari, SC, Saxena SN and Vyas KK (eds), *Biofertilizers.* Scientific Publishers, Jodhpur, 39-45.

Singh RN (1961) *Role of blue-green algae in nitrogen economy of Indian agriculture.* Indian Council of Agric Res, New Delhi.

Spiller H, Lattore HC, Hassan ME and Shanmugam KT (1986) isolation and characterization of nitrogenase depressed mutants of the cyanobacterium *Anabaena variabilis. J Bacteriol* 165: 412-419.

Stewart WDP (1980) Some aspects of structure and function of nitrogen-fixing cyanobacteria. *Annual Rev Microbiol* 34: 497-536.

Sutherland IW (1994) Structure function relationships in microbial exopolysaccharides. *Biotech Adv* 12: 393-448.

Teale FWJ and Dale RE (1970) Isolation and spectral characterization of phycobiliproteins. *Biochem J* 116: 161-169.

Thomas SP, Zaritsky A and Boussiba S (1990) Ammonium excretion by an L-methionine-DL-sulfoxime-resistant mutant of the rice field cyanobacterium *Anabaena siamensis*. Appl Environ Microbiol 56:3499-3504.

Tiwari DN, Kumar A and Mishra AK (1991) Use of cyanobacterial diazotrophic technology in rice agriculture. *Appl Biochem Biotechnol* 28/29: 387-396.

Tiwari DN, Pandey AK and Mishra AK (1981) Action of 2,4-D and rifampicin on growth and heterocyst formation in *Nostoc linckia. J Biol Sci* 3: 38-39.

Venkataraman GS (1972) *Algal biofertilizers and rice cultivation*. Today and Tomorrow's Publishers, Faridabad, India.

Venkataraman GS (1993) Blue-green algae (Cyanobacteria). In: Tata SN, Wadhwani AM and Mehdi MS (eds), *Biological Nitrogen Fixation*. Indian Council of Agric Res, New Delhi, 45-76.

Watanabe I (1982) *Azolla-Anabaena* symbiosis, its physiology and use in tropical agriculture. In: Dommergues Y and Dien HG (eds), *Microbiology of tropical soils and plant productivity*. M. Nijhoff, The Hague, 169-185.

Watanabe I (1987) Summary report of the *Azolla* programme of the International Network on Soil Fertility and Fertilizer Evaluationfor rice. In: Watanabe I (ed), *Azolla utilization*. International Rice Research Institute, Manila, 197-201.

Watanabe I and Liu CC (1992) Improving nitrogen fixing system and integrating them into successful rice farming. *Pl Soil* 141: 57-67.

Whitton BA and Potts M (2000) (eds) *The Ecology of Cyanobacteria*. Kluwer Academic Publisher, Dordrecht, The Netherlands.

Whitton BA and Roger PA (1989) Use of blue-green algae and *Azolla* in rice culture. *In*: Campbell R and Macdonald RM (eds), *Microbial Inoculation of Crop Plants*. 25: 89-100.

Witvrow M and De Clercq E (1997) Sulfated polysaccharides extracted from sea algae as potential antiviral drugs. *Gen Pharmacol* 29: 497-511.

Yokohama H, Hsu WJ, Poling SM and Hayman E (1982) Chemical regulation carotenoid biosynthesis. *In*: Britton G and Goodwin TW (eds), *Carotenoid Chemistry and Biochemistry*. Pergamon, Oxford, 371-385.

© R.K. Gupta & V.D. Pandey (*Editors*)
Advances in Applied Phycology, 2007
Daya Publishing House, New Delhi

155–168

Chapter 11

Cyanobacterial Hydrogen Metabolism with Special Reference to Transcriptional Regulation of Hydrogenases

D. Vyas[1] and R.K. Gupta[2*]

[1]*Lab of Microbial Technology, Department of Botany,*
Dr. Hari Singh Gour University, Sagar, M.P., India
[2]*Department of Botany, Govt. P.G. College, Rishikesh – 249 201 Dehra Dun, Uttranchal*

ABSTRACT

Hydrogen metabolism in cyanobacteria viz. nitrogenase(s), catalysing the production of molecular hydrogen concomitantly with the reduction of molecular nitrogen to ammonia, an uptake hydrogenase, catalysing the consumption of hydrogen produced by the nitrogenase, and a bidirectional hydrogenase, which has the capacity to take up and produce hydrogen. The three enzymes can be distributed in different ways. However, multiple copies of single hydrogenases have not been found so far in any cyanobacterium. Several nitrogen-fixing strains have all the three enzymes. Others have only an uptake hydrogenase and a nitrogenase, while only a bidirectional hydrogenase is present in the unicellular strain *Synechocystis* PCC 6803. An uptake hydrogenase is consistently present when a cyanobacterium has the capacity to fix atmospheric nitrogen. An exception might be the unicellular *Synechococcus* PCC 6301 (also *Anacystis nidulans*). It has been reported that it possesses an uptake hydrogenase although it is a non-nitrogen fixing species. The distribution of the bidirectional hydrogenase is not as clear as in the case of the uptake hydrogenase and it is missing in several strains. All hydrogenases known in cyanobacteria belong to [NiFe]-hydrogenases. Transcription of genes encoding various hydrogenases is discussed in the article.

* Corresponding Author: E-mail: rajankgupta1@rediffmail.com.

Introduction

Cyanobacteria are microorganisms that can be found in very different environments such as fresh and seawater, in the soil, deserts, polar regions hot water springs, and saline environments (Wynn-Williams 2000). They also have the capacity to form symbiotic relations with other organisms (Adams 2000).

The morphological variation of cyanobacteria is considerable. Both unicellular and filamentous forms are known, and variation within these morphological types occurs (Rippka *et al.*, 1979). Cyanobacteria are able to perform chlorophyll *a* based, oxygenic photosynthesis using a photosynthetic apparatus similar to that of chloroplasts of algae and higher plants. Some cyanobacteria are obligate photoautotrophs, while others can grow heterotrophically in the dark (Rippka *et al.*, 1979). Many cyanobacteria, both filamentous and unicellular species, have the capacity to fix atmospheric nitrogen, a capability only found in prokaryotic organisms. Some filamentous species can also develop a specialized cell, the heterocyst, where nitrogen fixation takes place. The heterocyst is slightly larger and rounded compared to the vegetative cell (Wolk 1996).

In 1896 Jackson and Ellms for the first time reported that *Anabaena* species from Massachusetts reservoir immediately produce hydrogen when placed in sealed bottled. There was a considerable gap in this area of research since the first report appeared. Studies on hydrogen metabolism by blue green algae (cyanobacteria) restarted only after the unequivocal report of nitrogen fixation by this group of algae. The real hydrogen production mediated by the nitrogenase enzyme complex was first of all reported in *Anabaena cylindrica* under *in vitro* conditions. Subsequently hydrogen production by intact filaments of *A. cylindrica* in the light under an atmosphere of argon and CO_2 was reported by Benemann and Weare. In the same year Russian workers have also reported H_2 production from cyanobacteria. Since then, extensive work has been done on various aspects of H_2 metabolism employing a variety of blue green algal strians (Vyas 1992, Vyas and Kumar 1995, Vyas and Gupta 2003, Vyas 2004).

Enzymes Involved in Cyanobacterial Hydrogen Metabolism

Three enzymes have been described to be directly involved in hydrogen metabolism in cyanobacteria; (1) nitrogenase(s), catalysing the production of molecular hydrogen concomitantly with the reduction of nitrogen to ammonia, (2) an uptake hydrogenase, catalysing the consumption of hydrogen produced by the nitrogenase, (3) a bidirectional hydrogenase, that has the capacity to take up and produce hydrogen. The three enzymes can be distributed in different ways. However, multiple copies of single hydrogenases have not been found so far in any cyanobacterium (Wunschiers and Lindblad 2003).

Several nitrogen-fixing strains have all three enzymes, the two filamentous, heterocystous strains *Anabaena* PCC 7120 (Houchins and Burns 1981b, Kaneko *et al.*, 2001) and *Anabaena variabilis* ATCC 29413 (Happe *et al.*, 2000, Schmitz *et al.*, 1995). Others have only an uptake hydrogenase and a nitrogenase, *e.g.* the filamentous, heterocystous strain *Nostoc punctiforme* (also *Nostoc* PCC 73102, *Nostoc* ATCC 29133) (Tamagnini *et al.*, 1997, Meeks *et al.*, 2001) while only a bidirectional hydrogenase is present in the unicellular strain *Synechocystis* PCC 6803 (Kaneko *et al.*, 1996, Appel and Schulz 1998).

An uptake hydrogenase is consistently present when a cyanobacterium has the capacity to fix atmospheric nitrogen. An exception might be the unicellular *Synechococcus* PCC 6301 (also *Anacystis nidulans*). It has been reported that it may possess an uptake hvdrogenase despite it is a non-nitrogen fixing strain (Boison *et al.*, 1996). The distribution of the bidirectional hydrogenase is not as clear as in the case of the uptake hydrogenase and it is missing in several strains (Tamagnini *et al.*, 1997, Tamagnini *et al.*, 2000).

All hydrogenases identified in cyanobacteria belong to [NiFe]-hydrogenases. However, no cyanobacterial hydrogenase has yet been purified and crystallised.

Nitrogen Fixation and Nitrogenase

Nitrogenase is an enzyme that is directly involved in hydrogen metabolism as a consequence of its production of molecular hydrogen during nitrogen fixation.

The overall reaction of nitrogen fixation can be written as follows:

$$N_2 + 8H^{-} + 16\ ATP \rightarrow 2NH_3 + H_2 + 16\ ADP + 16P.$$

The above reaction is catalysed by nitrogenase, which consists of two separate protein components: dinitrogenase and dinitrogenase reductase. The dinitrogenase is a heterotetramer consisting of two subunits of NifK and two subunits of NifD. The dinitrogenase reductase is a homodimer of NifH and plays a role in transferring electrons from a ferredoxin, or a flavodoxin, to the dinitrogenase, where the actual reduction of N_2 occurs. The reaction requires ATP and low-potential electrons. ATP may be generated by either cyclic photophosporylarion or oxidative phosporylation and the low potential electrons can come from NADPH that may be generated from the degeneration of carbohydrates produced during photosynthesis (Haselkorn and Buikema 1992). In hetcrocystous cyanobacteria, carbohydrates are imported from vegetative cells and products of nitrogen fixation arc exported to the vegetative cells (Bohme1998). Nitrogen fixation is oxygcn sensitive since nitrogenase becomes inactivated by 0_2, (Picnkos *et al.*, 1983). Since oxygcn is produced by photosystcm II during photosynthesis, these two processes must be separated either temporally or spatially by cyanobacteria. Heterocystous cyanobacteria separate the oxygen evolution and nitrogen fixation spatially by performing photosynthesis in the vegetative cells and nitrogen fixation in the hetcrocysts. To provide an environment with low oxygen the heterocyst lacks photosystem II activity, is surrounded by a thickened cell wall to reduce the diffusion of oxygen, and has a higher respiration rate (Bohme 1998).

Nitrogenase is subject to strict regulatory controls. Nitrogen fixation is not only inhibited by O_2 but also by ammonium and nitrate (Halbleib and Ludden 2000). A key protein in the control of nitrogen metabolism in cyanobacteria is NtcA, a transcriptional regulator that belongs to the CAP family (Herrero *et al.*, 2001). NtcA is present in both unicellular and filamentous, hetcrocystous cyanobacteria (Frias *et al.*, 1993, Herrero *et al.*, 2001), has been demonstrated NtcAto be essential for heterocyst development (Wei *et al.*, 1994). NtcA binds to a palindromic target motif GTA (N8) TAG upstream of the target gene. Studies of the motif in Synechococcus PCC 7942 demonstrated that some variation in the sequence was possible while still maintaining the binding capacity. However, GT(N10)AC was found to be essential for binding NtcA (Vazquez-Bermudez *et al.*, 2002). Examples of known genes regulated by NtcA are *gln*A (glutamine synthetase), nir (nitrate assimilation), NtcA (autoregulatory), and genes important ' for hetcrocyst development, *e.g.* het C and devBC*A* (Herrero *et al.*, 2001). In addition, NtcA has also been demonstrated to bind upstream of xis*A* which is a site-specific recombinase responsible for the excision of an 11.5 kb DNA element located within nifD in *Anabaena* PCC 7120 (Ramasubramanian *et al.*, 1994).

Cyanobacterial Uptake Hydrogenase

All known cyanobacterial uptake hydrogenases consist of two subunits, encoded by hupS and hupL, respectively (Figure 11.1). The small subunit, HupS, contains the iron-sulphur [Fe-S] cluster necessary for electron transfer to the active site, which is located in the large subunit, HupL. HupS also

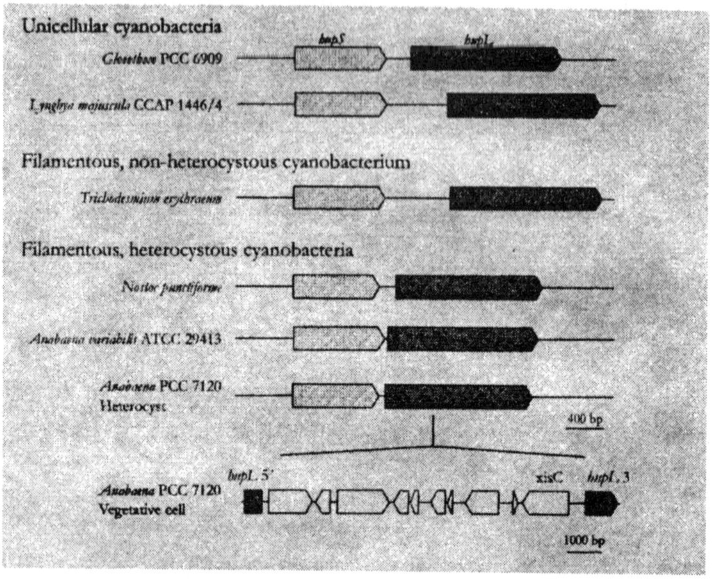

Figure 11.1: Physical Map of the Genes Encoding the Uptake Hydrogenase (hupSL) in Unicellular. Filamentous non heterocystous, and filamentous heterocystous cyanobacteria. All genes are contigous except hupL in *Anabaena* PCC 7120. In vegetative cells, a 10.5 kb DNA fragment is located–within hupL. However, this DNA fragment is erased by the site specific recombinase which is XisC located within the fragment. *Gloeothece* PCC 6909 (GenBank AY260103), *Lyngbya majuscula* CCAP 1446/4 (GenBank AF368526), *Trichodesmium erythraeum* (http://spider.jgi-psf.org/JGI_microbial/html/trichodesmium/trichod-homepage.html), *Nostoc punctiforme* (Oxelfelt *et al.*, 1998), *Anabnena variabilis* ATCC 29413 (Happe *et al.*, 2001), and *Anabaena* PCC 7120 (Carrasco *et al.*, 1995).

contains cystcins that are involved in the coordination of the [Fe-S] clusters. The small subunit in cyanobacteria differs from other small subunits from other microorganisms since the signal peptide in the N-terminus part of HupS is missing (Oxelfelt *et al.*, 1998). HupL contains two putative nickel-binding sites (R x CG x C) necessary for the coordination of the nickel in the active site (Carrasco *et al.*, 1995, Oxelfelt *et al.*, 1998, Happe *et al.*, 2000, Wunschiers and Lindblad 2003).

The physiological function of the uptake hydrogenase in cyanobactoria is to catalysc the consumption of hydrogen produced by nitrogenase (Happe *et al.*, 2000, Lindberg *et al.*, 2002, Houchins 1984, Oxelfelt *et al.*, 1995, Troshina *et al.*, 1996). A strong correlation between the activities of uptake hydrogenase and nitrogenase has been demonstrated in filamentous cyanobacteria. It is believed that the electrons derived from the hydrogen oxidation catalysed by the uptake hydrogenase recombine, through the respiratory chain, with oxygen in the oxyhydrogen reaction to form water. The advantage would be that energy from the H_2 produced by nitrogenase cań be recaptured and oxygen will be consumed and protect nitrogenase from oxygen. The recycling of hydrogen would also supply reducing equivalents to nitrogenase and other cell functions (Wunschiers and Lindblad 2003). However, the uptake hydrogenase is not essential for diazotrophic growth (Happe *et al.*, 2000, Lindberg *et al.*, 2002, Masukawa *et al.*, 2002). Inactivation of the uptake hydrogenase in *Anabaena variabilis* resulted in a lower rate of nitrogen fixation and slightly reduced growth rate compared to the wild type (Happe *et al.*, 2000).

In hetcrocystous cyanobacteria the uptake hydrogenase is located in the hetcrocyst with no activity in the vegetative cells (Peterson and Wolk 1978, Houchins and Burris 1981a, Houchins 1984, Carrasco *et al.*, 1995). The enzyme has been suggested to be membrane bound, located in the thylakoid (Eisbrenner *et al.*, 1978) or in the cytoplasmic membrane (Houchins and Burris 1981b). Since all identified cyanobacterial HupS are missing the N-terminal signal peptide important for membrane translocation, it has been suggested that the uptake hydrogenase is located on the cytoplasmic side of the thylakoid or cytoplasm membrane (Appel and Schultz 1998).

The activity of the uptake hydrogenase has been demonstrated to be influenced by different external factors, such as nickel, molecular hydrogen, carbon and nitrogen. Addition of extra nickel resulted in an increased hydrogen uptake activity' in *Nostoc punctiforme, Anabaena varaiabilis Oscillatorria subbrevis,*and *Anabaena* strains CA and 1 F (Xiankong *et al.*, 1984, Daday *et al.*, 1985, Kumar and Polasa 1991, Oxelfet *et al.*, 1995). However, a nickel concentration above 10 um does not stimulate the hydrogen uptake (Kumar and Polasa 1991, Oxelfelt *et al.*, 1995). A direct dependence of nickel was demonstrated for the induction of the uptake hydrogenase in *Anabaena* strains Cl and IF (Xiankong *et al.*, 1984) and *Anabaena cylindrica* (Daday *et al.*, 1985).

Molecular hydrogen has also been demonstrated to induce higher uptake activities in cyanobacteria. Studies on Anabaena PCC 7120 (Houchins and Burn's 1981a), *Nostoc punctiforme* (Oxelfelt *et al.*, 1995), Anabaena *cylindrica* and *Nostoc muscorum* (Eisbrenner *et al.*, 1978) demonstrated an increase of the hydrogen uptake activity when a fraction of the air was replaced by molecular hydrogen. However, in *Anabaena variabilis* only a slight stimulatory effect on the hydrogen uptake activity' was observed when exogenous hydrogen was added. It was suggested that the hydrogen produced from the nitrogenase is sufficient for hydrogenase induction (Troshina *et al.*, 1996).

A stimulation of the hydrogen uptake activity, together with nitrogenasc activity, could be observed in *Nostoc punctiforme* when organic carbon was added to the medium (Oxelfelt *et al.*, 1995). However, no effect could be observed in *Anabaena variabilis* after addition of carbon (Troshina *et al.*, 1996). Addition of ammonium decreased the activity' of both the uptake hydrogenase and the nitrogenase in *Nostoc punctiforme* (Oxelfelt *et al.*, 1995). A similar observation was made in *Anabaena variabilis* (Troshina *et al.*, 1996).

Transcription of the Genes Encoding the Uptake Hydrogenase

In the two cyanobacteria examined, hupS and hupL are located on a single transcript containing no additional ORFs (Happe *et al.*, 2000, Lindberg *et al.*, 2000). The size of the transcript was determined by Northern blotting in *Anabaena variabilis* ATCC 29413 to be approximately 2.7 kb (Happe *et al.*, 2000). In *Nostoc punctiforme*, hupSL was shown to be a transcript unit and a putative transcriptional terminator could be identified downstream of hupL. The intergenic region contains 7 bp repeats putatively forming a hairpin structure. The function of a hairpin formation in cyanobacterial hupSL is unknown but may be involved in transcript stability' or translational coupling between the structural genes (Lindberg *et al.*, 2000). Studies on the localisation of the hupSL transcript in *Anabaena* PCC 7120 demonstrated the presence, in the heterocysts only (Carrasco *et al.*, 1995). No hupSL-transcript could be detected in vegetative cells of *Anabaena variabilis* using either Northern blot or RT-PCR (Happe *et al.*, 2000). However, another study was able to detect a low level of hupSL,transcript in vegetative cells from a nitrogen-fixing culture of *Anabaena variabilis*. The authors suggested that this could be a result of a basal activity' of the hupSI~ promotor not necessarily resulting in translation (Boison *et al.*, 2000). A low H_2 uptake in ammonia grown cells of *Anabaena variabilis*, an activity thought to be due to the bidirectional hydrogenase has been reported (Troshina *et al.*, 1996). In addition, *Anabaena variabilis* contains an alternative nitrogenase expressed in vegetative cells under anaerobic conditions (Thiel *et al.*, 1995, 2001). In the study where a hupSL transcript was detected in vegetative cells (Boison *et al.*, 2000), no investigations of the presence of an alternative nitrogenasc or uptake hydrogenase were performed. Moreover, there is no data on the regulation of the transcription of hupSL in the vegetative cells of *Anabaena variabilis* during anaerobic conditions.

Transfer of non-nitrogen fixing vegetative cells to nitrogen-fixing conditions induces a hupL transcript in *Anabama* PCC 7120 and *Anabaena variabilis* (Carrasco *et al.*, 1995, Happe *et al.*, 2000, Boison *et al.*, 2000). Prior to expression of hupSL in Anabaena PCC 7120, a 10.5 kb DMA fragment is excised from within hupL. In *Anabaena* PCC 7120, two additional gene rearrangements occur (Golden *et al.*, 1985). Each excision requires a site-specific recombinase. XisC is responsible for the excision of the 10.5 kb in hupL and its gene is located on the excised fragment. Studies of the upstream region of another site-specific recombinase, xisA, revealed a binding site of the global nitrogen regulator NtcA. However, no obvious NtcA binding site could be detected upstream of xisC (Carrasco *et al.*, 1995). In contrast to *Anabaena* PCC 7120, no programmed rearrangement occurs in *Anabaena variabilis* ATCC 29413 or *Nostoc punctiforme* (Happe *et al.*, 2000 and Oxelfelt *et al.*, 1998). As in *Anabaena* PCC 7120, a hupL transcript was not detected in non-nitrogen fixing cells but was induced during nitrogen fixing conditions (Happe *et al.*, 2000).

The transcription start site of hupSL has been determined in *Anabaena variabilis* ATCC 29413 and *Nostoc punctiforme* using primer extension and 5 'RACE, respectively (Happe *et al.*, 2000, Lindberg *et al.*, 2000). The transcription start of hupS.L in *Anabaena variabilis* ATCC 29413 was located 103 bp upstream of the translation start site. One half of a putative Fnr binding site was found 144 bp upstream of the translational start site (Happe *et al.*, 2000). Fnr is a transcription factor playing a major role in altering the gene expression between aerobic and anaerobic conditions to facilitate changes in energy metabolism (Kiley and Beinert 1999). The transcription start in *Nostoc punctiforme* was located 259 hp upstrearn of the translation start. The promotor region has putative binding sites for NtcA and integration host factor (IHF). NtcA is the global nitrogen regulator in cyanobacteria and IHF is a DNA-binding protein consisting of two subunits that upon binding creates a sharp (more than 160°) bend. This allows proteins that bind further upstream to interact with the promotor region. In addition, IF-IP can also act directly as a repressor or activator (Wagner, 2000).

Cyanobacterial Bidirectional Hydrogenase

Initially, the bidirectional hydrogenase was suggested to be an enzyme with four subunit-s, consisting of a diaphorase part, HoxFU, and a hydrogenase part, HoxYH. These subunits are homologous to the heterotetrameric NAD'–reducing hvdrogenase of *Raslastonia eutropha* (Schmitz *et al.*, 1995). Recently, it was suggested that a fifth subunit, HoxE, belongs to the diaphorase part of the bidirectional hydrogenase. Thus, the cyanobacterial bidirectional hydrogenase is encoded by hoxEFUYH (Schmitz *et al.*, 2002).

The physiological function of the bidirectional hydrogenase in cyanobacteria is not completely clear. The bidirectional hydrogenase has been suggested to act as an electron valve during photosynthesis in *Synechocystis* PCC 6803. Inactivation of the bidirectional hydrogenase resulted in a higher fluorescence of photosystem II compared to the wild type (Appel *et al.*, 2000). The enzyme has also been proposed to play a role in fermentation by functioning as a mediator in the release of excess reducing power during anaerobic conditions (Stal and Moezelaar 1997, Troshina *et al.*, 2002). In addition, it has been suggested that the bidirectional hydrogenase is part of the respiratory complex I (Appel and Schulz 1996, Schmitz and Bothe 1996). In cyanobacteria, only 11 subunits out of 14 conserved subunits of a prokaryotic complex I have been identified. Some of the subunits of the bidirectional hydrogenase show sequence similarities with the missing subunits of the respiratory complex I (Schmitz *et al.*, 1995). However, the bidirectional hydrogenase has been demonstrated to be absent from several cyanobacterial strains (Tamagnini *et al.*, 1997, Tamagnini *et al.*, 2000) and studies of the respiration of *Nostoc punctiforme* a strain naturally lacking the bidirectional hydrogenase (Tamagnini *et al.*, 1997), demonstrated rates of respiration comparable to cyanobacteria having a bidirectional hydrogenase. In addition, mutants of hoxU in *Synchocystis* PCC 6301(Boison *et al.*, 1998) and hoxF in *Synechocystis* PCC 6803 showed non-impaired respirator 0_2 uptake whilst being affected in H_2 evolution (Howitt and Vermaas 1997). In general, it seems that the bidirectional hydrogenase does not play an essential role in those strains where it is present. lnactivation of hoxH in *Synechocystis* PCC 6803 and *Anabaena* PCC 7120 resulted in a small decrease in growth rate compared to the wild type (Appel *et al.*, 2000, Masukawa *et al.*, 2002).

The bidirectional hydrogenase is present in both vegetative cells and in heterocysts (Hallenbeck and Beneman 1978, Houchins and Burris 1981a). In *Anabaena* PCC 7120 the bidirectional hydrogenase appeared in the soluble fraction after cell disruption and it was suggested that the enzyme is soluble (Houchins and Burris 1981b, Hallenbeck and Beneman 1978). However, investigations in other cyanobacteria suggest an association with cell membranes. In *Anabaena variabilis* and *Synechocystis* PCC 6803, an association with the thylakoid membrane was demonstrated (Screbryakova *et al.*, 1994, Appel *et al.*, 2000). However, based on immunological data, an association with the cytoplasmic membrane in *Synechocystis* PCC 6301 has been suggested (Kentemich *et al.*, 1989).

The activity of the bidirectional hydrogenase has been examined in both unicellular and filamentous cyanobactcria. *In vivo*, NADH supports H_2 evolution in *Synechocystis* PCC 6301. NADPH also supports H_2 evolution but with less than 50 per cent of the activity obtained using NADH. For H_2 uptake, NAD–is the preferred electron acceptor (Schmitz and Bothe 1996).

The activity of the bidirectional hydrogenase has in several studies been demonstrated to be induced by anaerobic conditions (Schmitz and Bothe 1996, Screbryakova *et al.*, 1994, Houchins and Burris 1981a). The bidirectional hydrogenase in *Anabaena* PCC 7120 is active in both vegetative cells and in heterocysts in aerobically grown filaments, with a several-fold higher activity in hetcrocysts. Transferred to anaerobic conditions, the activity of the bidirectional hydrogenase increased with about two orders of magnitude with approximately the same activities in both cell types (Houchins and Burris 1981a). Similar results have been observed in *Anabaena variabilis* (Screbryakova *et al.*, 1994). In contrast to the filamentous cyanobacteria, the activity of the bidirectional hydrogenase in the unicellular *Gloeocapsa alpicola* is not directly dependent on oxygen. Higher activity is observed under nitrogen starvation and low light, and it was suggested that the bidirectional hydrogenase could act as an alternative electron donor to photosystem I after inactivation of photosystem II due to nitrogen starvation. Under dark anoxic conditions the unicellular cyanobacterium *Gloeocapsa alpicola* produces H_2 catalysed by the bidirectional hydrogenase (Troshina *et al.*, 2002). In addition, the unicellular strain *Chroocociidiopsis thermalis* contains a bidirectional hydrogenase with some catalytic properties more similar to an uptake hydrogenase. It is not inducible under anaerobic conditions or under nitrate starving conditions (Screbryakova *et al.*, 2000).

In contrast to the uptake hydrogenase, the bidirectional hydrogenase in Anabaena PCC 7120 did not respond to added H_2 in aerobically grown cells (Houchins and Burns 1981a).

Transcription of the Genes Encoding the Bidirectional Hydrogenase

The genes encoding the bidirectional hydrogenase, hoxEFUYH, in cyanobacteria are organised in a similar way in many strains. In some strains the genes are not adjacent and must thus be located on at least two operons. It is also possible to identify ORFs that are located between the hox-genes (Figure 11. 2).

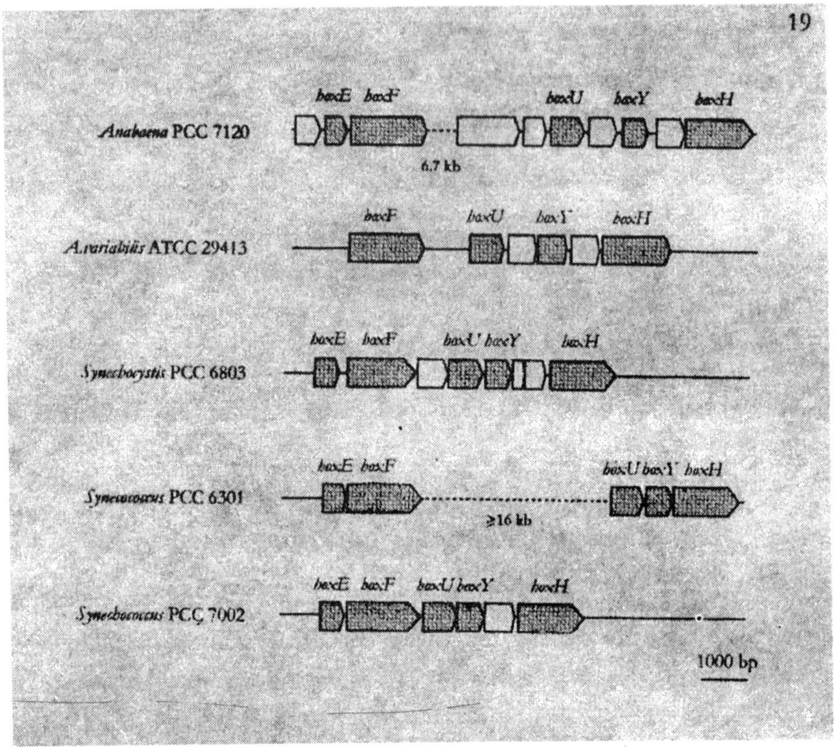

Figure 11.2: Physical Map of the Genes Encoding the Bidirectional Hydrogenase (hox E'FUYH in cyanobacteria.

***Anabaena* PCC 7120 (Kaneko *et al.*, 2001), *Anabaena variabilis* ATC(29413 (Schmitz *et al.*, 1995), *Synechocystis* PCC 6803 (Appel and Schulz 1996), *Synechococcus* PCC 6301 (also *Anacystis nidulans*) (Boison *et al.*, 1996), and *Synechocystis* PCC 700 (GcnBank AF381045).**

The information about the transcription and regulation of the hox-genes is limited in cyanobacteria. Transcript(s) of the bidirectional hydrogenase is present in both vegetative cells and heterocysts under nitrogen-fixing conditions and in vegetative cells during non-nitrogen fixing conditions *Anabaena variabilis* ATCC 29413. In addition, hoxFUYH were shown to be transcribed as a transcript unit together with two ORFs with unknown function. These experiments were performed using RT-PCR and do not exclude additional promotors within the operon (Boison *et al.*, 2000). The hox-genes in the unicellular *Synechocystis* PCC 6301 are located on two different transcripts. hoxEF form one transcript and at least 16 kb downstream of hoxF and hoxUYH is forming a second transcript together with hoxW, hypA and hypB (Boison *et al.*, 2000). In *Synechococcus* PCC 7942, hoxEFand hoxUYHW are located on two different transcripts. Using real time PCR and reporter gene constructs, it was suggested that a second promotor might be present between hoxH and hoxW. It was also demonstrated that the hox–genes had a circadian clock expression (Schmitz *et al.*, 2001).

Very few regulatory studies have been performed on the transcriptional regulation of the hox-genes in cyanobacteria. Studies of the transcription of hoxY and hoxW *Gloeocapsa alpicola* CALU 743 during nitrogen-limiting growth conditions demonstrated an increase in the enzyme activity but no regulation on the transcriptional level (Sheremetieva *et al.*, 2002). In contrast, transfer to a low level of oxygen in *Anabaena variabilis* induced both the enzyme activity as well a: the relative amount of hoxH (Sheremerieva *et al.*, 2002).

Maturation of Hydrogenases

The maturartion of hydrogenases is a complex process in which several proteins, encoded by the hyp-genes, are involved. The respective gene involved in the maturation of (NiFe)–hydrogenases have been identified sequenced, and characterized but the corresponding genes for the maturation of [Fe]-hydrogenases remain to be identified. Very little is known in cyanobacteria, most of the knowledge about hydrogenase maturation is from other microorganisms such as *Eschrichia coli*, *Ralastonia eutropha* (Wolf *et al.*, 1998), *Bradirhizobium* japonicum and *Rhizobium leguminosarum* (Rey *et al.*, 1993). Homologues of the hyp-genes have been identified in cyanobacteria but their role in maturation of the hydrogenases remains to be demonstrated. The Hyp-proteins are involved in the insertion of Ni, Fe, and the ligands, CO and CN, into the active site of the large, subunit. Other genes involved in the maturation of the hydrogenases, not belonging to the hyp-genes, encode endopeptidases, which are responsible for a specific cleavage of a C–terminal part of the large subunit of the hydrogenase.

One of the earliest steps in the maturation is the formation of the complex between HypC and the large subunit of the hydrogenase. HypC is assumed to act as a chaperone, maintaining the large subunit in a conformation accessible for metal insertion. A hypC mutant results in a metal free hydrogenase. The next step is the insertion of the ligands CN and CO into the large subunit. This is performed by HypF and of HypE (Reissman *et al.*, 2003). HypF has been shown bv mutational studies to be absolutely required for hydrogenase maturation, It has been demonstrated that the CN and CO ligands are derived from carbarnoylphosphate. Proteins catalysing O-carbamoylations contain a sequence motif (VxHHxAH) that is also found in HypF. In addition, HypF contains two zinc finger motifs. It is possible that HypF interacts with the large subunit and that the acyl-phosphatase and the carbonoylphosphate domains synthesise and insert the ligands in the active site. Using a two-hybrid method, it was shown that HypF and HypE interact in *Helocobacter pylori*. Inactivation of HypE results in a non-mature hydrogenasc, so HypE is also essential for hydrogenase maturation. HypB is suggested to be the main contributor of insertion of nickel.

Deletions of hypB leads to nickel free hydrogenase precursors and an inactive hydrogenase in *E.coli* (Maier *et al.*, 1993). In *R. leguminosarum* (Rey *et al.*, 1994), *B japonicum* (Olson *et al.*, 1997), HypB has also been shown to have a function in nickel storage, as a result of histidine rich domains in the amino terminus. The function of HypA is to cooperate with HypB during nickel insertion (Hube *et al.*, 2002). In *H. pylori* was demonstrated that HypA and HypB form a heterodimer (Mehta *et al.*, 2003). The role of HypD is unclear and remains to be identified. However, the protein has been demonstrated to form a complex with HupC in *E. coli* (Blokesch and Bock 2002).

The last identified step in the naturation of the large subunit is the proteolytic cleavage of the C-terminus. Recently, two ORFs putarively encoding hydrogenase specific endopeptidases, HoxW and HupW, were identified in *Anabaena* PCC 7120. It was suggested that they are responsible for the specific proteolytic cleavage of the C-terminal part of the bidirectional hydrogenase and uptake hydrogenase, respectively. Putative endopeptidases were also found in the unicellular *Synechocystis* PCC 6803, containing only the bidirectional hydrogenase, and the filamentous, hetcrocystous *Nostoc punctiforme*, containing the uptake hydrogenase only. Only one, though specific, putative endopeptidase was found in each of these two strains supporting the hypothesis that each hydrogenase has a specific endopeptidase (Wunschiers *et al.*, 2003).

Acknowledgements

The author (DV) is thankful to Department of Botany, Dr. H.S. Gaur University and UGC for the research Award. Author (RKG) is thankful to Principal, Govt. P.G. College, Rishikesh and to UGC, New Delhi for financial help.

References

Adams DG. 2000. Symbiotic interactions. In: The Ecology of Cyanobacteria. (Whitton BA and Potts M, eds) pp 523-561. Kluwer Academic Publishers, Dordrecht.

Appel J, and Schulz R. 1996. Sequence analysis of an operon of a NAD(P)-reducing nickel hydrogenase from the cyanobacterium *Synechocystis* sp. PCC 6803 gives additional evidence for direct coupling of the enzyme to NAD(P)H-dehydrogenase (complex 1). *Biochim Biophys Acta* 1298: 141- 147.

Appel, J, and Schulz R. 1998. Hydrogen metabolism with oxygenic photosynthesis: hydrogenases as important regulatory device for a proper redox poisoning? *J.Photochem Photobiol B: Biol* 47: 1-11.

Appel, J, Phupruch S, Steinmuller K, and Schulz R. 2000. The bidirectional hydrogenase of *Synechocystis* sp. works as an electron valve during photosynthesis. *Arch Microbiol* 173: 333-338.

Blokesch M, and Bock A. 2002. Maturation of (NiFe)-hydrogenases in *Escherichia coli.* the HypC cycle. *J Mol Biol* 324: 287-296.

Boison G, Schmitz O, Mikheeva L, Shestakov S, and Bothe H. 1996. Cloning, molecular analysis and insertional mutagenesis of the bidirectional hydrogenase genes from the cyanobacterium *Anacystis nidulans*. *FEBS Letters* 394: 153-158.

Boison G, Schmitz O, Schmitz B, and Bothe .H. 1998. Unusual gene arrangement of the bidirectional hydrogenase and functional analysis of its diaphorase subunit HoxU in respiration of the unicellular cyanobacterium *Anacystis nidulans. Curr Microbiol* 36: 253-258.

Boison G, Bothe H, and Schmitz O.2000. Transcriptional analysis of hydrogenase genes in the cyanobacteria *Anacystis nidulans* and *Anabaena variabilis* monitored by RT-PCR. *Curr Microbiol* 40: 315-321.

Bohme H. 1998. Regulation of nitrogen fixation in heterocyst-forming cyanobacteria. *Trends Plant Sci.* 9: 346-351.

Carrasco CD, Buettner JA, and Golden JW 1995. Programed DNA rearrangernent of a cyanobacterial hupL gene in heterocysts. *Proc Natl Acad Sci* 92: 791-795.

Daday A, Mackerras AH, and Smith GD. 1985. The effect of nickel on hydrogen metabolism and nitrogen fixation in the cyanobacterium *Anabaena cylindrica Journ of Gen Microbiol.* 131: 231-238.

Eisbrenner G, Distler E, Floener L, and Bothe H. 1978. The occurance of the hydrogenase in some blue-green algae. *Arch Microbiol* 118: 177-184.

Eiringer T, and Friedrich B.1991. Cloning, nucleotide sequence, and heterologous expression of a high-affinity' nickel transport gene from *Alcaligenes eutrophus. J Biological Chemistry* 266: 3222-3227.

Eiringer T, and Mandrand-Berthelot M-A. 2000, Nickel transport in microorganisms. *Arch Microbiol* 173: 1-9.

Frias JE, Merida A, Herrero A, Martin-Nieto J, and Flores E. 1993. General distribution of the nitrogen control gene *ntcA* in cyanobacteria. *J Bacteriol* 175: 5710-5713.

Golden JW, Robinson SJ, and Haselkorn R. 1985. Rearrangement of nitrogen fixation genes during heterocyst differentiation in the cyanobacterium *Anabaena. Nature*.314: 419-423.

Halbleib CM, and Ludden PW 2000. Regulation of biological nitrogen fixation. J Nutr 130: 1081-1084.

Hallenbeck PC, and Beneman JR. 1978. Characterization and partial purification of the reversible hydrogenase of *Anabaena cylindrica FEBS Lett* 94: 261-264.

Happe T, Schute K, and Bohme H. 2000. Transcriptional and mutational analysis of the uptake hydrogenase of the filamentous cyanobacterium *Anabaena variabilis*. ATCC 29413. *J Bacteriol* 182: 1624-1631.

Happe T, Hemschemeier A, Winkler M, and Kaminski A. 2002. Hydrogenases in green algae: do they save the algae's life and solve our energy problems? *Trends Plant Sci* 7: 246-50.

Haselkorn R, and Bulkema. WJ. 1992. Nitrogen fixation in cyanobacteria. In: Biological nitrogen fixation. (Staccy G, Burris RH, and Evans H), cds) pp 166-190. Chapman and Hall, London.

Haystead A, Robinson,R and Stewart,WDP 1970 Nitrogenase activity in extracts of heterocystous blue-green algae. *Arch Microbiol*. 82: 325-336.

Herrero A, Muro-Pastor AM, and Flores E. 2001. Nitrogen control in cyanobacteria. *J Bacteriol* .183: 411-425.

Houchins JP , and Burris RH. 1981a. Occurrence and localization of two distinct hydrogenases in the heterocystous cyanobacterium *Anabaena* sp. strain 7120. *J Bacteriol* 146: 209-204.

Houchins JP, and Burris RH. 1981b. Comparative characterization of two distinct hydrogenases *Anabaena* sp. strain 7120.*J Bacteriol* 146: 215-221.

Houchins JP 1984. The physiology and biochemistry of hydrogen metabolism in cyanobacteria. *Biochim Biophys Acta* 768: 227-255.

Howitt CA, and Vermaas WFJ. 1997. Analysis of respiratory mutants of *Synechocystis* 6803. In: Abstract of the IX International Symposium on Photosynthetic Prokaryotes (Peschek G A, Loffelhardt W, and Schmetterer G A, Eds, p 36, Vienna, Austria.

Hube M, Blokesch M, and Bock A. 2002. Network of hydrogenase expression in *Escherichia coli.* Role of accessory proteins HypA and HybF. *J Bacteriol* 14: 3879-3885.

Jackson DD and Ellms JW 1896 On odors of tastes of surface water, with special reference to *Anabaena* a microscopical organism found in certain water supplies of Massachusetts. In the 1896 Report of the Massachusetts State Board of Health,pp410-420.

Kaneko T, Sato S, Kotani H, Tanaka A, Asamizu E, Nakamura Y, Miyajima N, Hirosawa M, Sugiura M, Sasamoto S, Kimura T, Hosouchi T, Matsuno A, Muraki A, Nakazaki N, Naruo K, Okumura S, Shimpo S, Takeuchi C, Wada T, Watanabe A. Yamada M, Yasuda M, 'and Tabata S. 1996. Sequence analysis of the genome of the unicellular cyanobacterium *Synechocystis* sp. strain PCC 6803. II. Sequence determination of the entire genome and assignment of potential protein-coding regions. *DNA Res* 3: 185-209.

Kaneko T, Nakamura Y Wolk CP, Kuritz T, Sasamoto S, Watanabe A, Iriguchi M Ishikawa A, Kawashima K, Kimura T, Kishida Y, Kohara M, Matsumoto M, Matsumo A, Muraki A, Nakazaki N, Shimpo S, Sugimoto M, Takazawa M, Yamada M, Yasuda M, and Tabata S. 2001. Complete genomic sequence of the filamentous nitrogen-fixing cyanobacterium *Anabaena* sp. strain PCC 7120. *DNA Res* 8: 205-213.

Kentemich T, Bahnweg M, Mayer F, and Bothe H. 1989. Localization of the reversible hydrogenase in cyanobacteria. *Z Naturforch* 44c: 384-391.

Kiley PJ, and Beinert H. 1999. Oxygen sensing by the global regulator FNR: the role of the iron-sulfur cluster. *FEMS Microbiol Rev* 22: 341-352.

Kumar S, and Polasa H. 1991. Influence of nickel and copper on photobiological hydrogen production and uptake in *Oscillatoria subbrevis* strain 111. *Proc Indian Natn Sci.Acad* .281-285.

Lindberg P, Hansel A, and Lindblad P. 2000. hupS and bufL constitute a transcription unit in the cyanobacterium *Nostoc* sp. PCC 73102. *Arch Microbiol* 174: 129-133.

Lindberg P, Schutz K, Happe T, and Lindblad P. 2002. A hydrogen producing, hydrogenase-free mutant strain of *Nostoc punctiforme* ATCC 29133. *IntJ Hydrogen Energ* 27: 1291-1296.

Maier T, Jacobi A, Sauter M, and Bock A. 1993. The product of the hypB gene, which is required for nickel incorporation into hydrogenases, is a novel guanine nucleotide-binding protein. *J Bacteriol* 175: 630-635.

Masukawa H, Mochimaru M, and Sakurai H. 2002. Disruption of the uptake hydrogenase gene, but not of the bidirectional hydrogenase gene, leads to enhanced photobiological hydrogen production by the nitrogen-fixing cyanobacterium *Anabaena* sp. PCC 7121. *Appl Microbiol Biotechnol* 58: 618-624.

Meeks JC, Elhai J, Thiel T, Potts M, Larimer F, Lamerdin J, Prcd.ki P, and Atlas R. 2001. An overview of the genome of *Nostoc punctiforme*, a multicellular, symbiotic cyanobacterium. *Photosynth Res* 70: 85-106.

Mehta N, Olson JW and Maier RJ. 2003. Characterization of *Helicobacter pylori* nickel metabolism accessory proteins needed for maturation of both urease and hydrogenase. *J Bacteriol* 185: 726-734.

Olson JW, and Maier RJ. 1997. The sequences of hypF, hypC and hypD complete the hyp gene cluster required for hydrogenase activity in *Bradirhyzobium japonicum*. *Gene* 199: 93-99.

Oxelfelt F, Tamagnini P, Salema R, and Lindblad P. 1995, Hydrogen uptake in *Nostoc* strain PCC 73102; Effects of nickel, hydrogen, carbon and nitrogen. *Plant Physiol Biochem* 33: 617-623.

Oxelfelt F, Tainaginini P, and Lindblad P. 1998. Hydrogen uptake in *Nostoc* sp. strain PCC 73102. Cloning and characterization of a hypSL homologue. *Arch Microbiol* 169: 267-274.

Peterson RB, and Wolk PC. 1978. Localization of an uptake hydrogenase in *Anabaena*. *Plant Physiol* 61: 688-691.

Pienkos PT, Bodmer S, and Tabita FR. 1983. Oxygen inactivation and recovery of nitrogenase activity in cyanobacteria. *J Bacteriol* 153: 182-90.

Ramasubramanian TS, Wci TF, and Golden JW 1994. Two *Anabaena* sp. Strain PCC 7120 DNA-binding factors interact with vegetative cell and heterocyst specific genes. *J Bacteriol* 176: 1214-1223.

Reissniann S, Hochleitner E, Wang H, Paschos A, Lottspeich F, Glass RS, and Bock A. 2003. Taming of a Poison: Biosynthesis of the NiFe-Hydrogenase Cyanide Ligands. *Science* 299: 1067-1070.

Rey L, Murillo J, Hernando Y, Hidalgo E, Cabrera E, Imperial J, and Ruiz-Argueso T. 1993. Molecular analysis of a microaerobically induced operon required for hydrogenase synthesis in *Rhizobium leguminosarum* biovar viciae. *Mol. Microbiol* 8: 471-481.

Rey L, Imperial J, Palacios JM, and Ruiz-Argueso T. 1994. Purification of *Rhizobium leguminosarum* HypB, a nickel-binding protein required for hydrogenase synthesis. *J Bacteriol* 176: 6066-6073.

RippkaR, Deruelles J, Waterbury JB, Herdman M, and Stanier RY. 1979. Generic assignments, strain histories and properties of pure cultures of cyanobacteria. *J Gen Microbiol* 111: 1-61.

Schmitz O, Boison G, Hilscher R, Hundeshagen B, Zimmer W, Lottspeich F, and Bothe H. 1995. Molecular bio!ogical analysis of a bidirectional hydrogenase from cyanobacteria. *Eur J Biochem* 233: 266-276.

Schmitz O, and Bothe H. 1996. NAD(P)+-dependent hydrogenase activity in extracts from the cyanobacterium *Anacystis nidulans FEMS Microbiol Lett* 135: 97-101.

Schmitz O, Boison. G, and Bothe H. 2001. Quantitative analysis of expression. of two circadian clock controlled gene clusters coding for the bidirectional hydrogenase in the cyanobacterium Synechococcus sp. PCC 7942. *Mol Microbiol.* 41: 1409-1417.

Schmitz, O, Boison G, Salzman H, Bothe H,, Schutz K., and Happe T. 2002. HoxE-a subunit specific for the petameric bidirection hydrogenase coplex (HoxEFUYH)) of cyanobacteria. *Biochim Biophys Acta* 1554: 66-74.

Serebryakova LT, Zorin NA, and Lindblad P. 1994. Reversible hydrogenase in *Anabaena variabilis* ATCC 29413: presence and localization in non-N_2, fixing cells. *Arch Microbiol* 161: 140-144.

Serehryakova LT, Sheremetieva M, and Lindblad P. 2000. H_2-uptake and evolution in the unicellular cyanobacterium *Chroococcidiopsis* thermalis CALTJ 758. *Plant Physiol Biochem* 38: 525-530.

Sheremetieva M, Troshina O, Serebryakova LT, and Lindblad P. 2002. Identification of box genes and analysis of their transcription in the unicellular cyanobacterium *Gleocapsa alpicola* CALU 743 growing under nitrate-limiting conditions. *FEMS Microbiol Lett* 214: 229-233.

Stal LJ and Moczclaar R. 1997. Fermentation in cyanobacteria. *FEMS Microbiol Rev* 21: 179-211.

Tamagnini P, Troshina O, Oxelfelt F, Salema R, and Lindblad P. 1997. Hydrogenases in *Nostoc* sp. PCC 73102, a strain lacking a bi-directional hydrogenase. *Appl Environ Microbiol* 63: 1801-1807.

Tamagnini P, Costa, L, Almeida L, Oliveria M-J, Salema R, and. Lindblad P. 2000. Diversity of cyanobacterial hydrogenases, a molecular approach. *Curr Microbiol* 40: 356-360.

Thiel T, Lyons EM, Erkcr C, and Ernst A. 1995. A second nitrogenase in vegetative cells of a heterocyst.-forming cyanobacterium. *Proc Natl Acad Sci* 92: 9358-9362.

Thiel T, and Pratte B. 2001. Effect on heterocyst diffrentiation of nitrogen fixation in vegetative cells of the cyanobacterium *Anabaena variabilis* ATCC 29413. *J Bacteriol* 183: 280-286.

Troshina OY, Screbryakova LT, and Lindblad P. 1996. Induction of H_2-uptake and nitrogenase activities in the cyanobacterium *Anabaena varaiabilis* ATCC 29413: effects of hydrogen and organic substrate, *Curr Microbiol* 33: 1 1-15.

Troshina O, Serebryakova L, Sheremetieva M, and Lindblad P. 2002. Production of H_2 by the unicellular cyanobacterium *Gleocapsa apicola* CALU 743 during fermentation. *Int J Hydrogen Energ* 27: 1283-1289.

Vasquez-Bermudez MF, Flores E, Herrero A. 2002. Analysis of binding sites for the nitrogen-control transcription factor NtcA in the promoters of *Synechococcus* nitrogen-regulated genes. *Biochim Biophys Acta* 1578: 95-98.

Vyas D. 1992 Physiological studies on cyanobacteria with special reference to hydrogen uptake. Ph.D thesis BHU ,Varanasi.

Vyas D 2004 Cyanobacterial diversity and their hydrogen production potential *J.Bot.Soc.*Univ. of Sagar 39: 55-65.

Vyas D and Gupta, R.K. 2003. Impact of ultraviolet-B radiation on hydrogen production, and nitrogen fixation in cyanobacterium *Nostoc* spp. *J. Bot. Soc.* Univ. of Sagar 36-38: 89-97.

Vyas D and Kumar,HD 1995 Nitrogen fixation and hydrogen uptake in four cyanobacteria. *Int. J. Hydro. Engr* 20: 108-110.

Wagner R. 2000. The master-bender integration host factor (IHF) and its family of DNA-binding proteins. In: Transcription regulation in prokaryotes pp 227-232. Oxford University Press.

Wei TF, Ramasubramanian TS and Golden JW 1994. *Anabaena* sp. strain PCC 7120 ntcA gene required for growth on nitrate and hererocyst development. *J Bacteriol* 176: 4473-4482.

Wolf I, Buhrke T, DemeddeJ, Pohlmann A, and Friedrich B. 1998, Duplication of hyp genes involved in maturation of (NiFe) hydrogenases in *Alicaligenes eutrophus* H16. *Arch Microbiol* 170: 451-459.

Wolk CP. 1996. Heterocyst formation . *Annu Rev Genet* 30: 59-78.

Wunschicrs R, Scngcr H, and Schuiz R. 2001. Electron pathways involved in .H,-rnctabolisiTi in the green alga Scenedesnuu obliquns. Biochim Biophys Acta 1503: 271-278.

Wunschiers R, Batur M, and Lindblad P. 2003. Presence and expression of hydrogenase specific C-terminal endopeptidases in cyanobacteria. BMC *Microbiol* 3: 8

Wunschiers R and Lindblad P. 2003. Light-dependent hydrogen uptake and generation by cyanobacteria. Handbook of Photochemistry and Photobiology, in press

Wynn-WIlliams DD. 2000. Cyanobacteria in deserts - Life at the limit? In: The Ecology of Cyanobacteria (Whitton BA and Potts M, eds) pp 341-366. Kluwer Academic Publishers, Dordrecht.

Xiankong Z, Tabita, and van Baalen C. 1984. Nickel control of hydrogen production and uptake in *Anabaena* spp. strains CA and I F. *J Gen. Microbiol* 130: 1815-1818.

© R.K. Gupta & V.D. Pandey (*Editors*)
Advances in Applied Phycology, 2007
Daya Publishing House, New Delhi

Chapter 12

Algae and Cyanobacteria as the Living Resources for Sustainable Environmental Management

Chandra Bhushan Singh*

Department of Botany, Kurukshetra University, Kurukshetra – 136 119, Haryana, India

ABSTRACT

Environmental management for ecological sustainability has gradually attained a stage from where the issues and its major research needs may be defined on a global scale. The indiscriminate harvesting of natural resources causing loss of biodiversity, is to be regulated by enumerating that how much we have; and out of which how much we are losing and at which pace? The environmental degradation is not only to be prevented but is to also reversed, so that our limited *'non-living'* and *'living resources'* could be conserved in the wider perspectives of sustainable eco-managements by strengthening the areas of ecosystem functioning and restoration ecology. The present study has been aimed to focus on the long-term interdisciplinary ecological explorations in their problem solving mode by giving its rightful place in education, awareness, research and at all levels of human interactions with the Mother Nature. There are innumerable numbers of chemicals we use in our day-to-day life, industries, and agriculture etc.; and majority of them *per se* contain *'Heavy Metals'*. Therefore, these are usually present as the integral and prominent proportion of most of the pollutants; their bioaccumulation through adsorption &/or absorption by algae, microalgae, cyanobacteria and other microbes occurs via complexation, coordination, chelation, ion exchange, inorganic precipitation and/or a combination of all these. The core of such mechanisms is the predominant exchange of counter ions prevailing on the biomass. Algae and Cyanobacteria have not only been used as the early warning system for predicting pollution hazards and as the indicators

* E-mail: cbsinghbhu@gmail.com and cbsinghbhu@sancharnet.in.

of new substances entering into the aquatic environment and for eco-epidemiology, but also as the bioremediators and potential restoration agents of pollution load. Their efficient biotechnological exploitation as *biosorbants* of pollutants and toxic heavy metals through bioreactors, mass culture systems, strain improvement and genetic manipulations, appears to be more promising for the developing countries like India. Therefore, careful ecological information has become essential for developing conservation and restoration strategies for such useful plants. We need *'clean environment'* and this paper discusses that on the basis of varied responses of algae and cyanobacteria towards many of these stresses particularly in aquatic environments and wetlands, a suitable assemblage of *'most sensitive species'* could be identified for their use as tools for modern sustainable environmental management.

Introduction

Aquatic Macrophytes

Aquatic ecosystems, the key components of natural environments are subjected to continuous contamination due to increased generation of domestic, industrial and agrochemical wastes. The common aquatic macrophytes prevalent in the tropical country like India are: *Eicchornia cressipes, Hydrocotyl umbellatae, Lemna minor, Azolla pinnata, Salvinia pistia, Hydrillla sp., Typha latifolia, Altermanthera periloxirides,* etc. Bioremediation is biological based pollution treatment technology which bears potential to clean up a variety of toxic substances at lower costs. Aquatic Macrophytes and algae bear enough prospective in this regard, as they are the O_2-evolving photosynthetic organisms which grow at faster rates. All such plants accumulate heavy metals in ionic form from aquatic environments (*e.g.*, Pb, Cu, Cd, Fe, Hg, etc.). Aquatic vegetation, in general, can trap (filter) some of the nutrient-laden sediments flushed from the catchments in surface runoff as well as reduce the amount of nutrients entering the aquatic ecological systems or the wetlands. The US Environmental Protection Agency has projected algae as the indicators of wetland ecological conditions. Wetlands are complex, diverse ecosystems either directly linked to the landscape or to an area of seasonally, intermittently or permanently waterlogged soil or inundated land, whether it is fresh or saline, lentic (flowing) or lotic (static). *'Living wetland'* is a concept primarily concerned with basin, while the examples of flat wetlands are lakes, sumplands (swamps), damplands and palusplains (seasonally waterlogged flats). Wetland vegetation can remove some nutrients either directly from the water (*e.g.* uptake by algae and other aquatic vegetation) or by assimilation from the sediments via emergent rooted plants and microbes growing in the root zone. Plants that are useful in nutrient stripping in wetlands include the pithy sword sedge (*Lepidosperma longitudinale*), articulated sedge (*Baumea articulata*), duckweed (*Lemna minor*), thin duckweed (*Spirodela*), bulrush (*Typha orientalis, T. domingensis*) and fennel pondweed (*Potamogeton pectinatus*). Such plants can be easily harvested to recover and dispose-off the toxicants accumulated therein.

Heavy Metals

The term *'Heavy Metals'* is mainly used by Biologists and Environmentalists, which generally refers to those metals in the *Periodic Table* having atomic number >20 or alternatively: density > 5 g/cm^3. It is rather a vague term, which also includes those metals, which do not strictly fall as per definition of the *Periodic Table*, but otherwise toxic when present in excess (Rai *et al.*, 1981). Heavy Metals are needed in small quantities by almost all living organisms for their various physiological and metabolic processes but there elevated levels are generally toxic. The main toxic heavy metals are: Hg, Cd, Cu, Ni, Zn, Mn, Pb, etc. Others to follow include: Sn, Cr, Fe, As, Au, Ag, etc. Among these, Hg remains highly mobile in the environment, because of its various forms like *metallic* mercury, *inorganic* mercury and *organic* mercury. Its metabolism in phytoplankton (Havlik *et al.*, 1979), algae (Wilkinson *et al.*, 1989), cyanobacteria (Singh and Singh, 1987) and biotransformation through bacteria (Ramamoorthy at al., 1983) has been well documented. Algal blooms are almost synonyms with eutrophication, organic loading and wetland ecological conditions. Some algal populations bear intrinsic abilities to rapidly adapt and develop mechanisms for tolerance against various abiotic and biotic stresses. There are certain harmful algal blooms, which disrupt the economies of coastal communities through the closure of fisheries affected by algal toxins. Such Biota are referred to as *'hyper accumulator'* plants and have been reported for their remediation abilities at the heavy metal pollution sites (see Singh and Yadava, 1985). For example, the heavy metals such as (1) Cr is one of the heavy metal entering the environment both through natural and anthropogenic sources like burning of oil and coal, production of ferrochromium, chromates, chromium steel, cement, fungicide etc.; and therefore, virtually always present in the environment while (2) Pb, usually found in abundance along the roadside ecological habitats and tend to bioaccumulate in the fatty tissue of animals. They have long-term effects on faunal populations by the reducing reproductive success and survival rates of offsprings.

Phytoplanktons and the Algae

Phytoplanktons are the chief flora of aquatic habitats and algae constitute the major proportion of entire phytoplankton community. They are *one of the Best indicators of Ecological Pollution in Nature,* owing to their responsiveness towards pollutants and the ability to tolerate the pollutants (Shehata and Whitton 1982; *see* Pandey *et al.*, 1992). Algae are also one of the easily available ecological and biological indicators of pollution. By definition, *"Algae are the photosynthetic organisms that occur in most habitats, ranging from marine and freshwater to desert sands and from hot boiling springs to snow and ice. They vary from small, single-celled forms to complex multicellular forms, such as the giant kelps of the eastern Pacific that grow to more than 60 meters in length and form dense marine forests. Algae are found in the fossil record dating back to approximately 3 billion years in the Precambrian"*. They are the base organisms for the entire life on this planet earth, where water constitutes nearly $3/4^{th}$ of its surface *i.e.*, aquatic environments, which is dominated mainly by them. Algae being one of the important producers in aquatic environments play significant role in the structure and function of these ecosystems. They are the largest contributors of Gross Productivity on the planet. Recent reports have emphasized the potential use of selected heavy metal-sequestering algae for the purpose of phytoremediation, environmental clean-up and restoration (Singh, 2004). Their trophic independence for Carbon and even Nitrogen (*e.g.*, N_2-fixing blue-green algae) enables greater flexibility to them towards to adopting to altered/ varied environmental conditions. In addition, their oxygenation and mineralization capabilities in the treatment systems points towards their potential use as the bioremediators of aquatic ecosystems.

The Blue-green algae

The bluish appearing algae were earlier recognized as Blue-green Algae (BGA), but later they were further recognized as *'Cyanobacteria'*, owing to their resemblance with the prokaryotic nature conforming to that of bacteria. As they do not possess any flagella, they have developed a characteristic way of motion called *gliding movement*. Their photosynthetic pigments include the *'Biloprotein'* and *'Carotenoids'*, which can trap sunlight in the wave length range the chlorophylls fail to do. N_2 in the stratosphere is nearly $3/4^{th}$, which can be biologically fixed; and Carbon is also in abundance on this planet, which too can be fixed. Cyanobacteria are the simplest microorganisms which *can carry out both* at ease. They make use of both of these elements in an efficient way at the expense of sunlight, which is also in abundance in a tropical country like India (Singh and Singh, 1987). C-fixation

('Photosynthesis') sustains all life forms on earth, while N_2-fixation is a process for fixing the elemental nitrogen into usable form like ammonia. For their this ability, the BGA have been renowned as the *'Nitrogen Economy Builders of the Nation'*. Though, bacteria too can fix elemental N_2 but they are not photoautotrophs; it is only the cyanobacteria, which are booned to carry out oxygenic photosynthesis along with N_2-fixation. Theretofore, from past few decades, the cyanbacteria have become the emerging tools for environmental management. The ability of such microorganisms to selectively take-up the toxicants from their ambient medium enable them as the potential ecological indicators of pollution (Singh and Singh 1992a). For example, the effluents rich in NO_3^-& PO_4^{3-}, the promising bioindicator cyanobacteria are the species of O*scillatoria, Lyngbya, Phormidium, Anabaena, Nostoc, Aulosira, Aphanocapsa and Anacystis* (see Singh and Singh, 1992b).

Algae and Photosynthetic Prokaryotes

Algae as Ecological Indicators of Pollution

Algal blooms are almost synonyms with eutrophication, organic loading and wetland ecological conditions. Some of the algal populations bear intrinsic abilities to rapidly adapt and develop mechanisms for tolerance against various abiotic and biotic stresses. There are certain harmful algal blooms, which disrupt the economies of coastal communities through the closure of fisheries affected by algal toxins. The algal flora and other aquatic plants working as the hyperaccumulators of heavy metals have developed a detoxification mechanism consisting mainly of chelation and sequestering toxic metals by small metal binding proteins and peptides (Kratochvil and Volesky, 1998). There are two *metal-binding peptides, phytochelatins* and *metallothioneins,* which play vital role(s) in metal-detoxification homeostasis against heavy metals, however, the best known metal sequestering molecules are enzymatically produced small peptides the *'phytochelatins'*, which are derived from *glutathione* (see Asthana *et al.*, 1993). Some of the examples of the algae, which have been reported as the indicators for Pb pollution in (1) the mine waste discharge are the Diatom genera like *Synedra, Navicula, Cymbella,* (2) in the *Nainital Lake*: *Spirogyra adnata, Mougeotia scalaris* and *Oedogonium sp.*, etc, (3) for Mn and Fe: the genera are like *Hydrodictyon reticulatum* and (4) for organic wastes: *Capsosiphon fluvescens* have been reported. However, the algae should be considered as member of the total community only then, the absence or presence of certain species can be considered as reliable indicators of pollution.

Algae as Hyper-Accumulators of Pollution Load

The eukaryotic algae in general, and prokaryotic algae in particular, can selectively take-up toxicants like, heavy metals, radioactive nuclides, pesticides, herbicides, carcinogenic substances and other biocides etc., leading to their intracellular bioaccumulation at a level several fold higher than that of the their ambient microclimate (Singh *et al.*, 1992). Organic compounds and various heavy metals are the inevitable result of many industrial activities threatening potential hazards to the river ecosystems. Reddy and Venkateswarlu (1985) have made an assessment of the water quality and pollution load in the river *Tungbhadra* at Hyderabad by taking several physicochemical factors, heavy metal input and phycological characteristics. There is a general increase in the contents of bicarbamates, chlorides, phosphates, sulphates, total hardness and dissolved solids. The algal blooms are generally found increased after the application of insecticides due to temporary suppression of their grazing by aquatic invertebrates; also there is a shift in community composition from large filamentous Chlorophytes to smaller Blue-green algae and diatoms after a similar application of herbicides. Though the richness of algal species declines but the growth of filamentous algae, typically show proportionate increase due to acidification of aquatic ecosystems; and heavy metals, if present in small quantities play stimulatory actions. Acidification of aquatic ecosystems results in the decline of *Algal Biodiversity*. The nitrogen is turned into nitric acid by bacteria, which then flows into the water body and lowers the pH. They have been reported as the Indicators of Acidification (*e.g.*, occurrence of large populations of *Mougeotia sp.* in some naturally acidic Wetlands; Singh, 2004). Cd, Zn, Hg or Cr drastically reduced the structural and functional variables of algal populations in the phytoplankton community of river *Ganga* at Varanasi (Singh and Rai 1990). Barring the exception of *Anorthoneis excentrica*, the filamentous algal species were in general, showed greater tolerance against heavy metals than the unicellular forms.

Cyanobacteria as *'Ecotoxicological Tools'* for Sustainable Environmental Management

The photosynthetic prokaryotic algae (*e.g.*, Cyanobacteria) often dominate ecological niches, enriched with metal contaminants possibly by developing tolerance/resistance against them, as they are known to produce thiol and exopolysaccharide under the stress of heavy metals like Hg, Cu and Ni (Singh *et al.*, 1999). Such exopolymers are the effective and natural metal chelators like that of siderophores (Kaplan *et al.*, 1987). It involves the intricacies of heavy metal toxicity to cyanobacteria. This area of research encompasses the toxicological impacts of toxicants (*e.g.*, heavy metals) affecting the ecological viability of cyanobacteria and their resultant impact on the Environment. Cyanobacteria are also the bioremediators of rice grown acid soils (*e.g., Nostoc, Anabaena, Oscillatoria, Westiellopsis, etc.*). Cyanobacterial strains have been reported to reveal metabolism dependent uptake of mercury (Pant et al. (1992). They are pollutant tolerant *'hyper accumulators'* and can be safely raised up to the rank of *'Extremophiles'*, because of their ability to withstand the *'pollution load'* (Asthana *et al.*, 1996). It is now widely accepted that the immobilized cyanobacterial populations are one of the efficient bioaccumulators of heavy metals like Cu (Singh *et al.*, 1989). The *Algal and Cyanobacterial Biotechnologies* are the eco-friendly technologies for the management of polluted soil and aquatic environments. It employs various microalgae and cyanobacteria to treat the waste in Activated Sludge in various Pollution Treatment Plants and Sewage Treatment Plants. We need *'clean environment'* and this paper discusses that on the basis of the varied algal and cyanobacterial responses towards many of these stresses in wetlands and aquatic ecosystems, a suitable assemblage of *'most-sensitive species'* could be identified for their use as tools for modern environmental management, phytoremediation through such photosynthetic prokaryotes appears to be a viable method for cleaning large areas of soil and aquatic environments/ecosystems.

Sustainable Environmental Management

Environmental Problems

Environment is a complex set of physical, geographic, biological, social, cultural and political conditions which surround an individual or organism, and ultimately determines its form and nature of survival. Environmental problems result from complex interactions between its physical components (*non-living resources*) and all other living organisms (*living resources*) on hand, and the human beings, which alter both, on the other. The rapidly growing human populations and highly diverse human activities have impacted upon the natural environment, causing degradation of both physical and biological components. The depletion of natural resources, the loss of biodiversity as well as pollution of soil, air and water, has raised the alarm for environmental management by either (1) preventing/minimizing further degradation or/& (2) to take remedial measures to restore/rehabilitate the natural environment. Though, Ecology has been successful in analyzing the major environmental issues but it has almost failed to provide concrete solutions to the management of such environmental problems at lower costs. In this context, the press and media have played their important roles in increasing the public awareness about environmental problems throughout the country, yet, the degradation of the environment has continued unabated and at an accelerated rate. Therefore, there is a need to analyze the causes behind the present state of environment and to comprehend our failure to prevent degradation.

Environmental Management

Environmental management involves the effective and active measures, which are taken for the protection, conservation and presentation of the environment, heritage and natural resources for which a government, organization or individual is responsible. There exist various dynamic interrelationships between the organisms and their environment (both abiotic and biotic) at different levels of biological organization. The components of the earth include: (1) air, land and water, (2) the layers of the atmosphere, (3) organic and inorganic matter and living organisms, (4) the interacting systems, including their integral parts and (5) the cultural, socio-economic, environmental health and other items of environmental impacts. The major environmental issues include the non-living as well as biological components *and their interactions*, besides the political factors, which influence these components and their interactions. The complex of physical, chemical and biotic factors which act upon an organism or an ecological community ultimately determine its fate, form and survival. All environmental problems arise from the disruption of natural organization of organisms, the cycles of materials and energy flow. Another aspect of environmental problems concerns with the alterations in the biotic communities either due to harvesting of only useful organisms (plants and animals) from natural ecosystems or alternatively, through their elimination by the use of *biocides* or by causing indirect changes in their habitats. Other environmental issues such as *global climate change* and the depletion of ozone layer are a combined result of alterations in both physical and biological environments caused by modern way of humans life syles.

Environmental Management through *'Non-living Resources'* and *'Living Resources'*

For the assessment of management performance and adaptive management strategies, the ecological indicators should relate to community structure, biodiversity, health, growth and the reproductive potential of individuals in various ecosystems. Wetlands are complex communities where a diversity of species interacts with each other and with the non-living environment. The hydrological regime of many wetlands has been altered by development, commercialization and globalization. The increase in the stressor factor(s) does not necessarily result in the same response per unit of stress all along its gradient. For example, the response of an organism of a community to heavy metal contamination in soil can not be the same if the heavy metal concentration is raised from say, 0 to 5 ppm or 20 to 25 ppm or 100 to 105 ppm. The trophic interactions and networks of competitive interactions are needed to be understood at >1 spatial and temporal scale. Recent reports have emphasized the potential use of specially selected heavy metal-sequestering algal species for the purpose of Phytoremediation and environmental clean-up or restoration (Kohli *et al.*, 2005). Phytochelatins and metallothioneins are the two metal binding peptides, which play vital role in metal detoxification and homeostasis against heavy metals. Such metal binding proteins have been reported for *Chlorella ellipsoida C. pyrenoidosa, Dunaliella bioculata, euglena gracilis* and *Scenedesmus quadricauda* (Singh and Singh, 1990).

Sustainable Environmental Management

Environmental management for sustainable use of the goods and services provided by the natural ecosystems requires both preventive and corrective measures as well as reversal of degradation (*i.e.*, restoration and rehabilitation). The short-term fragmented studies and research makes it generally inapplicable for log-term planning and management. Thus, in order to predict or anticipate the long-term responses of biota and environmental changes, we need holistic and long-term interdisciplinary studies of various ecosystems, involving all components and functional aspects at the same time (Kohli *et al.*, 2005). The mobilization of resources and their use through of government controlled measures, conservation or improvement of both natural and economic goods and services in such a way so that their conflicts may be minimized towards their conservation and improvement. *Synechococcus*, a cyanobacterium has been demonstrated to be vital in the bioremediation of Hg long with some hetrotrophic bacteria. The studies on *Anabaena dolium* and *Chlorella vulgaris* have suggested that their alginate immobilized state is more suitable for bioaccumulation of Cu and Fe. The efficient biotechnological exploitation of algae and cyanobacteria as biosorbants of pollutants and toxic heavy metals through mass culture systems, bioreactors, strain improvement and genetic manipulations, appears to more attractive for the developing countries like India.

Biotechnological Explorations Using Cyanobacteria

Immobilized Cyanobacteria Reducing Heavy Metals from Aquatic Habitats

Alginate Spheres

The exponential phase cyanobacterial cells (400 μg protein/ml culture) were concentrated through centrifugation and repeated washings with distilled water to obtain the desired cell protein value in 2.0–5.0 per cent (w/v) sodium alginate (Pant *et al.*, 1992, Singh *et al.*, 1992) The suspension was mixed thoroughly and pumped drop-wise in cool 0.2M calcium chloride solution to give the alginate spheres (beads) with an average diameter of 5.0 mm, containing the cyanobacterial cells in the amount of 55 μg protein/bead and the allowed to harden for 2h. This was followed by repeated washings with sterile distilled water before inoculation into fresh growth medium and/or Sorption Medium for photoautotrophic growth under culture room conditions (Asthana *et al.*, 1995). The chemicals like $Ca(NO_3)_2$, NaCl, $CaCl_2$ (12.0 mM) were used as the desorbing agents to release the Cyanobacterial cells from alginate matrix. For the uptake of heavy metal experiments 6 to 30 days old spheres were used.

Alginate Biofilms

Pant *et al.* (1992) have reported that the cyanobacterial cells mixed with a lower concentration of alginate (0.25 per cent, w/v) were slowly spread over calcium chloride solution (0.2M) to a thickness, which allowed the formation of floating films containing algal cells of 1.0 mm average thickness. Adequate precautions were taken during the repeated washings of such *'Biofilms'* because of the problems related to their breakage. Such preparations were allowed to grow in the diazotrophic medium for 6 to 30 days, so that they could be used as and when needed for the experiments related to adsorption &/or uptake of desired heavy metals.

Polyurethane Foam

Entrapment of cyanobacteria has also been reported to be a successful measure for the reduction of heavy metals from aquatic environments either through adsorption or intracellular uptake. For this the polyurethane foam were cut into a size of 5.0 mm cubes and pre-sterilized before being added to the cyanobacterial cultures (Pant *et al.*, 1992). Visual observations revealed that 4-6 days were required for the established of cyanobacterial cells in the foam matrices. During its subsequent growth, a minor fraction of the cell population generally leaks out of the foam preparations, however, the experiments on growth &/or uptake of heavy metals, were based on the amount of cyanobacterial cells retained in the foam cubes.

Recovery of Precious Metals from Cyanobacterial Spheres

Bioremediation of metals from wastewaters is of major importance, as it offers a potential alternative over the conventional processes for the removal and recovery of valuable/toxic metals; and the feasibility of using inert microorganisms as potential biosorbants, has been studied widely (Singleton and Simmons, 1996; Sar *et al.*, 1999), The hyper accumulating immobilized cyanobacterial strains contain very large amounts of accumulated heavy metals from the ambient media. Such spheres may contain precious metals like Hg, Cu, Ag, Au etc. Chemical and metallurgical techniques have been reported for the successful recovery of these metals; however, their commercial viability remains to be explored. Such methodologies offer flexibility in developing the non-destructive desorption processes to recover the metals of interest along with the regeneration and reuse of biosorbants as well as several advantages like low operating cost, minimum volume of disposable chemicals and/or biological sludge, high detoxifying efficiency of very dilute effluents and no nutrient requirement (Tsezos *et al.*, 1995; Kratochvil and Volesky, 1998)

Conclusions

Algae in general and Cyanobacteria in particular, have skills either to tolerate or they have developed mechanisms for tolerance against such contaminant toxicants. The pollutants in general and heavy metals in particular, inhibit both structural and functional variables of phytoplankton in field microcosms. They can selectively take-up toxicants like heavy metals, radioactive nuclides, pesticides, herbicides, carcinogenic substances, etc., leading to their intracellular bioaccumulation at a level several fold high than their microclimatic habitats. Therefore, they have been now accepted as ecological indicators for biological monitoring of pollutants in various ecosystems, especially the aquatic ecosystems. They sequester heavy metals, consequently attributing their role towards phytoremediation leading to environmental clean up and finally restoration of degraded ecosystems. Algae are the one of the major contributors for sustainable development and stability of various ecosystems. Cyanobacteria are of paramount importance whenever algae and prokaryotic photoautotrophs are considered. Therefore, serious and careful steps should be taken to assess and conserve their biodiversity, so that the sustainable utilization of microorganisms could be ensured, as they are our *'living resources'*; and if such resources are managed properly, the non-living resources will automatically be conserved. Restoration of the environment, which has already been degraded, must be undertaken at a faster rate, especially because at least some continuing degradation can not be ruled out as long as the basic needs of the ever growing and huge human population have to be met.

References

Asthana, RK, AL Singh and SP Singh 1993. Comparison of Ni-sensitive and Ni-resistant strains of *Nostoc muscorum*. *World Journal of Microbiology and Biotechnology* 9: 323-327.

Asthana RK, A Pant, S Chatterjee and SP Singh 1996. Compartmentalization of Ni and Hg in cyanobacteria: Metal sorption and desorption. *Current Researches in Plant Sciences*, Vol-II, pp 51-57.

Asthana RK, S Chatterjee and SP Singh 1995. Investigations on nickel biosorption and its remobilization. *Process Biochemistry* 30: 729-734.

Helik B, J Stray, J Prasilova, K Kratzer and J Hanusova 1979. Mercury circulation in aquatic environment. Part-2: metabolism of methyl and phenyl mercury in phytoplankton. *Acta Hydrochim Hydrobiol*. 4: 401-408.

Kaplan D, D Christiaen and SM Arad 1987. Chelating properties of extracellular polysaccharides from *Chlorella* sp. *Appl. Environm. Microbiol.* 53: 253-256.

Kohli RK, DR Batish, and HP Singh (2005. Need to establish long-term ecological research network in India. Bulletin of the National Institute of Ecology 15: 271-276.

Kratochvil D and B Volesky 1998. Advances in the biosorption of heavy metals. *Trends in Biotechnology* 16: 291-300.

Ramamoorthy S, TC Cheng and DJ Kushner 1983. Mercury speciation in water. *Can. Fish Aqua. Sci.* 40: 85-89.

Pandey P, CB Singh and SP Singh 1992. Cu uptake in a cyanobacterium: Fate of selected photochemical reactions. *Current Microbiology* 24: 35-39.

Pant A, SC Srivastava and SP Singh 1992. Methyl mercury uptake by free and immobilized a cyanobacterium. *BioMetals* 5: 229-234.

Rai LC, JP Gaur and HD Kumar 1981. Phycology and heavy metal pollution. *Biological Reviews* 56: 99-151.

Reddy PM and Venkateswarlu V 1985. The assessment of water quality and pollution in the river *Tungbhadra* (AP), India. In: *All India Symposium on Biology of Algae,* Pune, Feb. 8-10, 1985, pp. 10.

Tsezos, M, E Remoudaki and V Angelatou 1995. A systematic study on equilibrium and kinetics of biosorptive accumulation: The case of Ag and Ni. *International Biodeterioration Biodegradation* 35: 129-153.

Sar, P, SK Kazy, RK Asthana and SP Singh 1999. Metal adsorption and desorption by lyophilized *Pseudomonas aeruginosa. International Biodeterioration and Biodegradation* 44: 101-110.

Sehata FHA and Whitton BA 1982. Zinc tolerance in strains of the blue-green ala *Anacystis nidulans. British Phycological Journal* 17: 5-12.

Singh AK and LC Rai 1990. Use of *in situ* structural and functional variables of phytoplankton of the river *Ganga* for assessment of heavy metal toxicity. *Biomedical and Environmental Science* 3: 397-405.

Singh AL, RK Asthana, SC Srivastava and SP Singh 1992. Nickel uptake and its localization in a cyanobacterium. *FEMS Microbiology Letters* 99: 165-168.

Singh CB (2004. Algae and cyanobacteria as emerging tools for environmental management. *National Seminar on Ecology and Environmental Management: Issues and Res. Needs,* Kurukshetra pp.78-79.

Singh CB and SP Singh 1987. Effect of mercury on photosynthesis in *Nostoc calcicola*: Role of ATP and interacting cations. *Journal of Plant Physiology* 129: 49-58.

Singh CB and SP Singh 1990. Hg^{2+}-induced leakage of electrolytes and inhibition of NO^-_3-utilization in *Nostoc calcicola*. Biology of Metals: Role of interacting cations. *Biology of Metals* 3: 208-212.

Singh CB and SP Singh 1992a. Protective effects of Ca^{2+}, Mg^{2+}, Cu^{2+}, and Ni^{2+} on mercury and methyl-mercury toxicity to a cyanobacterium. *Ecotoxicology and Environmental Safety* 23: 1-10.

Singh CB and SP Singh 1992b. Assessment of Hg^{2+} toxicity to a N_2-fixing cyanobacterium in long–and short-term experiments. *BioMetals* 5: 149-156.

Singh CB, SK Verma and SP Singh 1987. Impact of heavy metals on glutamine synthetase and nitrogenase activity in *Nostoc calcicola. Journal of General and Applied Microbiology* 33: 87-91.

Singh, N, RK Asthana, AM Kayastha, S Pandey, AK Chaudhary and SP Singh 1999. Thiol and exopolysaccharide production in a cyanobacterium under heavy metal stress. *Process Biochemistry* 35: 63-68.

Singh SP, RK Singh, PK Pandey and A Pant 1992. Factors regulating cupper uptake in free and immobilized cyanobacterium. *Folia Microbiologia* 37: 315-320.

Singh SP, SK Verma, RK Singh and PK Pandey 1989. Copper uptake by free and immobilized cyanobacterium. *FEMS Microbilol. Lett.* 60: 193-196.

Singh SP and V Yadava 1985. Cadmium uptake in *Anacystis nidulans*: Effect of modifying factors. *Journal of General and Applied Microbiology* 31: 39-48.

Singleton, I and P Simmons 1996. Factors affecting silver biosorption by an industrial strain of *Saccharomyces cerevisiae. J. Chem. Tech. Biotech.* 26: 973-981.

Wlikinson SC, KH Gouldingand PK Robinson 1989. Mercury accumulation and volatilization in immobilized algal cell systems. *Biotechnology Letters* 11: 861-864.

© R.K. Gupta & V.D. Pandey (*Editors*)
Advances in Applied Phycology, 2007
Daya Publishing House, New Delhi

179–185

Chapter 13

Cyanobacterial Remediation of Heavy Metal Pollution

M.R. Suseela* and Kiran Toppo

Algology Laboratory, National Botanical Research Institute, Lucknow – 226 001, U.P., India

ABSTRACT

Continuous shrinking of natural resources such as drinking water and land are ringing the bell not to take nature for granted. An integrated approach, aiming at reduced production of water, judicious use of the available water, recycling of the used water and treatment of the polluted soil have to be adopted to ensure the sustained availability of natural resources. The mechanical methods of treating wastewater and polluted land have generally been adopted because of space and treatment capacity but they are energy intensive and lead to loss of plant nutrients. The biological way of reclamation is eco-friendly, cost effective and resource generating. Since algae are considered to be the primary producers of aquatic bodies, deleterious effects of metals on algae may have serious consequences for ecosystem equilibrium. The primary productivity of system decreases and causes species diversity leading to dominance of a few tolerant algal forms as indicators of the levels of contamination due to toxic metals. This has been mainly due to their inherent potential to assimilate nutrients, accumulate and concentrate pollutants. However, besides their potential in biomonitoring they have not been utilized in indicating the assessing metal pollution loading in closed and flowing water system.

Keywords: *Bio-remediation, Cyanobacteria, Environment, Heavy metal, Pollution.*

---

* Corresponding Author: Tel: 0522-2205831 Ext. 238 (O); E-mail: mrsuseela25@rediffmail.com.

Introduction

The term "heavy metal" is somewhat imprecise, but includes most metals with an atomic number greater than 20 excluding lanthanide and actinides. Heavy metals are widespread pollutants of great environmental concern as they are non-degradable and thus persistent. It is well perceived that there is a permissible limit of each metal and above which they are generally toxic. Some of the heavy metals are essential to living organisms at trace level (Lee *et al.*, 2002). However, high concentrations can be deleterious or even lethal to organism. So, widespread contamination of land, water and air by heavy metals with specific gravity is a subject of global concern. Heavy metals enter into the water system by two ways. Natural sources include corrosion, weathering and volcanic activities, anthropogenic source includes direct addition of heavy metals by means of effluents from different industries. The following metals are released from different industries:

1. Tannery industry release Cr, Fe, Mn, Zn, Cu, Pb, Cd etc.
2. Steel industry release As, Cu, Fe, Pb, Zn, Cr, Al etc.
3. Paper and pulp industry release Cd, Cu, Fe, Pb, Zn, etc.
4. Zinc smelter plants release Cd, As, Cu, Fe, Pb, Zn, Mn etc.
5. Ore mining industries release Cr, Cu, Mn, Fe, Ni, Zn etc.

Toxic effluents from industries release into rivers and cause problems to people living in vicinity and the organisms living in water and disturb the aquatic as well as terrestrial ecosystem. There is continuous release of heavy metals by industries in Indian rivers (Table 13.1).

Table 13.1: Amount of Metals in Different Rivers of India

Water Bodies	*Cu*	*Zn*	*Ni*	*Co*	*Pb*	*Cr*	*Cd*	*Refeeence*
River Moosi (A.P.)	0.005–0.1	0.03–0.44	0.005–0.035	0.02–0.006	0.01–0.38	0.01	—	Venkateswarlu *et al.*, 1994
River Tungabhadra (A.P.)	0.06–1.10	0.09–2.31	0.07–0.20	0.05–0.16	0.16–0.44	0.07	—	Venkateswarlu *et al.*, 1994
River Godawari (A.P.)	0.02–0.08	0.06–1.5	0.02–0.04	0.02	0.02–0.08	0.02–0.1	—	Venkateswarlu *et al.*, 1994
River Cooum (Madras)	0.10–0.65	0.02–0.40	—	—	0.12–0.048	—	—	Jebanesan *et al.*, 1994
Lentic waters of Gwalior (M,P.)	0.017–0.034	0.065–0.120	0.001–0.004	0.003–0.009	0.002–0.009	0.20–0.048	0.009–0.19	Kaushik *et al.*, 1999
River Ganga Kanpur (U.P.)	—	.0.031–0.036	—	—	—	0.0027–0.0028	—	Singh *et al.*, 1999
River Yamuna, Allahabad (U.P.)	0.028–0.030	—	—	—	—	0.002–0.0026	—	Singh *et al.*, 1999

* Amount of heavy metals in mg/L

In another survey the concentration of heavy metals in ground water in Faridabad district of Uttar Pradesh was generally high and exceeded the maximum limit prescribed by ISI (1983) for drinking purpose.

Effect of Heavy Metal Pollution on Different Organisms

Metal pollution can adversely affect almost all living creatures on earth. The effect of heavy metal pollution on different organisms are as follows:

On Human Beings

Pollution by heavy metal is instantly recognized with the Minimata disease in Japan, when several thousands of people suffered mercury poisoning by consuming the fish caught in Minimata bay, which was the recipient of mercury released from a vinyl chloride plant. Similarly high level of cadmium in local food stuff in parts of Japan, attributed to irrigation water from the soil heaps of an abandoned mine, caused Itai-Itai Byo disease in 1955, mainly in women over forty. Chronic exposure of heavy metals can cause adverse health effects, especially in fetus and young children, even at low level. In human milk from industrialized countries, average background levels of lead is probably between 5-20µg/L while in the heavily polluted area this may be up to 20 times higher (Frkovic *et al.*, 1997). Lead (Pb) enters into young children through the bloodstream initially and then rapidly attached itself to the red blood cells (Harrison and Laxen, 1980) and cause circulatory disorder. Symptoms of chronic lead poisoning are subtle and manifestations include facial pollar, anemia, pain, headache, wrist drop, foot drop, restlessness, fatigue, convulsion and coma. It is estimated that more than half the under 12 years living in Indian metros have tremendously elevated blood lead level. About 3.5 crore people in West Bengal are at the risk of chronic arsenic poisoning from drinking contaminated ground water. Arsenic causes several disorders including hypertension and neurological effects. The pregnant women who had relatively low concentration of mercury in their bodies (hair concentration measured at 10-20 ppm) exposed their developing fetus to enough mercury to cause noticeable neurological problems such as delayed walking and learning deficits.

On Plants and Crops

Plants and crops are on direct target of heavy metals because metals get enter into the plant body through irrigation with metal contaminated water. These metals reduce the yield and cause several deformities in plants and crops. For example, cadmium effects on different plant material *i.e.* rice seedlings, germination, cell division, root growth, chlorophyll, water uptake and the loss of sugar into potato tuber. One µm chromium resulted into reduced activities of amino levulinic acid (ALAD) and nitrate reductase and there was decrease in total chlorophyll and protein content in *Nymphaea alba* (Vajpayee *et al.*, 2000). When *Azolla pinnata* was treated with 5-15ppm of different heavy metals for 1-14 days it was found that there was a marked reduction in chlorophyll content and nitrogen fixation activity. In *Vigna mungo*, all the growth parameters like germination, shoot length, root length, number of lateral root/shoot ratio, speed of germination index and vigour index showed a slight increase in 5mg/L concentration of chromium and cadmium solution but no such growth promoting effect was observed in base of mercury treatment. The 250mg/L concentration of mercury proved to be lethal for *Vigna mungo*. In case of *Oryza sativa*, Cd, Cu, Hg at concentrations 10, 100, 1000µg decrease the biochemical constituents and nitrate reductase activity in both etiolated and light grown seedling. But toxic effects of those heavy metals overcome by the supply of metabolites such as sucrose, glycine and citric acid. In *Pisum sativum* different concentration of mercuric chloride, cadmium chloride and lead nitrate altered nucleolar shape, size and number and reduced activity of r-RNA and protein synthesis. Tannery effluent a major source of heavy metal affected the mitotic indices and induced mitotic aberrations in root meristems of *Allium cepa*. Similarly 50,100,150, 200, 250, 300µg/mL of cupric chloride, cadmium chloride and mercuric chloride reduced the seedling growth of wheat in terms of root length, shoot length, fresh weight, dry weight and carbohydrate metabolism which includes a-amylase, B-amylase, invertase activities and reducing and non-reducing sugar contents were altered by heavy metals.

Bioremediation

Though plants are able to accumulate the heavy metals but there are many reports that these metals adversely affect the yield of plant and cause various abnormalities in plants and one of the major limitation of using plants for bioaccumulation and bioremediation of metals is that these plants take long time to grow and whole process is time taking. There should be a rapid method for bioremediation.

Cyanobacteria: A Better Option for Bioremediation

The current practice for the removal of soluble metal ions present in the industrial effluents includes simple chemical precipitation, ion exchange resins, solvent extraction etc. Low bulk metal concentration, ion specificity, narrow range of pH and poor settling colloidal properties may limit the effectiveness of these conventional processes. In the last decade, considerable research has been carried out on the materials of biological origin for metal removal. The microbial biomass can passively bind large amount of metal(s)–a phenomenon commonly referred to as biosorption (Raveender *et al.*, 2002).

Among photo autotrophs, algae became the candidate of interest due to bulk availability of their biomass from water bodies. Response of several green algae to toxic metals have been investigated and some of these taxa have been found to display tolerance to toxic metals. The macro algae most extensively used for monitoring heavy metal contamination of water belong to the genera *Fucus, Enteromorpha, Laminaria* and *Ulva* (Fytianos *et al.*, 1999). Bioaccumulation of heavy metals were greater in the alga *Enteromorpha* species than in *Ulva* species (Fytianos *et al.*, 1999). *Chlamydomonas reinhardtii* could tolerate 100 and 150μM copper sulfate (Boswell *et al.*, 2002). But cyanobacteria are known to be relatively more tolerant to heavy metals (Fiore and Trevores, 1994) and are easily available in nature.

Cyanobacteria (also known as blue green algae) are the largest and most diverse group of photosynthetic prokaryotes. Their habitats vary from fresh and marine water to terrestrial environment. They are oxygen-evolving organisms that respond to stress conditions such as light deprivation (Gardea-Torresdey *et al.*, 2002) and salt stress. Cyanobacteria have developed natural methods of responding to metals such as copper, lead and cadmium through surface binding of various functional groups (Gardea-Torresdey *et al.*, 2002). Cyanobacteria can also be used for reclamation of saline soils (Rai and Abraham, 1993).

Studies on cyanobacterial strains metal tolerance appear to have been initiated in 1982 when Zn-tolerent strain of *Anacystis nidulans* was isolated. Zinc tolerant strain of *Anacystis nidulans* displaying a zinc uptake comparable to Zn-sensitive wild type. The metal tolerance in the above strain was attributed to the intracellular detoxification mechanisms (Verma and Singh, 1995). Some reports suggested that removal of metal by cyanobacteria was quick *e.g.* uptake of metal in immobilized cells of *Aphanocapsa pulchara* showed 75-94 per cent removal in a retention time of 5 min. Increasing the retention time increased the removal percentage for cadmium and zinc. Copper was taken by *Aphanocapsa pulchara* at a faster rate (2 fold) when compared to cadmium and zinc both in light and dark. Several reports are present which are suggested that cyanobacteria are tolerant to high concentration of heavy metals. For instance under low density conditions, chronic treatment with chromium failed to produce much effect on growth of *Synechococcus* and *Nostoc* species (Thompson *et al.*, 2002). The mesophillic cyanobacteria *Anabaena flos-aquae* and *Synechococcus cerdrorum* were grown in various concentrations of carpet industry effluent. The effluent was stimulatory for the growth of both the cyanobacteria when used between 5 per cent to 50 per cent concentration. *Anabaena flas-aquae* showed maximum stimulation of growth at 25 per cent concentration while *Synechococcus cerdrorum* the concentration was 50 per cent. These two species of cyanobacteria could be grown even in 100 per cent concentration but the growth was very poor. The growth of *Anacystis nidulans* at 10mg/L selenium oxide was significantly greater than the control in terms of cell number (Lee *et al.*, 2002). Cyanobacteria used as single cell protein, *Spirulina plantensis,* was shown to contain detectable level of mercury and lead when grown under contaminated conditions (Slotton *et al.*, 1989). *Aphanocapsa pulchra* was tolerant to high concentration of zinc. At concentration 7.2mg/L there was only 20 per cent inhibition on chlorophyll-a content but about 90 per cent inhibition in phycocyanin and about 80 per cent inhibition in allophycocyanin content but decrease in phycocyanin and allophycocyanin was compensated by increasing the phycoerythrin content. Gardea-Torresdey *et al.* (2002) tested *Synechococcus* for its potential to bind metal ions from solution. This cyanobacterial strain to bind 11.3mg of Cu/gm of biomass, 30.4mg Pb/gm of biomass, 3.2mg of Ni/gm of biomass, 7.2mg of Cd/gm of biomass. More than 98 per cent of Cu, Pb and Ni metal ions were recovered, while over 50 per cent of Cd was recovered when treated with 0.1M HCl.

Mechanism of Metal Uptake

There are two phases of uptake of heavy metal ions by cyanobacteria. First a passive rapid phase, in which the ions bind to the cell wall and second a slower and metabolically dependent uptake in cytosol (Lee *et al.*, 2002). In case of resistant strain in the presence of metals, both phases can occur and can be augmented or obscured by various external as well as internal factors. Reports indicated that carboxyl group on algal cell biomass was responsible for binding to various ions (Gardea-Torresdey *et al.*, 1990). Intracellular polyphosphates found in algae, participate in metal sequestration and algal extracellular polysaccharides serve to chelate or bind metal ions (Zhang and Majidi, 1994). Certain archaebacteria and cyanobacteria may show a structure called sheath. When present, it appears as a loosely arranged homo–or heteropolymer of proteins with a small component of lipid and carbohydrates, or as fibrils of acidic polysaccharides. The sheath not only shows certain selectivity for metals but also participate in oxidation and precipitation of Fe and Mg from solution. A common metal induced response in many micro-organisms is the synthesis of intracellular metal binding proteins, metallothioneins (MTs). Through more frequently found in eukaryotes, MT-like proteins are reported from prokaryotes also. Metallothioneins are low molecular weight proteins or polypeptides (6,000-8,000amu), which bind metal ions in metal-thiolate cluster. These polypeptides are abundant in cystien residues (cys) and posses a characteristic pattern of sulfur containing amino acids. These polypeptides are commonly found in association with essential metal ions such as zinc, copper and also bind to toxic metals like cadmium, mercury and lead. These metal binding properties are mediated via the abundant cystein residue and their characteristic organization into–cys-cys,–cys-x-cys-, or–cys-x-x-cys–sequences (x corresponds to any other amino acid in protein sequence) (Gardea-Torresdey *et al.*, 2002). Pepi and Baldi, (1995) observed that cystein at 5μg/mL levels increased tolerance up to 10μg/mL for chromium, where as methionine only reduced the chromium toxicity in strain *Rhodosporidium torulodies.*

The genes for these proteins are widespread among the various genera and species. R2C cells, a rat testicular Leyding cell line, are sensitive to Cd-induced genotoxicity and that this sensitivity is associated with minimal expression of these genes can be increased the metal tolerance of cyanobacteria (Gardea-Torresdey *et al.*, 2002).

Conclusion

In India nearly 175 million hectares or about 50 per cent of our land area suffers from one type of degradation or the other and comes under the definition of waste land, which means that the productivity from this area is much less than what it should be. It is estimated that our national soil loss is about 6,000 million tons of nutrients. This is more than the quantity of fertilizers applied to our soil every year. Ground water availability for sustaining the life constitutes only 0.57 per cent of total water existing on this planet. The world population is increasing in an alarming manner. As per a report, it took 123 years for the world pollution to increases from one billion to two billion. The succeeding billion additions took only 33 years, 14 years, 13 years and 11 years. Now the world population is more than 6 billion, which has to depend on the same water and land resources what was being used by one billion population about two centuries back. Unless these precious materials are used judiciously and conserved properly, the increasing demand of water and food can't be met with. Heavy metal pollution of soil and water significantly contribute to degradation of soil and water. If we do not take adequate timely care of our soil and water resources most important constituent of our environment the time is not far when there will be famine and hunger stalking us and the next world war may be fought for water.

However, conventional treatment techniques are expensive and the increasing demand of eco-friendly technologies has lead to the search of low cost alternatives, which could be considered as single use materials. In such scenario, the cyanobacterial biomass has become a pet in terms of ecological and economic consideration.

Acknowledgements

We thank the Director, National Botanical Research Institute, Lucknow for his constant encouragement and providing all necessary facilities.

References

Lee LH, Lustigman BK and Murray SR (2002) Combined Effect of Mercuric Chloride and Selenium Dioxide on the Growth of the Cyanobacteria, *Anacystis nidulans*. *Bull. Env. Cont. and Toxi* 69:900-907.

Raveender V, Scaria J and Verma SK (2002) Application of Mutant Strains of Cyanobacteria for Cd Removal. *Bull. Env. Cont. Toxi* 69: 752-759.

Fytianos K, Evgenidou E and Zachariadis G. (1999) Use of Macroalgae as Biological Indicators of Heavy Metal Pollution in Thermaikos Gulf, Greece. *Bull. Env. Cont. Toxi* 62: 630-637.

Rai AK and Abraham G (1993) Salinity tolerance and growth analysis of the cyanobacterium *Anabaena doliolum*. *Bull. Env. Cont. Toxi* 51: 724-731.

Thompson SL, Manning FCR and McColl SM (2002) Comparison of the Toxicity of Chromium III and Chromium VI to Cyanobacteria. *Bull. Env. Cont. Toxi* 69: 286-293.

Boswell C, Sharma NC and Sahi SV (2002) Copper Tolerance and Accumulation Potential of *Chlamydomonas reinhardtii*. *Bull. Env. Cont. Toxi* 69: 630-637.

Frkovic A, Kras M and Alebic-Juretic A (1997) Lead and Cadmium Content in Human Milk from the Northern Adriatic Area of Croatia *Bull. Env. Cont. Toxi* 58:16-21.

Verma SK and Singh SP (1995) Multiple metal resistance in the cyanobacterium *Nostoc muscorum*. *Bull. Env. Cont. Toxi* 54:614-619.

Vajpayee P, Tripathi RD, Rai UN, Ali MB, and Singh SN (2000) Chromium (VI) accumulation reduces chlorophyll biosynthesis, nitrate reductase activity and protein content in *Nymphaea alba* L. *Chemosphere*. 41(7): 1075-82.

Pepi M and Baldi, F (1995) Chromate tolerance in strains of *Rhodosporidium toruloides* modulated by thiosulfate and sulfur amino acids. *Biometals* 8: 99-104.

Slotton DG, Goldman CR and Franke A. (1989) Commercially grown *Spirulina* found to contain low levels of mercury and lead. *Nutr Rep Int* 40:1165-72.

Fiore MF and Trevors JT (1994) Cell composition and metal tolerance in cyanobacteria. *Biometals*, 7: 83-103.

Harrioson RM and Laxen DPH (1980) The Physico-chemical speciation of Lead in drinking water. *Nature*, 286:791-793.

Gardea-Torresdey JL, Hejazi M, Tiemann KJ, Parsons JG, Duarte-Gardea M and Henning J (2002) "Use of Hop (Humulus lupulus) Agricultural Byproducts for the Reduction of Aqueous Lead(II) Environmental Health Hazards", J. Hazard. *Mater*. 91: 95.

Gardea-Torresdey JL, Becker-Hapak MK, Hosea JM and Darnall DM (1990) Effect of Chemical Modification of Algal Carboxyl Groups on Metal Ion Binding, Eviron. *Sci. Technol.*, 24:1372.

Zhang W and Majidi (1994) Monitoring the Cellular Response of *Stichococcus bacillaris* to Exposure of Several Different Metals Using in Vivo ^{31}P NMR and Other Spectroscopic Techniques, *Environ. Sci. Technol.*, 28, pp. 1577-1581.

Venkateswarlu V, Manikya Reddy P and Rajkumar B (1994) Heavy metal pollution in the rivers of Andira Pradesh, India. *J Environ Bio*, 15(4):275-282.

Singh GS and Singh AS (1999) Variation and correlation of dissolved oxygen with effluent quantity and stage of river Ganga at Varanasi (India). *Indian J Environ Hlth*, 36(2): 79-83.

Kaushik S, Sahu BK, Lawania RK and Tiwari RK (1999). Occurrence of heavy metals in lentic water of Gwalior region. *Pollution research* 18(2): 137-140.

Jebanesan A, Selvanayagam M (1994) Population dynamics of aquatic hemipterans in the river Cooum, Madras. *J Environ Bio*, 15 (3): 213-220.

© R.K. Gupta & V.D. Pandey (*Editors*)
Advances in Applied Phycology, 2007
Daya Publishing House, New Delhi

Chapter 14

Heavy Metal Toxicity to Rice Field Cyanobacteria

Mihir Kumar Das*

P.G. Department of Botany, G.M.College (Autonomous), Sambalpur – 768 004 (Orissa)

ABSTRACT

Living organisms require trace amount of some heavy metals, including cobalt, copper, iron, manganese, molybdenum, vanadium, stromium and zinc. Excessive levels of essential metals, however, can be detrimental to living organisms. Human activities have drastically altered the biochemical and geochemical cycles and balance of some heavy metals. Heavy metals are stable and persistent environmental contaminants since they can not be degraded or destroyed. Therefore, they tend to accumulate in the soil and sediments. The heavy metals released into the environment are recalcitrant and are released gradually into the water bodies or terrestrial ecosystem causing serious threat to organisms including algae. Cyanobacteria are subjected to several stresses including heavy metals. Heavy metals are generally components of several pesticides, extensively used in the rice fields. Excessive use of heavy metals affects the growth of these prokaryotic algae, thereby reducing the fertility of the soil. These organisms are recognized as biological indicators of pollution and efficient tools for the removal of heavy metals from the aqueous system. Heavy metals are removed by biomineralization and by passive accumulation in cells through surface binding with chemical functional groups. These chemical functional groups evolved most likely as a response mechanism to the metal toxicity.

Keywords: *Cyanobacteria, Heavy metal, Toxicity.*

* Tel: 0663-2548197, (M) 94373-93301; E-mail: mihir01@sify.com.

Introduction

While man's fascination with the precious metal is legendary, the chemistry of toxic metals is not a subject which has inspired poets, and literary references to metal stressed ecosystems are rare (Stokes, 1983). The poet Bottomly has lamented in his poem for the legacy of the metal mining and processing industries, which is perceived as a lifeless wasteland, inhabitable by man.

"When you destroy a blade of grass

You poison England at her roots

Remember no man's foot can pass

Where evermore no green life shoots".

Less predictable and of more relevance in the present context, is the poet's observation in a later stanza:

"The grass, forerunner of life, has gone,

But plants that spring in rains and shards

Attend until your dream is done,

I have been hemlock in your yards".

The poet has identified one of the most intriguing aspects of metal pollution. The accelerating frenzy of industrialization to meet the demands for comfort of man has overstretched the self-purification capacity of biosphere to a point, where its delicate equilibrium is threatened. Since our independence, considerable progress has been made and today, we are among the first ten industrialized nations of the world, thanks to the vision of Pandit Jawaharlal Nehru. Associated with any development, there is always some amount of environmental degradation. More than 60,000 chemicals are being used by the human population, all over the world. Most of them are manufactured by the industries, which discharge huge quantities of heavy metals into each and every imaginary lines passing between two poles on the earth's surface, resulting serious health hazard through contamination of food chains, ultimately reaching human being.

Forstner and Wittmann (1979) have listed some of the sources of heavy metals as: geological weathering, industrial processing of ores and metals, the everyday use of metals and items with metal components, leaching of metals from garbage and solid waste dumps and animal and human excreta. A number of heavy metals, including zinc, copper, iron, manganese and cobalt under normal conditions are required as micro-nutrients in biological systems to act as co-factors and/or as part of the prosthetic groups of enzymes in a variety of metabolic and developmental pathways. But, however, excessive levels of essential metals, however, can be detrimental to living organisms.

The term 'heavy metal' is often used to describe 'trace element". The term is ill-defined, unsatisfactory of even unacceptable to some chemists, although most ecologists use it. The heavy metals are those having a density at least five times that of water *i.e.* a density greater than 5gm Cm^{-3}.

In unperturbed natural systems, the two processes of (a) weathering and release and (b) accumulation of heavy metals in sediments is farley balanced. But this balance is sometimes disturbed, by various anthropogenic inputs of heavy metals into the surrounding aquatic ecosystems, both by altering the aquatic bioproductivity and through direct toxic action on aquatic organisms. Two factors contribute to the deleterious effects of heavy metals as environmental pollutants: (i) heavy metals can not be destroyed or degraded by biotic or abiotic processes (ii) therefore they tend to accumulate in the soils and sediments of rivers and lakes by association with organic and inorganic matters by process of absorption, complexation and chemical co-ordination (Rai and Dubey, 1989).

Hazards of Heavy Metal Pollution

The hazards of heavy metal pollution evidenced by such episodes as the Minamata and Nigata Bay incidents and by the occurrence of itai-itai (ouch-ouch) disease, caused by mercury and cadmium poisoning respectively have received much attention. Ingestion of metals such as lead cadmium mercury, arsenic and chromium may pose great risks to human health.

Lead

Because of size and charge similarities lead can substitute for calcium and be included in bone. Children are especially susceptible to lead because developing skeletal muscles require high calcium levels. Lead that is stored in bone is not harmful, but if high levels of calcium are ingested later, the lead in the bone may be replaced by calcium and mobilized. Once free in the system, lead may cause nephrotoxicity, neurotoxicity and hypertension.

Cadmium

Cadmium may interfere with the metallothionein's ability to regulate zinc and copper concentration in the body. Metallothionein is a protein that binds to excess essential metals to render them unavailable. When cadmium induces metallothionein activity, it binds to copper and zinc, disrupting the homeostasis level.

Mercury

Mercury poses a great risk to humans. Mercury occurs naturally in the environment and it can also be released into the air through industrial pollution. Bacteria in the water cause chemical changes that transform mercury into methlmercury, which binds tightly to the protein in fish tissue. Mercury tends to accumulate in the food chain, so predatory fish species (eg. Shark) tend to have higher levels than non-predatory fish or species at lower level on the food chain. High amounts of mercury can damage the nervous system of people, triggering such health problems as memory loss, slurred speech, hearing loss, lack of co-ordination, loss of sensation in fingers and toes, reproductive problems, coma and possibly death.

Arsenic

Arsenic ingestion can cause severe toxicity through ingestion of contaminated food and water. Ingestion causes vomiting, diarrhea and cardiac abnormalities.

Chromium

The presence of abundant chromium anions in the water is generally a result of industrial waste. The chronic adverse health effects are respiratory and dermatologic.

A lot of information concerning effects of heavy metals on animals and higher plants and their fate in various ecosystems has accumulated during the last few decades. The algae and higher plants of contaminated localities tend to accumulate heavy metals to a dangerous extent. Algae incorporate solar energy into biomass, produce oxygen that is dissolved in water and used by aquatic organisms, function critically by the cycling and mineralization of chemical elements and serve as food for herbivorous and omnivorous animal.

Cyanobacteria

Cyanobacteria (blue-green algae) are the fascinating group of prokaryotic photosynthetic nitrogen-fixers, endowed with the tremendous capacity of their distribution in diverse ecological habitats. Rice is perhaps the only major cereal crop to whose nitrogen economy; nitrogen-fixating cyanobacteria make a significant contribution (Roger and Kulasooriya, 1980). Agronomic potential of cyanobacteria in rice cultivation was first recognized by De (1939), who attributed the natural fertility of tropical rice fields to biological nitrogen fixation by these organisms. Since then, long term fertility experiments and nitrogen fixation measurements have confirmed the importance of cyanobacteria in maintaining a moderate, but constant rice production in field receiving no fertilizer (Roger, 1988). However, cyanobacteria are subjected to several stresses including heavy metals both on aquatic as well as terrestrial system. The high metal concentration resulting from anthropological and agricultural activities affect their growth, ultimately leading to loss of fertility in the rice-fields. Many cyanobacteria are capable of accumulating and concentrating metals several folds higher than in the medium. Within the cells of the algae, metals are regulated by homeostatic control. At high metal concentrations, however, the homeostatic regulation ceases and metals act as toxicants severely damaging the entire system. There are several reports on the adverse effects of heavy metals on micro-organisms including cyanobacteria (Henriksson and Da Silva, 1978; Stratton and Corke, 1979a,1979b; Allen *et al.*, 1980; Singh and Pandey, 1981). Heavy metals are generally components of several pesticides and are used extensively as an algicide in controlling cyanobacterial blooms in lakes and ponds (Edmondson, 1969; Hodson *et al.*, 1979). Further Cyanobacteria isolated from metal polluted habitats are often more tolerant to metals than those from non-metal habitats. The accumulated metals are chemically detoxified by the algae rendering them inactive. The toxic effects of metals, their uptake and accumulation and detoxification mechanisms operating in the tolerant algae vary from species to species. It is, therefore, essential to study the toxicity of metals to cyanobacteria, their EC_{50} and the influence of various factors on their toxicity, uptake and accumulation. Some work has been done to assess the impact of different kinds of heavy metals such as Hg, Cd, Ni, Cu, Zn, Pb, Ag, Cr, Fe etc. on the growth, pigment composition heterocyst frequency and nutrient uptake on different species of cyanobacteria (Stratton *et al.*, 1979; Rai and Raizada, 1985, 1986; Campbelt and Smith, 1986; Rai and Dubey, 1988; Rai *et al.*, 1990; Poonguzhali and Rao, 1999; Maheswari *et al.*, 2001)

Effect of Heavy Metals on Cyanobacteria

Growth of *Anabaena inaequalis* was inhibited significantly above 2ppb Hg^{2+} and completely inhibited at 8 ppb. Low levels of Hg^{2+} stimulated acetylene reduction and photosynthesis. The lysis of vegetative cells was the primary action of mercury ions, resulting in the inhibition of growth, photosynthesis and nitrogenase activity (Stratton *et al.*, 1979). *Anabaena inaequalis* was sensitive to nickel ion (0.025, 0.05, 0.075, 0.1 and 0.125 ppm) in order of decreasing sensitivity of growth photosynthesis and acetylene reduction. The growth was inhibited at 0.125 ppm. Nickel toxicity was proposed be due to poisoning of intracellular enzyme systems by nickel ions (Stratton and Corke, 1979a). However, the responses of *Anabaena inaequalis* towards combination of mercuric, cadmium and nickel ion was dependent upon the order of metal addition and actual metal concentration involved. Mercuric and cadmium ions interacted synergistically towards photosynthesis and nitrogenase activity, but resulted in mixed synergism and antagonism towards growth at 0.003 ppm concentration (Stratton and Corke, 1979). Rai and Raizada (1985) observed that $NiCl_2$ stimulated carbon fixation and nitrogenase activity of *Nostoc muscorum* at 2.1 μM. But silver ($AgCl_2$) at 0.013, 0.026, 0.052 μM seems to be much more toxic than nickel at 2.1, 4.2, 5.04 μM. EDTA protected carbon fixation and calcium ameliorated the nitrogenase activity in the test alga. But when the same organism was treated with $K_2Cr_2O_7$ (34, 68, 102μM) and $PbCl_2$ (36, 72, 108 μM), approximately 50 per cent algae survived at 68.00 and 72.0 μM concentration of Cr and Pb respectively. Chromium was proved to be more toxic than lead because its sub-lethal concentration (68 μM) inhibited approximately 82 per cent CO_2 uptake and 64 per cent acetylene reduction. Toxicity was found to be reversed by calcium, but not by EDTA (Raizada and Rai, 1985). Metal toxicity may be appreciably counteracted by reducing substances and amino acids which are likely to occur in natural habitats (Rai and Raizada, 1988).

Nickel transport was dependent on the membrane potential of the cells of *Anabaena cylindrica* and the rate of uptake was decreased in the dark or by the inhibitors. The cells transported nickel ions by a carrier facilitated transport process with the concentration factor for the ions being determined by the cell membrane potential (Campbell and Smith, 1986; Singh and Singh,1987), while working with Cu^{2+}, Cd^{2+} and Zn^{2+}, found that 5 μM Cu^{2+}, Cd^{2+} and Zn^{2+} were inhibitory to 10 μM H_2O_2 supported Hill activity and oxygen evolution in membrane preparation from *Anacystis nidulans.* But when they treated Hg^{2+} (0.20 and 0.25μM) against *Nostoc calcicola,* the photosynthesis was inhibited at the very initial stage. Photosynthetic O_2 evolution was more sensitive to Hg^{2+} stress than $^{14}CO_2$ uptake.Hg^{2+} ion predominantly attack the action sites of Mn^{2+} and ATP generating steps of photosynthesis in the cyanobacterium.

Sodium arsenite at concentration above 50 μM inhibited the growth of *Synechococcus leopoliensis* UTEX 625 in a defined culture medium. Inhibition was transitory, with growth resuming after a lag period, the duration of which depended on the arsenite concentration. Cells grown for several hours in presence of 10 μM arsenite became tolerant to concentrations of arsenite that inhibited the growth of untreated cells. At a concentration of 200 μM, aresenite temporarily halted the growth of sensitive cells, but did not affect that of tolerant cells. This concentration of arsenite inhibited net photosynthesis in both sensitive and tolerant cells. At the same time, it selectively decreased the incorporation of carbon in the light into **α**–amino acids especially glutamate, in sensitive, but not in tolerant cells. Simultaneously, incorporation of carbon into pyruvic acid markedly increased. The activity of the partially purified pyruvate dehydrogense complex of *S.leopoliensis* was abolished by 45 μM arsenite. Therefore, it is concluded that inhibition of pyruvate dehydrogensase by arsenite is sufficient to explain its inhibition of growth in this organism (Budd *et al.*, 1986).

Exposure of *Nostoc muscorum* to different concentration of $NiCl_2$ and $AgCl_2$ brought about reduction in growth, carbon fixation, heterocyst production and nitrogenase activity and increase in loss of ions (K^+, Na^+). It was found that level of protection of ascorbic acid and glutathione was more for Ag^{2+} and Ni^{2+}. However, metal induced inhibition of growth and carbon fixation was equally ameliorated by methionine. Protection of metal toxicity by amino acids lends further support to self–detoxifying ability of cyanobacteria, because they are known to synthesize all essential amino acids (Rai and Raizada, 1987). Jampani (1988) found that growth and survival of *Synechococcus aeruginosus* were completely inhibited by lead nitrate ($PbNO_3$) at 100 and 200 μg/ml respectively. The starred (N_2, C, PO_4^-deficient) cells expressed more sensitivity to lead toxicity. All the nutrients offered protection against lead. Fructose (0.5 per cent) significantly stimulated the process, like growth, heterocyst differentiation, nitrate reductase, nitrogenase, glutamine synthetase and $^{14}CO_2$ uptake, treated with 0.4mM chromium, followed by glucose (1.0 per cent), sucrose (0.5 per cent) and acetate (0.1 per cent) (Rai and Dubey, 1988).

Rueter (1988) observed that increased iron nutrition ($FeCl_3$) in the range of 10^{-8}M to 10^{-6}M resulted approximately four fold increase in carbon and nitrogen fixation rates. Ahluwalia and Kaur (1989) while working with $NiSO_4$ against *Anabaena variabilis* found that the nickel ion was found to be algicidal where as at 0.5ppm it enhanced the growth over control. Heterocyst was differentiated in pairs or in small chains at 2.0-5.0 ppm. The response to metal combination (Cu+Ni, Cu+Fe, Ni + Fe) was dependent upon the order in which the metals were added. The Cu + Ni combination resulted in synergistic interaction in contrast to the antagonism of Cu+Fe and Ni +Fe. (Mallick and Rai, 1989). Addition of low concentration (3 and 6 µM) of $HgCl_2$ to intact Cells of *Spirulina platensis* caused an enhancement in the intensity of fluorescence emitted from phycocynin at room temperature and induced blue shifts in the emission peak, suggesting the energy transfer within the phycobilisomes. (Murthy *et al.*, 1989), where as Petterson and Bergmann (1989) found that addition of low concentration of $AlCl_3$ (3.6 to 36 µM) increased the ATP pool 20-40 per cent within 24 hours, the effect being more marked with time.

Exposure of *Nostoc muscorum* to combinations of Ag + Cr, Ag + Ni, Ag + Pb and Ag, Cr, Ni and Pb alone brought about a complex pattern of inhibition of growth, NO_3, NH_4 and $^{14}CO_2$ uptake and nitrogenase activity. The order of inhibition of growth and NO_3 uptake was AG + Cr > Ag + Ni > Ag + Pb. The combination of Ag + Pb was less toxic for growth, nitrogenase, NO_3 and NH_4 uptake (Rai, 1989). All these test metals, when used individually depicted toxicity against growth, nitrate uptake, ammonia, $^{14}CO_2$ fixation and nitrogenase activity of the test algae (Rai and Raizada, 1989). Chromium (40 µg/ml CrO_3) was proved to be much more toxic than tin(50 µg/ml $SnCl_2$) against *Anabaena doliolum* and the sequence of protective efficiency of synthetic and natural complexing ligands may be given as EDTA> nitrilotriacetate acid > citrate > Pyridine dicarboxylic acid (Rai and Dubey, 1989). The toxicity of the test metals (CrO_3 and $SnCl_2$) was lowered at alkaline p^H and increased at acidic p^H. Sodium Chloride at 20mM was found to decrease metal toxicity. Humic acid was the most effective in regulating metal toxicity (Dubey and Rai, 1990).Addition of Ni^{2+} (1 µM to 5 µM), Hg^{2+}(0.1 µM to 0.5 µM) and Cu^{2+} (0.1 mM to 0.5mM) inhibited the growth, O_2evolution and O_2 uptake in *Cylindrospermum* sp. (Hg^{2+} > Ni^{2+}> Cu^{2+}) (Singh *et al.*, 1989).

A general reduction in cell dimension, thylakoid surface area, number and volume of polyhedral bodies, polyphosphate bodies cyanophycean granules, lipid bodies, membrane limited crystalline inclusion, volume and number of wall layers and mesosome was observed when $CdCl_2$ (0.12 µM and 1.18 µM) was treated against *Anabaena flos-aquae* (Rai *et al.*, 1990). Husaini and Rai (1991), when exposed *Nostoc linckia* to same metal ($CdCl_2$) with concentrations ranging from 0.01 to 0.5 µg/ml, it showed a concentration dependent inhibition of growth, nutrient uptake and activities of enzymes responsible for assimilation of nutrients.

Mercury induces alternation of energy transfer in phycobilisome of *Spirulina platensis* by selectively affecting the pigment protein, and phycocyanin (Murthy and Mohanty, 1991). PS II was found to be more sensitive both to low and high concentration of chromium (34, 68, 102, 136, µM) and lead (36, 72, 108, 142 µM), used against *Nostoc muscorum*. A considerable inhibition of PS I was, however, observed at high concentration only (Prasad *et al.*, 1991). But *Nostoc calcicola*, while saturated with 40µM $CuSO_4$ for 1 hour, showed more than 50 per cent inhibition of PS II and 95.4 per cent of $^{14}CO_2$ fixation compared with only 15.5 per cent decrease in PS I activity (Pandey *et al.*, 1992).

The effects of combined nitrogen sources on cadmium toxicity in cyanophage N-1 resistant mutant strain of Cyanobacterium *Nostoc muscorum* were investigated (Pandey and Pandey, 1994). The study clearly indicated that the cyanobacterial sensitivity to cadmium can be modified by the availability of combined nitrogen sources. Nitrate and urea reduce cadmium toxicity either through increased intracellular amino acid pool or by increasing the p^H of the medium leading to a decrease in Cd^{2+}

availability. In case of NH_4^+, on the other hand, Cd^{2+} reduces NH_4^+ uptake leading to a N-limiting condition inside the cell. Further, as NH_4^+ do not bring alkalinization of the culture medium, free Cd^{2+} ion concentration will be high facilitating its availability and toxic effects. Such results may be of ecophysiological significance in Indian paddy fields, where nitrogenous fertilizes are frequently applied to improve the grain yield.

The effect of copper on a non-heterocystous and a heterocystous cyanobacterium *Oscillatoria foreaui* and *Calothrix elenkinii* respectively was investigated. The study revealed that lower concentration (5µM) of copper was stimulatory to growth and metabolic activities of *O. foreaui*, but not to *C. elenkinii. Calothrix elenkinii* was more tolerant to copper than *O. foreaui,* since it grew at concentrations greater than 30 µM(Uma *et al.*,2001). Choudhury (2003) investigated the interactive effects of 0.1 µM of mercury and nickel with seven chelating agents, *viz.* EDTA, ascorbic acid, citric acid casein hydrolysate, glucose, sucrose and sodium acetate on the growth, pigment contents and heterocyst frequency of *Nostoc linckia*. Sucrose (15mM) stimulated these process, followed by sodium acetate (1mM) and casein hydrolysate (10 µg/ml). It was found that level of protection of toxic effect by sucrose was more for mercury than nickel.

Mechanism of Toxicity Regulation

Metal ions are present in free or bound states in aquatic environment. The ions have to pass through or become adsorbed on to the cell wall. Exclusion can operate at this point of the cell wall or the sheath is able to take up to metal ion and thus prevent its encountering the cell membrane. Free ions will pass through the membrane with the help of carriers and permeability can be controlled so that ions can not pass through or even if they do will move slowly. On entry, the metal may cause changes in morphology, physiology and biochemistry of the cell. The metals may be adsorbed into the cell wall, may accumulate in the cytoplasm, mitochondria and microsomal fractions. It is also understood that lipids and polysaccharides are important constituents of cell-wall and membranes of various algae and seem to be the preferred sites of heavy metal interaction (Adhikary *et al.*, 1986). Inside the cell, binding of cations may take place which requires production of (i) organic matter or ligands, either for transporting the metals outside the cell or probably for binding the metals to non-sensitive sites within the cell and preventing their further action. At this stage, several mechanisms may operate such as binding of metals with (i) sulphydryl groups (ii) active sites of the enzymes either to induce or retard their activity (iii) organic acids (eg. Citrate, malate and succinate) (v) phytochelatin complexes and (vi) RNA and DNA inclusions (mitochondrial and microsomal fractions) (Poonguzhali and Rao, 1999).

Conclusion

Cyanobacteria seem to be endowed with a self-detoxifying mechanism against heavy metal toxicity. This could be one of the reasons why cyanobacteria, compared to other algae, flourish luxuriantly in polluted environments. Thus studies on the level of different heavy metals in the environment, their toxicity towards primary producers and the ameliorative effects of various chelating agents are necessary to identify metal pollution indicator species and assess pollution status of aquatic bodies.

Acknowledgement

I express my deep appreciation to Dr. Shiba Prasad Adhikary, Professor of Botany, Utkal University, Bhubanerwar for his interest, inspiring guidance and criticisms during the course of investigation and preparation of this manuscript. I would like to thank Mrs. Ratna Choudhury, who assisted me in providing information on this subject.

References

Adhikary SP, Weckessor J, Jurgens UJ, Golecki JR and Borowiak D (1986) Isolation and chemical characterization of the sheath from the cyanobacterium *Chroococcus minutus* SAG. B.41.76. *J Gen Microbiol* 132: 2595-2599.

Ahluwalia AS and Kaur M (1989) Nickel toxicity on growth and heterocyst formation in nitrogen–fixing blue green algae. *Phykos* 28:196-200.

Allen HE, Hall RH and Brisbin TD (1980) Metal speciation: Effects of aquatic toxicity. *Environ Sci Technol* 14:441-443.

Budd K, Casey JR and Mac Arthur JD (1986) Arsenite toxicity and arsenite tolerance in the cyanobacterium *Synechococcus leopoliensis*. *Can J Bot* 64:2433-2440.

Campbell PM and Smith DG (1986) Transport and accumulation of nickel ions in the cyanobacterium *Anabaena cylindrica*. *Arch Biochem Biophysics* 244:470-477.

Choudhury R (1993) Toxicity of Mercuric Chloride and Nickel Sulphate on the rice field cyanobacterium *Nostoc linckia*. *M.Phil.Thesis*, Utkal University.

De PK (1939) The role of blue-green algae in nitrogen fixation in rice fields. *Proc Roy Soc London* 127:121-134.

Dubey SK and Rai LC (1990) Heavy metal toxicity in a nitrogen fixing cyanobacterium *Anabaena doliolum*: Regulation of toxicity by certain environmental factors. *Biomed Environ Sci* 3:240-249.

Edmondson WT (1969) Eutrophication in North America. In: *Eutrophication: Causes, Consequences, Correctives*. Natl Acad Sci, Washington D.C., 124-149.

Forstner U and Wittmann GTW (1979) *Metal pollution in the aquatic environment*. Springer-Verlag,Berlin, Heidelberg and New York.

Henriksson LE and Da Silva EL (1978) Effcet of some inorganic elements on nitrogen fixation in blue-green algae and some ecological aspects of pollution. *Z Allegem Microbiol* 18:487-494.

Hodson PV, Borgmann V and Shear H (1979) Toxicity of copper to aquatic biota. In: Nriaqu JO (eds) *Copper in the environment*. John Wiley and Sons, New York, 302-372.

Husaini Y and Rai LC (1991) Studies on nitrogen and phosphorous metabolism and the photosynthetic electron transport system of *Nostoc linckia* under cadmium stress. *J Plant Physiol* 138:429-435.

Jampani CSR (1988) Lead toxicity of alga *Synecococcus aeruginosa* and its recovery by nutrients. *J Environ Bio* 88:261.

Maheswari RU, Elango K, Gayathri V and Anand N (2001) Differential response of cyanobacteria to copper. *Phykos* 40(182):155-161.

Mallick N and Rai LC (1989) Response of *Anabaena doliolum* to bimetallic combinations of copper, nickel and iron with special reference to sequential addition. *J Appl Phycol* 1:301-306.

Murthy SDS and Mohanty P (1991) Mercury induced alternation of energy transfer in phycobilisome by selectively affecting the pigment protein, phycocyanin in the cyanobacterium, *Spirulina platensis*. *Plant Cell Physiol* 32:231-237.

Murthy SDS, Sabat SC and Mohanty P. (1989) Mercury induced inhibition of photosystem-II: Activity and changes in the emission of fluorescence from phycobilisomes in intact cells of the cyanobacterium *Spirulina platensis*. *Pl Cell Physiol* 30:1153-1157.

Pandey PK, Singh CB and Singh SP (1992) Copper uptake in a cyanobacterium: Fate of selected photochemical reactions. *Curr Microbiol* 24:35-39.

Pandey U and Pandey J (1994) Effect of cadmium on growth, photosynthesis and nitrogen fixation of *Nostoc muscorum* and cyanophage N-1 resistant mutant. *Phykos* 33(1 &2):19-23.

Pettersson A and Bergman B (1989) Effect of aluminium on ATP pools and utilization in the cyanobacterium *Anabaena cylindrica*: a model for the *in vivo* toxicity. *Physiol Plant* 76:527-534.

Poonguzhali TV and Rao VNR (1999) Uptake and accumulation of metals by microalgae. *Phykos* 38(182):135-140.

Prasad SM, Singh JB, Rai LC and Kumar HD (1991) Metal induced inhibition of photosynthetic electron transport chain of the cyanobacterium *Nostoc muscorum*. *FEMS Microbial Lett* 82:95-100.

Rai LC (1989) Silver toxicity in nitrogen fixing cyanobacterium: Interaction with chromium, nickel and lead. *Bio metals* 2:122-128.

Rai LC and Dubey SK (1988) Chromium toxicity to a cyanobacterium: Possible role of carbon sources in toxicity amelioration. *Microbios* 55:193-203.

Rai LC and Dubey SK (1989) Protective effects of certain natural and synthetic complexans on the toxicity of chromium and tin to a nitrogen-fixing cyanobacterium *Anabaena doliolum*. *Biomed Environ Sci* 2:271-278.

Rai LC, Jensen TE and Rachlin JW (1990) A morphometric and X-ray energy dispersive approach to monitoring p^H altered cadmium toxicity in *Anabaena flos-aquae*. Arch Environ Contam Toxicol 19:479-487.

Rai LC and Raizada M (1985) Effect of nickel and silver ions on survival, growth, carbon fixation and nitrogenase activity of *Nostoc muscorum*: Regulation of toxicity by EDTA and calcium. *J Gen Appl Microbiol* 31:329-337.

Rai LC and to Raizada M (1986) Nickel induced stimulation of growth, heterocyst differentiation, $^{14}CO2$ uptake and nitrogenase activity in *Nostoc muscorum*. *New Phytol* 104:111-114.

Rai LC and Raizada M (1987) Toxicity of nickel and silver to *Nostoc muscorum*: Interaction with ascorbic acid, glutathione, sulfur containing amino acids. *Ecotox Environ Safety* 14:12-21.

Rai LC and Raizada M (1988) Impact of chromium and lead in *Nostoc muscorum*: Regulation of toxicity by ascorbic acid, glutathione and sulfur containing amino acids. *Ecotox Environ Safety* 15:195-205.

Rai LC and Raizada M (1989) Effect of biometallic combinations of nickel, chromium and lead on growth, uptake of nitrate and ammonia, $^{14}CO2$ fixation and nitrogenase activity of *Nostoc muscorum*. *Ecotox Environ Safety* 17:75-85.

Raizada M and Rai LC (1985) Metal induced inhibition of growth, heterocyst differentiation, carbon fixation and nitrogenase activity of *Nostoc muscorum*: Interaction with EDTA and calcium. *Microbios Lett* 30:153-161.

Roger PA (1988) Reconsidering the utilization of blue-green algae in wetland rice cultivation. In: Dutt SK and Sloger C (eds) *Biological N_2-fixation associated with rice production*. Oxford and IBH Publ Co Pvt Ltd. New Delhi, 119-137.

Roger PA and Kulasooriya SA (1980) *Blue-green algae and Rice*, IRRI, Manila, Los Banos.

Rueter JG (1988) Iron stimulation of photosynthesis and nitrogen fixation in *Anabaena* 7120 and *Trichodesmium* (cyanophyceae). *J Phycol* 24:249-254.

Singh SP and Pandey AK(1981) Cadmium toxicity in a cyanobacterium: Effect of modifying factors. *Environ Exp Bot* 21:257-265.

Singh DP and Singh SP (1987) Action of heavy metals on Hill activity and O_2 evolution in *Anacystis nidulans*. *J Pl Physiol* 83:12-14.

Singh HN, Singh AK and Katiyar S (1989) Genetic construction of strains of diazotrophic cyanobacteria with potentials in biofertilizer technology. In: Trivedi ML, Gill BS and Saini SS (eds) *Plant Science Research in India*. Today and Tomorrows, New Delhi, 565-571.

Stokes PM (1983) Response of freshwater algae to metals. *In*: Round FE and Chapman DJ (eds) *Progress in Phycological Research* Vol.II Elsevier Science Publishers, UK, 87-109.

Stratton GW and Corke CT (1979a) The effect of nickel on growth, photosynthersis and nitrogenase activity of *Anabaena inaequalis*. *Can J Microbiol* 25:1094-1099.

Stratton GW and Corke CT (1979b) The effect of mercuric, cadmium and nickel ion combinations on a blue-green alga. *Chemosphere* 10:731-740.

Stratton GW, Huber AL and Corke CT(1979) Effect of mercuric ion on the growth, photosynthesis and nitrogenase activity of *Anabaena inaequalis*. *Appl Environ Microbiol* 38:537-543.

© R.K. Gupta & V.D. Pandey (*Editors*)
Advances in Applied Phycology, 2007 196–208
Daya Publishing House, New Delhi

Chapter 15

Cyanobacterial Toxins: Risks and Impacts on Human Health

B.V. Singh* and Shalini Singh

Centre for Conservation and Utilization of Blue Green Algae, Indian Agricultural Research Institute, Pusa, New Delhi - 110 012, New Delhi

ABSTRACT

The cyanobacteria also provide an extraordinarily wide-ranging contribution to human affairs in everyday life and are of economic importance. Both the beneficial and detrimental features of the cyanobacteria are of considerable significance. They are important primary producers and their general nutritive value is high. The nitrogen-fixing species contribute globally to soil and water fertility. The use of cyanobacteria in food production and in solar energy conversion holds promising potential for the future. However, cyanobacteria may also be a source of considerable nuisance in many situations. Abundant growth of cyanobacteria in water reservoirs creates severe practical problems for water supplies. The development of strains containing toxins is a common experience in polluted inland water systems all over the world, as well as in some coastal waters. Thus cyanobacterial toxins, or "cyanotoxins", have become a concern for human health.

Introduction:

Cyanobacteria are members of a group known as eubacteria or true bacteria. For a long time they were not recognized as bacteria, more often being referred to as blue-green algae. All bacteria belong to a group of organisms known as prokaryotes, a Latin word meaning 'before nucleus'. Bacteria have no organized nucleus. Cyanobacteria are classified as bacteria, not algae, since their genetic material is not organized in a membrane-bound nucleus. Unlike other bacteria, they have chlorophyll and use

* Corresponding Author: E-mail: bvsingh10@yahoo.co.in.

the sun as an energy source. They are often referred to as 'blue-greens', since the first cyanobacteria identified were blueish-green in colour. Some are olive or dark green, and others are even purplish in colour.

Toxic cyanobacterial blooms occur because of favorable conditions including hot, sunny days and warm, nutrient-rich water. The bloom is most abundant during late summer and early autumn. Under favorable conditions, bacterial numbers multiply rapidly, doubling in one day or less. Blooms usually do not last long. Rain, heavy winds or cooler temperatures often inhibit growth or break up the bloom, mixing it into the water body within a few days. However, under continuing favorable conditions blooms may last for several weeks. The formation of toxic blooms is unpredictable. The presence of bacteria may often be determined by a bluish tinge to the water. Concentrations of bacteria are often bluish green but may vary in color from dark green to brownish green depending on the total bacterial population. Also, not all cyanobacteria are poisonous, and the cyanobacteria which can generate poisonous toxins do not always do so.

Mass Occurrences of Toxic Cyanobacteria

The toxicity of cyanobacterial mass occurrences (blooms) was originally brought to the attention of scientists through reports of animal poisonings by farmers and veterinarians, with the first well documented case being reported from Australia in 1878 (Francis 1878). In most, if not all, reported cases since that time, afflicted animals consumed water from water bodies where there was an obvious presence of a cyanobacterial scum on the water surface (Yoo *et al.*, 1995). More recent measurements of cyanobacterial toxins using sensitive modern analytical methods have often revealed high frequencies of toxic blooms even when animal poisonings have not been reported. Throughout the world, it appears that liver-toxic (hepatotoxic, microcystin-containing) freshwater blooms of cyanobacteria are more commonly found than neurotoxic blooms. Liver-toxic blooms have been reported from all continents and almost every part of the world where samples have been collected for analysis. Nevertheless, mass occurrences of neurotoxic cyanobacteria are common in some countries and these have been reported from North America, Europe and Australia, where they have caused several animal poisonings. Blooms which have caused both liver and kidney damage due to the toxin cylindrospermopsin (and possibly related cyanotoxins) have been reported in Australia, Japan, Israel and Hungary. In recent years, surveys have been carried out in a number of countries in South America, Africa, Australia, Asia and Europe. The conclusion that can be drawn from these surveys is that toxic cyanobacteria are internationally ubiquitous, and that as further surveys are carried out more toxic cyanobacterial blooms and new toxic species will be discovered. This is particularly true of tropical and subtropical regions that are currently under-represented in the literature. It seems likely that every country in the world will have water bodies which support blooms of toxic cyanobacteria at some time or another. It is also important to note that mass occurrences of toxic cyanobacteria are not always associated with human activities causing pollution or "cultural eutrophication". For example, massive blooms of toxic cyanobacteria have been reported in Australian reservoirs with pristine or near-pristine catchments (watersheds), and toxic benthic cyanobacteria have killed cattle drinking from oligotrophic, high-alpine waters in Switzerland. The organisms can grow rapidly in favourable conditions, such as calm nutrient-rich fresh or marine waters in warm climates or during the late summer months in cooler parts of the world. Blooms of cyanobacteria tend to occur repeatedly in the same water, posing a risk of repeated exposure to some human populations. Cyanobacterial toxins in lakes, ponds, and dugouts in various parts of the world have long been known to cause poisoning in animals and humans; one of the earliest reports of their toxic effects was in China 1000 years ago (Chorus and Bartram 1999).

The Cause

Cyanobacteria or blue-green algae occur worldwide especially in calm, nutrient-rich waters. Some species of cyanobacteria produce toxins that affect animals and humans. People may be exposed to cyanobacterial toxins by drinking or bathing in contaminated water. The most frequent and serious health effects are caused by drinking water containing the toxins (cyanobacteria), or by ingestion during recreational water contact. Toxins from these bacteria are poisonous to cattle, horses, sheep, pigs, chickens, ducks (domestic and wild), pigeons, geese, herons, songbirds, dogs, rabbits, small wild and domestic animals, and even frogs, fish and snakes. Cyanobacterial toxins are primarily neurotoxic (affect the nervous system) and hepatotoxic (affect the liver). Clinical signs in cyanobacterial poisoning include nervous derangement, staggering, tremors and severe abdominal pain. The toxins are also poisonous to humans. People are mainly exposed to cyanobacterial toxins by drinking or bathing in contaminated water. Other sources include algal food tablets. Some species form a scum on the water, but high concentrations may also be present throughout the affected water. Surface scums, where they occur, represent a specific hazard to human health because of their particularly high toxin contact. Contact, especially by children, should be avoided.

Cyanobacterial toxins are classified by how they affect the human body. Hepatotoxins (which affect the liver) are produced by some strains of the cyanobacteria *Microcystis, Anabaena, Oscillatoria, Nodularia, Nostoc, Cylindrospermopsis* and *Umezakia*. Neurotoxins (which affect the nervous system) are produced by some strains of *Aphanizomenon* and *Oscillatoria*. The structural diversity of known cyanotoxins includes many variants of alkaloids and cyclic peptides. A variety of cyanobacteria produce one or more cyanotoxins (Li *et al.*, 2001) (Table 15.1).

Table 15.1: Cyanotoxins and their Producers
A range of structural variants has been identified for the various toxins.
For each genus listed, toxic and nontoxic strains are known to exist.

Cyanotoxin	*Known Toxin Producers*
Hepatotoxins	
Microcystins	*Microcystis, Planktothrix, Nostoc, Anabaena, Anabaenopsis*
Nodularins	*Nodularia*
Cylindrospermopsin	*Cylindrospermopsis, Aphanizomenon, Umezakia, Raphidiopsis*
Neurotoxins	
Anatoxin-a	*Anabaena, Planktothrix, Aphanizomenon*
Anatoxin-a(S)	*Anabaena*
Saxitoxins	*Anabaena, Aphanizomenon, Cylindrospermopsis, Lyngbya, Planktothrix*
Dermatoxins	
Lyngbyatoxin-a	*Lyngbya*
Aplysiatoxins	*Lyngbya, Schizothrix, Planktothrix*

Cyanobacterial Blooms and Surface Scums

Toxic cyanobacterial blooms have occurred throughout recorded history in Australia, as causes of livestock deaths and water unpalatability. Human injury from cyanobacterial toxins is rarely recorded, but members of two local populations in Australia have suffered from an acute toxicity believed to be a consequence of copper sulphate treatment of water blooms. Minor health effects have been correlated with cyanobacterial contamination of drinking water drawn from rivers. Recreational exposure to toxic cyanobacteria is comparatively common, with demonstrated minor adverse effects on health. The World Health Organization has drawn up guideline values for microcystin in drinking water and recommendations for recreational waters. The major unknown is the potential for cancer stimulation by cyanobacterial toxins, particularly gastrointestinal cancers in Australia and other affluent countries and liver cancer in poorer nations.Effective assessment of the risk of cyanobacterial toxins to human health requires data that relate the dose of toxin to the clinical effects in a population. In general, when an adverse health effect has been suspected to have been caused by a cyanobacterial bloom in a water supply, no measurements of toxins have been undertaken. Even in the recent case of the deaths of more than 50 dialysis patients in Brazil from cyanobacterial toxicity, the best assessment of toxic dose that could be achieved was the retrospective analysis of post-mortem samples of human liver and blood and of the filters from the treatment plant itself (Jochimsen *et al.*, 1998; Pouria *et al.*, 1998). As a result of the lack of adequate human exposure and toxicity data, animal toxicity data are often used for risk assessment. Approaches from human epidemiological data, relating illness to cyanobacterial contamination of drinking water, would provide effective measures of risk, if the problem of measuring toxin exposure could be overcome. In Australia, the major problem with toxic cyanobacteria in drinking water is with blooms of M. aeruginosa and the microcystin toxins they produce in supply reservoirs located in temperate areas. Although cyanobacterial cell numbers are regularly monitored in reservoirs, toxin concentrations may only rarely be measured in water supplies, even on occasions of confirmed toxic blooms in the source water. The other major potential hazards in water supplies in Australia are cylindrospermopsin, from the tropical *Cylindrospermopsis raciborskii* and saxitoxins from *A. circinalis*. The toxic cyanobacteria occur as intermittent blooms in water supply reservoirs, and they are often controlled by the supply authority with applications of copper sulphate. This lyses the bloom cyanobacteria, liberating toxins into the water, making their removal in conventional water filtration plants difficult (Falconer *et al.*, 1989).

The formation of water blooms results from the redistribution and often rapid accumulation of buoyant planktonic populations. When such populations are subjected to suboptimal conditions, they respond by increasing their buoyancy and move upward nearer to the water surface. Water turbulence usually prevents them reaching the surface. If, however, turbulence suddenly weakens on a calm summer day, the buoyant population may 'over-float' and may become lodged right at the water surface. There the cells are exposed to most unfavourable and dangerous conditions, like O_2 supersaturation, rapidly diminishing CO_2 concentrations and intense solar radiation, which are inhibitory to photosynthesis and N_2-fixation, causing photo-oxidation of pigments and inflicting irreversible damage to the genetic constitution of cells. A frequent outcome of surface bloom formation is massive cell lysis and rapid disintegration of large planktonic populations.

Most of these conditions are produced by just three blue-greens, informally referred to as Annie (*Anabaena flos-aquae*), Fannie (*Aphanizomenon flos-aquae*) and Mike (*Microcystis aeruginosa*). An oversupply of nutrients, especially phosphorus and possibly nitrogen, will often result in excessive growth of blue-greens because they possess certain adaptations that enable them to outcompete true algae. Perhaps the most important adaptation is their positive buoyancy, which is regulated by their gas vesicles which are absent in true algae.

A number of properties and reactions to environmental conditions are discussed below in order to describe these ecostrategists and to aid the understanding of their specific behaviour.

Light Intensity

Like algae, cyanobacteria contain chlorophyll *a* as a major pigment for harvesting light and conducting photosynthesis. They also contain other pigments such as the phycobiliproteins which include allophycocyanin (blue), phycocyanin (blue) and sometimes phycoerythrine (red). These pigments harvest light in the green, yellow and orange part of the spectrum (500-650 nm) which is hardly used by other phytoplankton species. The phycobiliproteins, together with chlorophyll *a*, enable cyanobacteria to harvest light energy efficiently and to live in an environment with only green light. Many cyanobacteria are sensitive to prolonged periods of high light intensities. The growth of *Planktothrix* (formerly *Oscillatoria*) *agardhii* is inhibited when exposed for extended periods to light intensities above 180 µE m-2 s-1. Long exposures at light intensities of 320 µE m-2 s-1 are lethal for many species (Van Liere and Mur 1980). However, if exposed intermittently to this high light intensity, cyanobacteria grow at their approximate maximal rate (Loogman 1982). This light intensity amounts to less than half of the light intensity at the surface of a lake, which can reach 700-1,000 µE m-2 s-1. Cyanobacteria which form surface blooms seem to have a higher tolerance for high light intensities. Whereas the green alga (*Scenedesmus protuberans*) grew faster at high light intensities, growth of the cyanobacterium (*Planktothrix agardhii*) was faster at low light intensities. If both organisms were grown in the same continuous culture at low light intensity, *Planktothrix* could out-compete *Scenedesmus*. At high light intensities, the biomass of the green alga increased rapidly, causing an increase in turbidity and a decrease in light availability. This increased the growth rate of the cyanobacterium, which then became dominant after 20 days. Although cyanobacteria cannot reach the maximum growth rates of green algae, at very low light intensities their growth rate is higher. Therefore, in waters with high turbidity they have better chances of out-competing other species. This can explain why cyanobacteria which can grow under very poor nutritional conditions often develop blooms in nutrient-rich eutrophic waters.

Gas Vesicles

Many planktonic cyanobacteria contain gas vacuoles. These structures are aggregates of gas-filled vesicles, which are hollow chambers with a hydrophilic outer surface and a hydrophobic inner surface (Walsby 1978). A gas vesicle has a density of about one tenth that of water and thus gas vesicles can give cyanobacterial cells a lower density than water.

Growth Rate

The growth rate of cyanobacteria is usually much lower than that of many algal species (Reynolds 1984). At 20 °C and light saturation, most common planktonic cyanobacteria achieve growth rates of 0.3-1.4 doublings per day, while diatoms reach 0.8-1.9 doublings per day and growth rates of up to 1.3-2.3 doublings per day have been observed for single-celled green algae (Van Liere and Walsby 1982). Slow growth rates require long water retention times to enable a bloom of cyanobacteria to form. Therefore cyanobacteria do not bloom in water with short retention times.

Phosphorus and Nitrogen

Because cyanobacterial blooms often develop in eutrophic lakes, it was originally assumed that they required high phosphorus and nitrogen concentrations. This assumption was maintained even though cyanobacterial blooms often occurred when concentrations of dissolved phosphate were lowest. Experimental data have shown that the affinity of many cyanobacteria for nitrogen or phosphorus is higher than for many other photosynthetic organisms. This means that they can out-compete other phytoplankton organisms under conditions of phosphorus or nitrogen limitation. In addition to their high nutrient affinity, cyanobacteria have a substantial storage capacity for phosphorus. They can store enough phosphorus to perform two to four cell divisions, which corresponds to a 4-32 fold increase in biomass. However, if total phosphate rather than only dissolved phosphate is considered, high concentrations indirectly support cyanobacteria because they provide a high carrying capacity for phytoplankton. High phytoplankton density leads to high turbidity and low light availability, and cyanobacteria are the group of phytoplankton organisms which can grow best under these conditions. A low ratio between nitrogen and phosphorus concentrations may favour thedevelopment of cyanobacterial blooms. A comparison between the optimum N:P ratios for eukaryotic algae (16-23 molecules N:1 molecule of P) with the optimum rates for bloom-forming cyanobacteria (10-16 molecules N: 1 molecule P), shows that the ratio is lower for cyanobacteria (Schreurs 1992).

Population Stability

While many planktonic algae are grazed by copepods, daphnids and protozoa, cyanobacteria are not grazed to the same extent, and the impact of grazing by some specialised ciliates and rhizopod protozoans is usually not substantial. Cyanobacteria are attacked by viruses, bacteria and actino-mycetes, but the importance of these natural enemies for the breakdown of populations is not well understood. Because they have few natural enemies, and their capacity for buoyancy regulation prevents sedimentation, the loss rates of cyanobacterial populations are generally low. Thus, their slow growth rates are compensated by the high prevalence of populations once they have been established.

Temperature

Maximum growth rates are attained by most cyanobacteria at temperatures above 25 °C (Robarts and Zohary 1987). These optimum temperatures are higher than for green algae and diatoms. This can explain why in temperate and boreal water bodies most cyanobacteria bloom during summer.

Toxic Cyanobacteria and Other Water Related Health Problems

The contamination of water resources and drinking water supplies by human excreta remains a major human health concern, just as it has been for centuries. By contrast, the importance of toxic substances, such as metals and synthetic organic compounds, has only emerged in the latter half of the twentieth century. Although eutrophication has been recognised as a growing concern since the 1950s, only recently have cyanobacterial toxins become widely recognised as a human health problem arising as a consequence of eutrophication. The importance of such toxins, relative to other water-health issues, can currently only be estimated. A significant proportion of cyanobacteria produce one or more of a range of potent toxins. If water containing high concentrations of toxic cyanobacteria or their toxins is ingested, they present a risk to human health. Some cyanobacterial substances may cause skin irritation on contact. The relationship between water resources and health is complex. The most well recognised relationship is the transmission of infectious and toxic agents through consumption of water. Drinking water has therefore played a prominent role in concerns for water and human health. Diseases arising from the consumption of contaminated water are generally referred to as "waterborne". Globally, the waterborne diseases of greatest importance are those caused by bacteria, viruses and parasites, such as cholera, typhoid, hepatitis A, cryptosporidiosis and giardiasis. Most of the pathogens involved are derived from human faeces and the resulting diseases are generally referred to as "faecal-oral" diseases; however they can also be spread by means other than contaminated water, such as by contaminated food. Waterborne diseases also include some caused by toxic chemicals, although many of these may only cause health effects some time after exposure has occurred and may therefore be difficult to associate directly with the cause.

Cyanobacterial toxins may enter the body through oral consumption, inhalation or skin absorption, although the proportions can be expected to be widely different between toxins. Although the relative proportions have not been experimentally evaluated, nonrecreational exposure can be expected to be almost entirely via the oral route. Cyanobacterial toxins comprise a very diverse group of organic molecules, with microcystins the only examples for which uptake data are available. Oral administration studies have shown that uptake of microcystin into the blood occurs through the gastrointestinal lining, whereas excretion is almost entirely through the faeces. Thus, oral consumption of microcystin will result in the whole gastrointestinal tract lining being exposed to the toxin, with the potential for cell injury. Experimental assessment of the possible carcinogenic or tumour-promoting effects of exposure to this toxin have been carried out in our laboratory over the last six years (Falconer and Humpage 1996). Recent data have shown that the growth of aberrant crypt foci in the mouse colon is stimulated by Microcystis extracts in the drinking water (Humpage *et al.*, 2000).

Disease due to cyanobacterial toxins varies according to the type of toxin and the type of water or water-related exposure (drinking, skin contact, etc.). Humans are affected with a range of symptoms including skin irritation, stomach cramps, vomiting, nausea, diarrhoea, fever, sore throat, headache, muscle and joint pain, blisters of the mouth and liver damage. Swimmers in water containing cyanobacterial toxins may suffer allergic reactions, such as asthma, eye irritation, rashes, and blisters around the mouth and nose. Animals, birds, and fish can also be poisoned by high levels of toxin-producing cyanobacteria. Cyanobacteria from the species *Cylindroapermopsis raciborski* may also produce toxic alkaloids, causing gastrointestinal symptoms or kidney disease in humans. People are mainly exposed to cyanobacterial toxins by drinking or bathing in contaminated water. Other sources include algal food tablets. Some species form a scum on the water, but high concentrations may also be present throughout the affected water. Surface scums, where they occur, represent a specific hazard to human health because of their particularly high toxin contact. Contact, especially by children, should be avoided. Drinking stagnant pond water during hot, dry weather can cause death in animals. Intoxication with cyanobacteria is characterized by convulsions, ataxia (in–coordination), bloody diarrhea and sudden death. Affected animals rarely range far from the water source. Freshwater toxins are diverse and result in a variety of symptoms that depend on the toxin composition of the algae. Symptoms may range from gastroenteritis and neurological disorders to liver damage and respiratory failure. Effects of various toxins in humans are as follows:

Anatoxin-a

It is relatively stable in the dark, but in pure solution in the absence of pigments it undergoes rapid photochemical degradation in sunlight. Breakdown is further accelerated by alkaline conditions (Stevens and Krieger 1991). The half-life for photochemical breakdown is 1-2 hours. Under normal day and night light conditions at pH 8 or pH 10, and at low initial concentrations (10 µg l-1), the half-life for anatoxin-a breakdown was found to be approximately 14 days. Toxin affects muscle contraction by mimicking a neurotransmitter (*i.e.*, acetylcholine) that causes ion channels to remain open. Muscle cells contract until they fail from exhaustion. Symptoms include staggering, decreased movement, abnormal breathing, convulsions, and death due to respiratory paralysis.

Anatoxin-a(s):

Toxin binds to the enzyme responsible for deactivating acetylcholine, leaving the ion channel open. In domestic animals, symptoms include hypersalivation, mucous nasal discharge, diarrhea, and tremors.

Homoanatoxin-a

It is a potent neuromuscular blocking agent with an i.p. LD50 in mice of 250 µg kg-1 bw. Toxicosis in the lethal dose range leads to severe body paralysis, convulsions and death by respiratory arrest in 7-12 minutes. Experiments with rat phrenic nerve hemidiaphragm preparations demonstrated that the physiological effects of homoanatoxin-a are related to those observed for *d*-tubocurarine. Recent studies have shown that homoanatoxin-a enhances the influx of Ca^{+2} ions in the cholinergic nerve terminals (Aas *et al.*, 1996).

Saxitoxins

Saxitoxin and some of its analogues are produced by *Anabaena circinalis* in Australian freshwaters and *Aphanizomenon flos-aquae* in the USA (Humpage *et al.*, 1994). The saxitoxin group has been the cause of paralytic shellfish poisoning (PSP) in people. Several species of dinoflagellates produce PSP toxins that accumulate in molluscs which filter-feed on these organisms. People who have consumed shellfish containing high levels of PSP toxins may suffer from this acute illness. The signs and symptoms of PSP in humans may range from a slight tingling and numbness about the lips to complete paralysis and death from respiratory failure. Nearly all the systemic actions of saxitoxin can be explained by its pharmacological effect on nerve axon membranes. This involves a wide spread blockage of sodium ion channels of the excitable membranes of nerves, thereby affecting (partially or completely, depending on dose) impulse generation in peripheral nerves and skeletal muscles (Catterall 1980). In mammals, these effects lead to paralysis, respiratory depression and respiratory failure.

Cylindrospermopsin

This cyanotoxin was initially isolated from a culture of *Cylindrospermopsis raciborskii* obtained from a water supply reservoir in tropical northern Australia. The organism was identified as a result of an outbreak of acute hepato-enteritis and renal damage among an Aboriginal population on Palm Island, off the coast of North Queensland. Intraperitoneal injection of the lysed organism to mice resulted in widespread and progressive tissue injury, with cell necrosis in the liver, kidneys, adrenals, lung, heart, spleen and thymus (Hawkins *et al.*, 1997). *In vitro* studies with pure cylindrospermopsin have shown that it inhibits glutathione synthesis and protein synthesis in genera.

Other Bioactive Compounds in Cyanobacteria

Cyanobacteria produce a wide variety of bioactive compounds in addition to the cyanotoxins described in this chapter. They include anti-tumour (cytotoxic), anti-viral, and anti-fungal compounds, antibiotics and protease inhibitors (Namikoshi and Rinehart, 1996). Further screening of these biomedically interesting compounds is underway and is likely to lead to the discovery of many new compounds in the future, some of which may be toxic. Bioassays of cyanobacterial cell extracts have often revealed a higher toxicity than expected from the content of known toxins in the extract.

Treatment Options

The first treatment step should always be bloom prevention. Natural surface water will occasionally bloom regardless of the best efforts at prevention, but the frequency and severity of the bloom can be reduced by using good management practices. Runoff should be controlled to minimize fertilizer and/or waste inputs, and livestock should not be permitted to water directly from surface water sources. Water should be kept as nutrient-free as possible.

Aeration can also be a valuable tool in combating blooms. Good aeration keeps the water moving and maintains a more constant temperature from top to bottom. This helps to prevent extremely warm layers of water from forming at the surface during the hot summer months. Aeration also prevents severe oxygen depletion initiated by the death and decay of an algal bloom. Although algae may still bloom with aeration, cyanobacteria do not thrive in moving water. Cyanobacteria tend to bloom under warm, calm conditions. Proper aeration helps to prevent these conditions from occurring.

If a bloom begins to form in a surface water source, determine the size and type of the bloom. If it is a filamentous bloom, there are a few options. Small filamentous algal blooms close to shore can be removed with a rake or hoe and placed away from the watershed area to prevent re-entry of the dead bloom into the water. Filamentous algae decompose easily, and can be used for compost if combined with other materials to increase air circulation. Large blooms of planktonic or filamentous organisms are more difficult to handle. If the situation is severe, there are a number of chemical options available for treating surface water. Be aware that no chemical treatment is completely effective for long term control. Cyanobacteria can build up tolerance to repeated chemical applications. Chemical application should only be used as a last resort, not as routine maintenance.

Chemical treatment options commonly include one of the following four compounds: copper sulphate, lime (as quicklime, or calcium hydroxide), alum (as aluminum sulphate, ferric chloride Lime, alum and ferric chloride are all coagulants–they bind with suspended and dissolved particles to form clumps that settle to the bottom of the dugout. This includes binding with algae and cyanobacteria. Copper sulphate, also known as bluestone, kills cyanobacteria, yet is only marginally effective on algae. Following treatment, the dead cells settle to the bottom. There are advantages and disadvantages to each method of chemical control. If the dugout to be treated contains fish, great care should be taken when applying any chemical.

Safe Practices

The placing of barriers that reduce exposure to a cyanotoxin hazard is an important measure and involves identifying "critical control points" and implementing measures for their monitoring and control. In the case of cyanobacteria, critical control points might include, for example, noting the tendency of a water body to develop blooms, scums or mats. Monitoring schemes need to be developed that are capable of detecting proliferation of cyanobacteria (linked to a programme of appropriate actions) and drinking water treatment technology needs to be in existence that is capable of preventing human exposure if cyanobacteria occur in source waters.

A drinking water supply safe from cyanotoxins will either draw upon a resource which does not harbour cyanotoxins (*e.g.* groundwater or surface water which does not support cyanobacterial growth), or have treatment in place that is likely to remove cyanobacterial cells (without causing their disrupture) as well as removing cyanotoxins. However, in many circumstances a potential cyanotoxin hazard can be managed effectively without the necessity of advanced treatment processes, through water resource management techniques and removal of intact cells. Most of the reported incidents of human injury that have raised awareness of the importance of cyanotoxins in drinking water have involved the inappropriate treatment of water supplies, such as the use of copper sulphate in dealing with an established bloom of cyanobacteria. A very effective approach to safe practices may involve changing the drinking water source. In a number of regions, surface waters are used for reasons of easy access and tradition, although groundwater of high quality is available. Exploring options of improving practices of drinking water abstraction with low technological input (such as drilling wells, or using bankside filtration) may lead to health benefits. In China, a high prevalence of endemic primary liver cancer was related to several factors: hepatitis B, aflatoxins in the diet, and drinking surface water polluted with cyanobacteria likely to contain microcystins. Changing the drinking water source from shallow, eutrophic ponds and ditches to groundwater was a major element of a package of measures which showed some success in improving health.

Recreational water use is likely to be a major route of exposure to cyanotoxins in some parts of the world. Whereas similar approaches to resource protection apply as for drinking water, there are very few further management options available once cyanobacteria proliferate or accumulate in a recreational water. Because adequate surveillance is sometimes difficult and management options, except precluding or discouraging use, may be scarce, a large share of the responsibility for safe practices lies with the users of a bathing site. The provision of adequate information to the public becomes, therefore, a major responsibility of public authorities. The growth of cyanobacteria in lakes and rivers used for recreational purposes has been well recognised as a public nuisance. Water blooms of cyanobacteria may be associated with unpleasant odours and the offensive appearance of lake shores, especially when scums of the organisms accumulate and decay. Areas with extensive cyanobacterial scums or accumulated detached mats on bathing beaches may be avoided by swimmers and other water users because of the obviously unpleasant environment, particularly if locally anaerobic water conditions or cyanobacterial toxins cause fish-kills, further emphasising the unattractiveness of water contact. In temperate climates, cyanobacterial dominance is most pronounced during the summer months, when the demand for recreational water is highest. In some regions, cyanobacteria have been abundant for more than a generation and visitors have accepted this water quality as "normal" for their region. Multiple anecdotal observations of children playing with scum material have been reported.

Molecular Detection of Toxic Cyanobacteria

The morphological discrimination between toxic and nontoxic cyanobacteria can be difficult, as some genera contain both toxic and nontoxic members. Some cyanobacteria are known to be toxic, some may be genetically capable of producing toxins (toxigenic) but do not under all conditions, and some do not produce toxins at all. DNA-based detection methods have become popular because of their potential specificity (targeting genes involved in toxin biosynthesis), sensitivity, and speed. This review highlights some of these advances and the use of molecular tools to detect, identify, and study toxic cyanobacterial ecology.

The number of publications using molecular methods to detect toxic cyanobacteria is rapidly increasing. Before the sequences of the microcystin biosynthesis gene cluster were published, DNA-based methods for the detection and phylogenetic analysis of toxic cyanobacteria were investigated. Recently sequenced microcystin biosynthesis genes provide very specific molecular targets (Nishizawa *et al.*, 2000). These sequences are being used throughout the world for the design and construction of primer sets for PCR-based toxin gene detection (Tillett *et al.*, 2000).This approach is appealing as an early warning diagnostic, and is very sensitive because of the amplification achieved by PCR. Following the sequencing and publishing of genes involved in other cyanotoxin biosynthetic pathways, molecular methods will be developed to detect these genes. Potential identification of genes involved in the biosynthesis of the toxins nodularin and cylindrospermopsin (Schembri *et al.*, 2001) have already been reported PCR provides a rapid approach for the detection of toxic *Microcystis*, and is potentially a very valuable tool for understanding the geographical and temporal distribution of the organism. It can also be used in the analysis of the cyanobacterial communities with and without *Microcystis*, allowing us to assess the relationships within communities that influence its presence and/or proliferation. Comparing cyanobacterial communities from a single water body over time may reveal a temporal fluctuation of *Microcystis* that correlates with community structure changes. Both fingerprinting and quantitative approaches are being developed for this purpose.

Future Aspects

Research into developing further understanding of the human health significance of cyanobacteria and individual cyanotoxins, and into practical means for assessing and controlling exposure to cyanobacteria and to cyanotoxins, is a priority. A major gap also lies in the synthesis and dissemination of the available information. Information concerning the efficiency of cyanotoxin removal in drinking water treatment systems is limited. Especially, simple, low-cost techniques for cyanobacterial cell removal, such as slow sand filtration, should be investigated and developed further. More information is also needed on the capability of simple disinfection techniques, such as chlorine, for oxidising microcystins and cylindrospermopsin. If this is found to be applicable, or if "conventional" treatments are found to be effective if properly operated, these approaches would provide a practical tool for removing cyanotoxins in many situations. Whilst cyanobacterial blooms remain sporadic or occasional events, most emphasis is still placed upon the protection of drinking water supplies through the preparation of contingency plans and their activation when appropriate. Early warning systems and predictive models can facilitate this and should be based upon available information on the conditions leading to cyanobacterial bloom development and on occurrence, localisation and movement of scums.

Epidemiological evidence is of particular value in determining the true nature and severity of human health effects (and therefore the appropriate response) but is generally lacking in relation to human exposures to cyanobacteria. The limited studies undertaken to date in relation to recreational exposure require further substantiation. Opportunistic studies into exposures through drinking water may provide further valuable insights. Information from experimental toxicology also needs to be strengthened. In particular, long-term exposure studies (of at least one year or longer) should be carried out to assess the chronic toxicity of microcystins and cylindrospermopsins. Uptake routes (*e.g.* through nasal tissues and mucous membranes) require further investigation. Further systematic studies are also required into the suggested tumour-promoting effects of some cyanotoxins, particularly in the dose range of potential oral uptake with drinking or bathing water. Lipopolysaccharide (LPS) endotoxins from cyanobacteria pose a potential health risk for humans, but knowledge of the occurrence of individual LPS components, their toxicology, and their removal in drinking water treatment plants, is so poor that guidelines cannot be set at present. Further bioactive cyanobacterial metabolites are also identified frequently and the health significance of these requires investigation.

References

Aas P, Eriksen S, Kolderup J, Lundy P, Haugen J E., Skulberg O M and Fonnum F 1996. Enhancement of acetylcholine release by homoanatoxin-a from *Oscillatoria formosa. Environ. Toxicol. Pharmacol.* 2: 223-232.

Catterall W 1980. Neurotoxins that act on voltage-sensitive sodium channels in excitable membranes. *Ann. Rev. Pharmacol. Toxicol.* 20: 15-43.

Chorus I and Bartram J 1999. Toxic cyanobacteria in water: a guide to their public health consequences, monitoring, and management. London.

Falconer L R, Runnegar M T C, Buckley T, Huyn V L and Bradsha W P 1989. Use of powdered and granular activated carbon to remove toxicity from drinking water containing cyanobacterial toxins. *Journal of the American Water Works Association* 18: 102105.

Falconer I R and Humpage A R 1996. Tumour promotion by cyanobacterial toxins. *Phycologia* 35: 74-79.

Francis G 1878. Poisonous Australian lake. *Nature* 18: 11-12.

Hawkins P R, Chandrasena N R, Jones G J, Humpage A R and Falconer I R 1997. Isolation and toxicity of *Cylindrospermopsis raciborskii* from an ornamental lake. *Toxicon* 35: 341-346.

Humpage A R, Rositano J, Bretag A H, Brown R, Baker P D, Nicholson B C and Steffensen D A 1994. Paralytic shellfish poisons from Australian cyanobacterial blooms. *Aust. J. Mar. Freshwater Res.* 45: 761-771.

Humpage A R, Hardy S J, Moore E J, Froscio S M and Falconer L R 2000. Microcystins (cyanobacterial toxins) in drinking water enhance the growth of aberrant crypt foci in the mouse colon. *Journal of Toxicology and Environmental Health* 61: 155-165.

Jochimsen E M, Carmichael W W, Cardo D M, Cookson S T, Holmes C E M, Antunes M B, Filho D A, Lyra T M, Barreto V S T and Jarvis W R 1998. Liver failure and death after exposure to microcystins at a haemodialysis center in Brazil. *New Engl. J. Med.* 338: 873-878.

Li RH, Carmichael WW, Brittain S, *et al.*, 2001. First report of the cyanotoxins cylindrospermopsin and deoxycylindrospermopsin from *Raphidiopsis curvata* (Cyanobacteria). *J Phycol* 37: 1121–26.

Loogman J G 1982. Influence of photoperiodicity on algal growth kinetics. Thesis, University of Amsterdam.

Namikoshi, M. and Rinehart, K L 1996 Bioactive compounds produced by cyanobacteria. *J. Ind. Microbiol.* 17: 373-384.

Nishizawa T, Ueda A and Asayama M 2000. Polyketide synthase gene coupled to the peptide synthetase module involved in the biosynthesis of the cyclic heptapeptide microcystin. *J. Biochem.* 127: 779–89.

Pouria S A, Barbosa R L Cavalcanti V T S, Barreto C J, Ward W, Presier G K and Codd G A? 1998.? Fatal microcystin intoxication in haemodialysis unit in Caruaru, Brazil.? *The Lancet* 35: 2:21-26.

Reynolds C S 1984. *The Ecology of Freshwater Phytoplankton.* Cambridge University Press, Cambridge.

Robarts R D and Zohary T 1987. Temperature effects on photosynthetic capacity, respiration, and growth rates of bloom-forming cyanobacteria. *N.Z. J. Mar. Freshwat. Res.* 21: 391-399.

Schreurs H 1992. Cyanobacterial dominance, relation to eutrophication and lake morphology. Thesis, University of Amsterdam.

Schembri M A, Neilan BA, and Saint C P 2001. Identification of genes implicated in toxin production in the cyanobacterium *Cylindrospermopsis raciborskii. Environ Toxicol* 16: 413–21.

Stevens D K and Krieger R I 1991. Stability studies on the cyanobacterial nicotinic alkaloid anatoxin-a. *Toxicon* 29: 167–179.

Tillett D, Dittmann E and Erhard M 2000. Structural organization of microcystin biosynthesis in *Microcystis aeruginosa* PCC7806: an integrated peptide-polyketide synthetase system. *Chem Biol* 7: 753–64.

Van Liere L and Walsby A E 1982. Interactions of cyanobacteria with light. In: *The Biology of the Cyanobacteria* (Ed. Carr N G and Whitton B A) Blackwell Science Publications, Oxford, 9-45 pp.

Van Liere L and Mur L R 1980. Occurrence of *Oscillatoria agardhii* and some related species, a survey. *Dev. Hydrobiol.* 2: 67-77.

Walsby A E 1978. The gas vesicles of aquatic prokaryotes. In: *Relations Between Structure and Functioning in the Prokaryotic Cell* (Ed. Stanier R Y, Rogers H J and Ward J B), Cambridge University Press, 338 pp.

Yoo S, Carmichael W W, Hoehn R C and Hrudey S E 1995. *Cyanobacterial (Bluegreen Algal) Toxins: A Resource Guide.* American Water Works Association Research Foundation, Denver, Colorado.

© R.K. Gupta & V.D. Pandey (*Editors*)
Advances in Applied Phycology, 2007
Daya Publishing House, New Delhi

209–212

Chapter 16
Algae and Disease Prevention

U.K. Chaturvedi* and I. Habib

Botany Department, Bareilly College, Bareilly – 243 005

ABSTRACT

Algae are of great interest in several ways. They are a good sources of food and energy for the aquatic animals and man. They live in almost every conceivable habitat, dominantly growing in water. Recently, their antibiotic properly has come to light. They have been found to cure several diseases in man and animals. Their role in the prevention of human diseases has been given by several workers.

Introduction

Algae are ubiquitous in distribution. The are the bases of aquatic food chains for fishes and other animals which may lead to man. Whatever is written on the 'role of algae to man' would be less as they are the first organisms on this earth which were able carry out the oxygenic photosynthesis and thus converted carbon dioxide into oxygen. Thus, they undoubtedly played a major role in the oxygenation of the air-that we breathe. They, thus helped to create the conditions needed for the emergence of aerobic organisms. In fact, they are the jewels of the plant world. Here, in this article, we will emphasize on the role of certain algae in the prevention of diseases and also their antibiotic activities.

Pharmacological Importance of Algae

The algae have been found to produce substances that inhibit the growth of organisms particularly Gram+ve and Gram-ve bacteria and fungi which cause diseases and thus, they help in the prevention of diseases in man, animals and plants. R. Pratt (1944), perhaps for the first time showed that in cultures of *Chlorella pyrenoidosa* and *C. vulgaris*, a substance known as 'Chlorellin' accumulates, out in the cultures of the alga *Phormidium uncinatum*.

* Corresponding Author: Tel: 0581-2470758, 9412145933

A large number of marine algae belonging to the algal groups Phaeophyceae, Rhodophyceae, Chrysophyceae and Bacillariophyceae have also shown some antibiotic properties. Investigations by Gupta and Srivastava have brought about the broad spectrum antibiotic properties of certain fresh water algae, particularly *Oscillatoria princeps, Lyngbya majuscula* and *Hydrodictyon reticulatum* which are effective against the growth of *Salmonella aureus. Bacillus subtillis, Salmonella lutea, B. brachyseptica, B. typhosus* and *Escherichia coli* bacteria. Several species of Spirogyra are reported to reduce the number of pathogenic bacteria in drinking water. Similarly, a common fresh water red alga–*Compsopogon coeruleus* also shows antibacterial activity against these bacteria.

From times immemorial the people of China, Japan and other Asian countries have been using sea-weeds for medicinal purposes, particularly in the treatment of 'goitre' disease or as worm expellents. These diseases were unknown among the coastal people of Japan who were in the habit of eating Iodine containing sea weeds. Alga *Aisidium helminthocorthon*, that grows on the shores of Corsica is a well known vermifuge. Similarly, the extracts of *Corallina* and *Codium* algae are also used as effective vermifuge.

Studies have also shown the production of antibacterial substances from the extracts of several marine algae such as the species of *Ascophyllum, Rhodomela pelvetia, Polysiphonia* and *Laminaria digitalis.* Species of the alga *Laminaria*, produce a mixture of sodium-Laminarin sulphate which is used as blood anti-coagulants. The nature of antibiotic substances has been studied in some algae. They are found to be of phenolic compounds, fatty acids, carboxyl compounds or terpenes etc. Acrylic acid, which is a kind of volatile acid, has been reported from a brown alga–*Phacocystis pauchetii*, while the alga *Rhodomela larix* produces a brominated phenolic compound. The mechanism, as how these antibacterial substances act is very little known, however, it is believed that these antibiotic substances block one or more of the essential enzyme systems in the metabolic process.

Beside, these above mentioned algae producing antibiotic substances, there are several other seaweeds which are used in the treatment of lymphatic and glandular disorders in the body. Species of *Ulva* and *Fucus* are used in stomach disorders and burn injuries. For the treatment of urinary bladder and kidney troubles, the antibiotic substances produced from the alga *Acetabularia major* act as a effective cure. Algal species, such as *Chondrus crispus* and *Gigartina stelleta* have been used, by the natives of Great Britain who use to live in the coastal areas. for the treatment of stomach and chest diseases. It is believed that owing to the presence of Potassium Chloride, Pot. Iodide and high vitamin contents in the seaweeds, that their intake as food by the natives of Great Britain, China and Japan, keep them healthy and thus they lead a disease free life.

Carrageenin extract, which is another product of algal origin, acts as a blood coagulant, while Alginic acid has been found to stop bleeding. As per the reports are available, an effective vermifuge is obtained from the extracts of *Digenea, Codium, Alsidium* and *Durvillea.* Extracts of *Spirulina* has been found to prevent the development of oral tumor. According to Venkataraman (1989), *Spirulina* based ointment has brought about quick wound healing in rats. Gustafsan (1989) found *Spirulina-sulpholipids* and glycolipids to be remarkably acute against the AIDS virus. Yamane (1988) has reported *Spirulina* reduced nephrotoxicity caused by mercury and other pharmaceutical drugs in rats. Agar-agar is used as a good laxative. It is also used for clotting of blood. Also it is claimed that *Nitzschia* can reduce the number of *E. coli* in water.

Now a days as modern drugs are increasingly available and which are more effective, the use of seaweeds as medicine is gradually declining. However, seaweeds are still meeting the increasing demands of several pharmaceutical industries.

As per the report published from FAO few years ago, about 10 per cent of the world population is suffering from malnutriton owing to the shortage of food supply and increase in population. According to a survey carried out in India and elsewhere, have brought about the high incidence of nutritional related deficiency diseases mostly in children and pregnant women. The deficiency of Vit-A, protein, calcium, iron etc. may lead to blindness, anaemia and week bones. Overcome this problem, the nutritional values of a micro-alga *Spirulina* is recently gaining ground. Since it is the richest source of protein and Beta-carotene, therefore, it is excellent for eye sight and immunity against diseases. This alga is richest source of Iron, Vitamin-B12 and Gamma Lanoline Acid (GLA). Iron is essential for maintaining haemoglobin levels in the blood, Vitamin-B12 helps reduce stress and emotional fatigue while GLA prevents heart problems and reduces 'bad' cholesterol.

There was a report published in some journal that a patient who was suffering from prolapsed stomach when fed on diet containing dried agar-agar and was asked to drink a lot of water. The stomach got distended and later on regained its normal condition. Similarly there are few algae which are a good source of antibiotic and which inhibit the growth of other pathogenic bacteria. It has also been reported that extracts of *Cladophora* and *Lynbya* contains antiviral proprieties and kill the strains of certain bacteria *e.g. Pseudomonas* and *Mycobacterium*. On the other hand, member of the order Charales have been claimed to possess larvicidal properties. Caballero (1919) considered charales useful in the destruction of mosquito larvae. Agar-agar is an important algal product used in the manufacture of pills and ointments. Besides this, it forms a base for many kinds of medicines, which are used as laxatives.

A good amount of information is available on the antibiotic property of algae and thus help in the prevention of diseases in man and animals. An antibacterial substance isolated from alga (*Liguharia tussilyinea*) obtained after steam distillation proved to be very efficient against *Staphylococcus aureus* and *Trichophyton asteroides*. Also it is claimed that Nitschia can reduce the number of *Escherichia coli* in water. Kidney's play a viral role in eliminating unwanted metabolic substances and toxins from the body. The phycocyanin present in *Spirulina* gave a cleaning effect of the kidneys. Diet rich in *Spirulina* enhances the immune response by stimulating macrophage function. Extracts of *Spirulina* or a combination of *Spirulina* and Dunaliaella prevented the development of oral tumor. Mohan (1992) carried out feeding trials in a patient who was suffering from joints with a dose of *Spirulina*.

Many algae have also been an important source of folk medicine (Brorowitzba and Bomwitzka, 1986). Recently, the scientists from National Cancer Institute, USA, discovered 'cyanovirin' from Nostoc ellipsosporun (Boyli *et al.*, 1997), which inactivates HIV without adversely affecting the host cells. The pigment phycocyanin from *Spirulina* is not only used as a food colorant but also as phycofuor' in immunological diagnostic kits. In this context, the contribution of Venkataraman (1983, 1998) in the understanding of the importance of *Spirulina* and bringing it to the commercial market is noteworthy. The potential of cyanobacteria as sources of drug and pharmaceutically important products as well as the role of cyanobacteria in the environment as pollution indictors and scavengers are being assessed (Patterson, 1996).

So, it can be concluded in brief, that algae are capable to preventing or even curing of certain diseases and at the same time can fulfil the nutritional requirements of the masses.

References

Borowitzka and Borowitzka, L.J. (Eds.) 1986. *Microalgal Biotechnology*. Cambridge University Press, Cambridge.

Boyd, M.R., *et al.*, 1997. Discovery of cyanovirin-N, a novel human immunodeficiency virus inactivating protein. *Antimicrob. Agents Chemother.*, 41: 1521–1530.

Cabellero, A., 1919. *Bot. R. Soc. Espanola Hist. Nat. Madrid*, 19: 449.

Gupta, A.B. and Srivastava, G.S., 1963. Production of an antibacterial substance of hydrodicton. *Carr. Sci.*, 23: 19–20.

Gustafson, K., 1989. AIDS: Antivirus sulpholipids in cyanobacteria. *J. Nath. Cancer Institute*, p. 1254.

Mohan, Malathi, 1992. Case study-efficacy of *Spirulina* in case of rheumatoid arthritis. *EITA Nat. Sym., MCRC.*

Patterson, G.M.L., 1996. Biotechnological applications of cyanobacteria. *J. Sci. Ind. Res.*, 55: 669–684.

Pratt, R., 1944. *Limnol. Oceanogr.*, 11: 447.

Venkataraman, L.V., 1989. *Spirulina* state of art and emerging prospects. *Phykos*, 28(1&2): 231–250.

Vidyavati, 2002. Recent trends in algal biotechnology. *Algological Research in India*, p. 1–6.

Yamani, Y., 1988. The effect of *Spirulina* on nephrotoxicity in rats.

© R.K. Gupta & V.D. Pandey (*Editors*)
Advances in Applied Phycology, 2007
Daya Publishing House, New Delhi

213–228

Chapter 17

Molecular Approaches to Cyanobacteria

S. Singh*[1], B.V.Singh[1], D.W. Dhar[1] and Rajinder K. Gupta[2]

[1]*Centre for Conservation and Utilisation of Blue Green Algae, Indian Agricultural Research Institute, New Delhi – 110 012*

[2]*School of Biotechnology, Guru Gobind Singh Indraprastha University, Delhi – 110 006, India*

ABSTRACT

Evolutionary physiology represents an explicit fusion of two complementary approaches: evolution and physiology. Stimulated by four major intellectual and methodological developments (explicit consideration of diverse evolutionary mechanisms, phylogenetic approaches, incorporation of the perspectives and tools of evolutionary genetics and selection studies, and generalization of molecular techniques to exotic organisms), this field achieved prominence during the past decade. It addresses three major questions regarding physiological evolution: (*a*) What are the historical, ecological, and phylogenetic patterns of physiological evolution? (*b*) How important are and were each of the known evolutionary processes (natural selection, sexual selection, drift, constraint, genetic coupling/hitchhiking, and others) in engendering or limiting physiological evolution? and (*c*) How do the genotype, phenotype, physiological performance, and fitness interact in influencing one another's future values? To answer these questions, evolutionary physiology examines extant and historical variation and diversity, standing genetic and phenotypic variability in populations, and past and ongoing natural selection in the wild. Also, it manipulates genotypes, phenotypes, and environments of evolving populations in the laboratory and field. Thus, evolutionary physiology represents the infusion of paradigms, techniques, and approaches of evolutionary biology, genetics, and systematics into physiology. The reciprocal infusion of physiological approaches into evolutionary biology and systematics can likewise have great value and is a future goal.

* Corresponding Author: E-mail: shalini222003@yahoo.co.in.

Introduction

Cyanobacteria are aquatic and photosynthetic organisms, that is, they live in the water, and can manufacture their own food. Because they are bacteria, they are quite small and usually unicellular, though they often grow in colonies large enough to see. They have the distinction of being the oldest known fossils, more than 3.5 billion years old, in fact! It may surprise you then to know that the cyanobacteria are still around; they are one of the largest and most important groups of bacteria on earth. The cyanobacteria and prochloroaceae are the only prokaryotic groups that share the use of photosystems I and II and hence the ability to carry out oxygenic photosynthesis with all photosynthetic eukaryotic organisms (Stanier, 1977). The structure of the reaction center complexes seems to be evolutionarily conserved in all these organisms, but there is a large diversity in their antenna chlorophyll complexes (Glazer, 1983). Cyanobacteria are unique since they are also capable of anaerobic metabolism, and several groups among the cyanobacteria share with the prokaryotic anoxygenic phototrophic bacteria the ability to perform anoxygenic photosynthesis using reduced electron donors (Padan and Cohen, 1982; Cohen *et al.*, 1986). They also share with many archaebacteria the ability to use elemental sulfur for anaerobic dark respiration (Oren and Shilo, 1979).

The key to the genetic study and manipulation of cyanobacteria is the development of a transformation system (either natural or artificial) for DNA uptake. Such a system of cellular competence could be used to internalize exogenous DNA with or without a special treatment to the recipient cells prior to their incubation with the donor DNA. The majority of cyanobacterial strains capable of DNA-mediated transformation possess physiological, or natural, competence for DNA uptake. The mechanism involved in such a process is still unknown but the phenomenon has been extensively used to transform cyanobacterial cells either chromosomally or with plasmids.

Our present understanding of the ecology, physiology, and molecular genetics of cyanobacteria allows the exploration of the use of cyanobacteria for several biotechnological uses, such as the production of specific photosynthetic pigments and herbicides and the use of cyanobacteria for agricultural dinitrogen fixation.

Cyanobacterial DNA

Like other prokaryotes, cyanobacteria contain two types of DNA, chromosomal DNA and smaller, autonomous, extrachromosomal molecules. Both types of DNA can be used for gene transfer and manipulation, each having both advantages and disadvantages. In both instances, transformability of the cell is a prerequisite for gene manipulations. The introduction of foreign DNA can be monitored either by following a genetic marker or by observing some phenotypic alteration indicating a successful transformation. Targeting of the foreign DNA into specific sites in the acceptor DNA and the copy number of the introduced gene may both be of great importance for its expression. Mobilizing DNA sequences into or out of a cyanobacterial cell is not always easier when using extrachromosomal self-replicating molecules. So far, no specific function has been assigned to cyanobacterial plasmids, and in some instances, curing the cells of their plasmids did not have any deleterious effect on the cells, nor was their viability harmed. Nevertheless, various procedures for transforming cyanobacteria have been developed using both types of cyanobacterial DNA.

Genome Size

The genome size of 128 strains representing all major taxonomic groups of cyanobacteria has been measured from the kinetics of renaturation of DNA. The range of size is between 1.6×10^9 Da in the unicellular strains, which is comparable in size to those of other bacterial genomes and 8.6×10^9

Da, in the larger filamentous strains, which greatly exceeds the largest genome previously described in prokaryotes. Even though a larger genome might be anticipated in organisms capable of fixing nitrogen and exhibiting morphological differentiation, the great excess of DNA in these strains raises the possibility that some DNA sequences do not possess a coding capacity. Genetic mapping of the chromosome of cyanobacteria may shed light on such enigmas as well as establish a physical map of the chromosome. Wolk and coworkers have initiated such a study with *Anabaena variabilis* by producing a set of serially overlapping cosmid clones (Herrero and Wolk, 1986). The genome sizes of cyanobacteria are discontinuously distributed into four distinct groups with means of 2.2×10^9, 3.6×10^9, 5.0×10^9, and 7.4×10^9 Da. This suggests that genome evolution in cyanobacteria occurred by a series of duplications of a small ancestral genome (Herdman *et al.*, 1979).

Cyanobacteria contain several identical chromosomes in each cell, which suggests the possibility of interactions between them. Even though genome interaction seems very likely, no direct molecular evidence for recombination between chromosomes in cyanobacteria has been documented. An intrachromosomal gene conversion mechanism involving the *psb*A gene family in *Synechococcus* PCC 7942 was elucidated. Interchromosomal recombination was observed in *Synechocystis* PCC 6803. Thus, the general assumption favors the existence of such mechanisms in cyanobacteria. The *rec*A gene was identified in some cyanobacterial species, and in many instances, recombination between chromosomal DNA and plasmid DNA has been proposed. To minimize fortuitous results, this phenomenon should be taken into account whenever introduction of foreign or engineered genes into cyanobacteria is considered (Williams, 1988).

Plasmids

The existence of nonchromosomal DNA molecules in cyanobacteria was first observed in *Anacystis nidulans* (*Synechococcus* sp.). Ever since, it has been evident that nearly all cyanobacteria possess endogenous plasmids (Rebiere *et al.*, 1986), suggesting that it might be feasible to use these molecules as tools in molecular genetic studies. However, no function encoded by cyanobacterial plasmids has been identified so far, and the regulatory mechanism of their replication is still unknown. Several workers have tried to attribute various functions to plasmids, such as gas vacuolation, toxin production, resistance to high salt concentrations or heavy metals, resistance to antibiotics and synthesis of restriction/modification enzymes. It was observed that spontaneous loss of plasmids from different unicellular strains, such as *Synechococcus* sp. PCC 6301 and 73109 and *Synechocystis* sp. PCC 6803, could occur without causing any obvious phenotypic change (Whitehead and Brown, 1985).

Still, several research groups have produced various cloning vectors or shuttle vectors (*E. coli*-cyanobacterium) by utilizing these self-replicating molecules. Nevertheless, this approach is limited since the parameters related to mechanisms involved in plasmid replication and control of plasmid copy number in cyanobacteria remain obscure. The number of autonomous plasmids per cyanobacterial cell may vary from one up to eight, with sizes ranging from 1.3 kb to about 130 kb (Tandeau de Marsac and Houmard, 1987).

In a few instances, different strains of cyanobacteria possess plasmids of identical size and endonucleolytic digestion pattern (Hauman, 1981) or sequence homologies (Schwabe *et al.*, 1988). Since some of these strains are of different geographical origins, an implication can be made that interspecific or even intergeneric plasmid transfer may occur in nature.

In other cases, multimeric forms of a single plasmid species were elucidated by digestion with restriction enzymes. Reaston *et al.* (1980) found in *Nostoc* sp. PCC 7524, three plasmids, pDU1, pDU2, and pDU3 (6.1, 11.8, and 37.3 kb respectively), where pDU2 is a dimeric form of pDU1. This raises questions in regard to the function and preservation of such forms in the cell, particularly if an active recombination mechanism followed by segregation is considered. Even though the plasmids obtained from different cyanobacterial species are generally similar when purified by different researchers, there are exceptions that in few instances were attributed in part to the extraction protocol used in the various laboratories. However, the very extensively studied cyanobacterium *Calothrix* sp. PCC 7601 (*Fremyella diplosiphon*) is a real exception since different restriction patterns and different plasmid numbers were obtained in various laboratories (Bogorad *et al.*, 1983). These variations could not be explained or related to phenotypic changes of the cells. However, these results, together with previous observations, that spontaneous pigment mutants arise with high frequency (10^{-3} to 10^{-4}) in this organism led the de Marsac group (Tandeau de Marsac and Houmard 1987) to suggest the existence of mobile genetic elements in the *Calothrix* strain. So far, whether these elements might be transposons, insertion elements, or Mu-type phages has not been determined. If mobile DNA elements do exist, one would expect high frequencies of genome rearrangements in cyanobacteria. Nevertheless, the only rearrangement documented so far has been the well-analyzed rearrangement of the *nif* region in the filamentous dinitrogen-fixing strain *Anabaena* sp. PCC 7120, which accompanies the differentiation of vegetative cells into heterocysts (Golden *et al.*, 1985)

DNA-DNA hybridization

One important technique for comparing prokaryotes at the molecular level is DNA-DNA hybridization. DNA-DNA hybridization gives a measure of relatedness across the whole genome. In this test, the genomic DNA from one species is mixed with the DNA from a second species and the similarity of the DNAs is reflected in the extent to which strands of DNA from one organism anneal with strands of DNA from the other organism. The sensitivity of DNA-DNA hybridization declines rapidly as the organisms become more diverged, limiting the method to characterization of closely related strains, species and genera. In addition, testing the relationships of a new organism can require many hybridizations between different strains of *Pseudanabaena*, LPP group, *Anabaena* and unicellular forms have been elucidated using filter hybridization technique.

A polyphasic approach was used to clarify the taxonomy of the water bloom forming *Oscillatorioid* cyanobacteria. Seventy-five strains of *Oscillatorioid* cyanobacteria were characterized by DNA-DNA hybridization, 16S rDNA sequence analysis, fatty acid composition, phycobilin pigment composition, complementary chromatic adaptation, morphological characters, growth temperature and salinity tolerance. Genomic DNA homologies were examined from six *Microcystis* strains including five different species, *Microcystis aeruginosa*, *Microcystis ichthyoblabe*, *Microcystis novacekii*, *Microcystis virdis* and *Microcystis wesenbergii*. All DNA-DNA reassociation values between two strains of *M.aeruginosa* and the other four species exceeded 70 per cent, which is considered high enough for them to be classified within the same bacterial species. The current taxonomy of cyanobacteria depends too much upon morphological characteristics and much be reviewed by means of bacteriological methods as well as traditional botanical methods (Otsuka *et al.*, 2001).

The PII protein is encoded by a unique glnB gene in *Synechococcus* sp. strain PCC 7942. Its expression has been analysed in the wild type and in NtcA-null mutant cells grown under different conditions of nitrogen and carbon supply. DNA-DNA hybridization experiments revealed the presence of one transcript species 680 nucleotides long, whatever the nutrient conditions tested. A second transcript species, 620 nucleotides long, absent in the NtcA null mutant, was observed in wild-type cells that were nitrogen starved for 2 h under both high and low CO_2 and in the presence of nitrate under a high CO_2 concentration. Primer extension analysis indicated that the two transcript species are generated from two tandem promoters, a sigma 70 Escherichia coli-type promoter and an NtcA-dependent promoter, located 120 and 53 nucleotides, respectively, from the glnB initiation codon. The NtcA-dependent promoter is up-regulated under the conditions mentioned above, while the sigma 70 E. coli-type promoter displays constitutive levels of transcripts in the NtcA null mutant and slightly different levels in the wild-type cells, depending on the nitrogen and carbon supplies. In general, a good correlation between the amounts of the two transcript species and that of the PII protein was observed, as revealed by immunodetection with specific antibodies. The phosphorylation level of PII in the wild type is inversely correlated with nitrogen availability and directly correlated with higher CO_2 concentration. This regulation is correspondingly less stringent in the NtcA null mutant cells. In contrast, the dephosphorylation of PII is NtcA independent (Lee-Hyun *et al.*, 1999)

Random Amplified Polymorphic DNA (RAPD)

Random amplified Polymorphic DNA technique found immediate favour with molecular and population biologists due to its simple and straight forward protocol. RAPD allows the detection of multi-locus genetic variation using short primers of arbitrary sequence (Welsh amd McClelland 1990). In contrast to other standard molecular techniques, such as restriction fragment length polymorphism (RFLP), DNA sequencing and allozymes, the RAPD technique is very easy to perform and requires no prior knowledge of the genomes under investigation. Finally, a good interpretability and allow complexity in banding pattern are definitely desirable in RAPD technique. RAPDs are viewed as having several advantages over other DNA fingerprinting techniques; the methodology is relatively easy to apply, an unlimited number of loci can be examined and because the primers consist of random sequences that do not discriminate between coding and non-coding regions, it samples the genome more randomly than conventional methods (Weising *et al.*, 1995). Despite its advantages, RAPD analysis presents some practical problems, for example, it has been claimed that RAPD markers are rather difficult to reproduce. Nevertheless, if the conditions of DNA extraction and PCR amplification are correctly standardized, reproducibility of the patterns should not be a problem.

Nishihara and coworkers (1997) were the first to utilize RAPD analysis for cyanobacteria. This technique is advantageous as a taxonomic method for cyanobacteria as many strains exhibit poor/slow growth and yield low amounts of DNA. Their studies indicated a good divergence for discrimination of the affiliated groups of *Microcystis*. Growth of *Microcystis* in fresh waters presents a considerable threat to the health of humans. Random amplified polymorphic DNA (RAPD) analysis was used to discriminate genotypes in five species of *Microcystis* cyanobacteria. Strains of each group with the identical allozyme genotype gave similar RAPD patterns characterizing the respective group. On the other hand, no similarities in RAPD patterns were observed among strains of which allozyme genotypes were different. A good accordance between the RAPD analysis and allozyme divergence indicated a high reliability of both methods for discrimination of the affiliated groups of *Microcystis*. Several amplified DNA fragments, which were expected to be markers for a particular taxon with identical allozyme genotype, were also observed on the RAPD patterns. Genetic homogeneities of *M. novacekii*, *M. viridis* and *M. wesenbergii* were shown by RAPD analysis as well as the allozyme genotype. However, significant variations were observed in *M. aeruginosa* and *M. ichthyoblabe* in the levels of DNA and proteins.

As a part of the molecular taxonomic revision of the genus *Microcystis* (Cyanophyceae), genome DNA divergence within and between three *Microcystis* species, including seven strains from different areas, was analysed electrophoretically after RAPD analysis using 24 primers. The genus was divided into the following four taxonomic entities, differing in morphology and genetics: (1) *M. aeruginosa* 101,

M. aeruginosa 7820, *M. aeruginosa* 90 *and Microcystis sp.* (*Dianshan*) (2) *M. viridis* (3) *M. wesenbergii*; and (4) *M. aeruginosa* 41. M. aeruginosa comprises two heterogeneous taxonomic entities which could be viewed as separate species. Because of their unique genotypes and minor morphological variation, *M. viridis* and *M. wesenbergii* are regarded as well-established species. As anticipated, *Anabaena* sp. 7120, as the control, was genetically distinct from the *Microcystis* strains. The results from this RAPD analysis of the genus *Microcystis* indicated that it is possible to analyse intra–and inter-specific variation of Cyanophytes on the basis of the genotype (Pan-Hui *et al.*, 1999)

Nitrogen fixing *Anabaena azollae* strains isolated from four different *Azolla* cultures were characterized based on their total protein profile and RAPD profile to study the existing variation among them. As expected, the isolates showed almost similar protein banding patterns, but exhibited differences in 40-70 KDa protein subunits. Polymerase chain reaction of the DNA of the isolates, using four different primers, amplified specific sequences of DNA and showed clear polymorphism among the isolates. The RAPD profile generated the fingerprinting pattern characteristic of each strain based on the sequence of the primers used. Common band sharing observed between the strains *A. azollae*-RS-KK-SK-AM and *A. azollae*-RS-KK-SK-RP probably represents maternal inheritance of DNA to the progeny. The polymorphic bands were generated specifically for the isolates *A. azollae*-RS-KK-SK-RP and *A. azollae*-RS-KK-SK-AM with primers numbered 2 and 4, respectively, which could be developed as possible markers for these isolates (Subhashini *et al.*, 2002)

Restriction Fragment Length Polymorphism (RFLP)

The RFLP technique is generally used to identify and classify organisms at the population or the species level. It has been productivity used to study marine *Synechococcus* sp. strains and cyanobacterial symbionts. In the case of angiosperms symbiosis, twelve free living cultures derived from symbionts of eight *Gunnera* species growing in Sweden, New Zealand and USA were characterized by Zimmerman and Bergman (1990). Two methods were used: The RFLP analysis using heterologous probes from *Anabaena* sp. Strains PCC 7120 and the immuno-staining of the protein profiles. Twelve strains of phycoerythrin rich cyanobacteria of the *Synechococcus* type were isolated from the pelagial of lake Constance in different phases during the growth period of the year 1994. By analyzing the restriction fragment length polymorphism of the DNA three new genotypes were distinguished. Studies based on restriction fragment length polymorphism (RFLP) and PCR techniques have been used to examine the *Anabaena-Azolla* symbiosis for classification of the cyanobacterial symbionts from different *Azolla* species. Restriction fragment length polymorphism and the sequencing of the 16s rRNA gene are used for characterization and grouping of axenic, planktonic cyanobacteria (*Anabaena, Aphanizomenon, Microcystis, Nodularia, Nostoc* and *Oscillatoris*). Isolates from cycads and *Gunnera* have been studied with respect to genetic diversity by using protein profiles and the RFLP technique (Lupski and Weinstock, 1992).

Microcystins are harmful hepatotoxins produced by many, but not all strains of the cyanobacterial genera *Anabaena, Microcystis, Planktothrix,* and *Nostoc*. Waterbodies have to be monitored for the mass development of toxic cyanobacteria; however, because of the close genetic relationship of microcystin-producing and non-producing strains within a genus, identification of microcystin-producers by morphological criteria is not possible. The genomes of microcystin-producing cells contain mcy genes coding for the microcystin synthetase complex. Based on the sequence information of mcy genes from *Microcystis* and *Planktothrix*, a primer pair for PCR amplification of a mcyA gene fragment was designed. PCR with this primer pair is a powerful means to identify microcystin-producing strains of the genera *Anabaena, Microcystis* and *Planktothrix*. Moreover, subsequent RFLP analysis of the PCR products generated genus-specific fragments and allowed the genus of the toxin producer to be identified. The assay can be used with DNA from field samples (Hisbergues *et al.*, 2003)

Terminal restriction fragment length polymorphism (TRF or T-RFLP) analysis and 16S rDNA sequence analysis from clone libraries were used to examine cyanobacterial diversity in three types of predominant soil crusts in an arid grassland in Utah, USA. Total DNA was extracted from cyanobacteria–lichen or moss-dominated crusts that represent different success ional stages in crust development, and which contribute different amounts of carbon and nitrogen into the ecosystem. Cyanobacterial 16S rRNA genes were amplified by PCR using cyanobacteria-specific 16S rDNA primers. Both TRF and clone sequence analyses indicated that the cyanobacterial crust type is dominated by strains of *Microcoleus vaginatus*, but also contains other cyanobacterial genera. In the moss crust, *M. vaginatus*-related sequences were also the most abundant types, together with sequences from moss chloroplasts. In contrast, sequences obtained from the lichen crust were surprisingly diverse, representing numerous genera, but including only two from *M. vaginatus* relatives. By obtaining clone sequence information, we were able to infer the composition of many peaks observed in TRF profiles, and all peaks predicted for clone sequences were observed in TRF analysis. This study provides the first TRF analysis of biological soil crusts and the first DNA-based comparison of cyanobacterial diversity between lichen-, cyano–and moss-dominated crusts. Results indicate that for this phylogenetic group, TRF analysis, in conjunction with limited sequence analysis, can provide accurate information about the composition and relative abundance of cyanobacterial types in soil crust communities (Redfield *et al.*, 2002).

Amplified Fragment Length Polymporphism (AFLP)

Amplified fragment length polymorphisms are polymerase chain reaction based markers for the rapid screening of genetic diversity. AFLP methods rapidly generate hundreds of highly replicable markers from DNA of any organism; thus, they allow high-resolution genotyping of fingerprinting quality. The time and cost efficiency, replicability and resolution of AFLPs are superior or equal to those of other markers (allozymes, random amplified polymorphic DNA, restriction fragment length polymorphism, microsatellites) expect that AFLP methods primarily generate dominant rather than codominant markers. Because of their high replicability and ease of use, AFLP markers have emerged as a major new type of genetic marker with broad application in systematics, pathotyping, population genetics, DNA fingerprinting and quantitative trait loci (QTL) mapping (Avise 1994).

The advantage of AFLP over other techniques is that multiple bands are derived from all over the genome. This prevents over interpretation or misinterpretation due to point mutations or single-locus recombination, which may effect other genotypic characteristics. The main disadvantage of AFLP markers is that alleles are not easily recognized. PCR has proven to be successful in detecting genetic variation amongst plant-pathogenic fungi, as well as cyanobacteria. The utility, repeatability and efficiency of the AFLP technique are leading to broader application of this technique in the analysis of cyanobacteria populations. The purpose of this study was to investigate the usefulness of AFLP, which is based on the selective amplification of genomic restriction fragments by PCR, to differentiate between geographical unrelated strains of *Microcystis* spp. (Janssen *et al.*, 1996).

Molecular marker analysis is becoming increasingly capable of identifying informative genetic variation. Amplified fragment length polymorphism markers (AFLPs) are among the recent innovations in genetic marker technologies, and provide a greater capacity for genome coverage and more reproducible results than previous technologies. The usefulness of AFLP, which is based on the selective amplification of genomic restriction fragments by PCR, to differentiate between geographical unrelated *Microcystis* strains. In total 23 strains were subjected to the AFLP fingerprinting. After analysis of the data on the basis of the average linkage method, known as the unweighted pair group method using arithmetic averages (UPGMA), a dendrogram with four clusters was obtained. Cluster 1 consisted mainly of the NIES strains that originated from Japan, while in cluster 2 the European strains grouped together. The South African strains that originated from the northern part of the country group together in cluster 3, while the strains collected from the central and southern regions group together with the US strains in cluster 4. The study had revealed extensive evidence for the applicability of AFLP in cyanobacterial taxonomy, and furthermore clearly demonstrates the superior discriminative power of AFLP towards the differentiation of geographical unrelated *Microcystis aeruginosa* strains that belong to the same species, as well as highlighting the potential of this fingerprinting method in evolutionary studies (Oberholster *et al.*, 2005).

Microsatellite (STRR)

The presence of repeated DNA [short tandemly repeated repetitive (STRR) and long tandemly repeated repetitive (LTRR)] sequences in the genome of cyanobacteria was used to generate a fingerprint method for symbiotic and free-living isolates. Primers corresponding to the STRR and LTRR sequences were used in the PCR, resulting in a method which generate specific fingerprints for individual isolates. The method was useful both with purified DNA and with intact cyanobacterial filaments or cells as templates for the PCR. Twenty-three *Nostoc* isolates from a total of 35 were symbiotic isolates from the angiosperm *Gunnera* species, including isolates from the same *Gunnera* species as well as from different species. The results show a genetic similarity among isolates from different *Gunnera* species as well as a genetic heterogeneity among isolates from the same *Gunnera* species. Isolates which have been postulated to be closely related or identical revealed similar results by the PCR method, indicating that the technique is useful for clustering of even closely related strains. The method was applied to nonheterocystus cyanobacteria from which a fingerprint pattern was obtained (Rasmussen, 1998)

The cyanobacteria belonging to the genus *Nostoc* fix atmospheric nitrogen, both as free-living organisms and in symbiotic associations with a wide range of hosts, including bryophytes, gymnosperms (cycads), the small water fern *Azolla* (Pteridophyte), the angiosperm genus *Gunnera*, and fungi (lichens). The *Gunnera-Nostoc* symbiosis is the only one that involves a flowering plant. In Chile, 12 species of *Gunnera* have been described with a broad distribution in the temperate region. The genetic diversity of *Nostoc* symbionts from three populations of *Gunnera tinctoria* from Abtao, Chiloe Island, southern Chile, and microsymbionts from other two species of *Gunnera* from southern Chile, using PCR amplification of STRR (short tandemly repeated repetitive) sequences of the *Nostoc* infected tissue are examined. To our knowledge, this is the first report of PCR fingerprinting obtained directly from symbiotic tissue of *Gunnera*. Genetic analyses revealed that *Nostoc* symbionts exhibit important genetic diversity among host plants, both within and between *Gunnera* populations. It was also found that only one *Nostoc* strain, or closely related strains, established symbiosis with an individual plant host (Guevara *et al.*, 2002).

Symbiotically associated cyanobacteria from 18 accessions within all known species in the genus *Azolla* were examined and classified by the use of polymerase chain reaction (PCR)-fingerprinting. A repetitive sequence specific for cyanobacteria, the short tandemly repeated repetitive (STRR) sequence, was used as a primer in the reaction. Cyanobacterial filaments isolated directly from the *Azolla* leaf cavity or contained within homogenised symbiotic *Azolla* tissue were used as templates. Based on the fingerprint pattern, distinct differences were demonstrated between cyanobacteria isolated from the

Euazolla and *Rhizosperma* sections. In addition, individual fingerprints were obtained from all cyanobacteria isolated from the different *Azolla* species. The fingerprints were used to generate a phylogenetic tree. Three clusters were distinguished: one contained the four isolates from the section *Euazolla*, a second the isolate from *Azolla filiculoides*, and a third the three isolates from the section *Rhizosperma*. By the use of STRR-PCR fingerprinting, new data on the taxonomy of cyanobacteria in *Azolla* were obtained, which have been difficult to generate by other classification methods. PCR-fingerprinting may, therefore, be a valuable tool for diversity and classification studies of symbiotic cyanobacteria from *Azolla* and, as co-evolution between the cyanobacteria and its corresponding host exists the method may also be useful for the taxonomy of *Azolla* (Zheng *et al.*, 1999).

Cylindrospermopsis raciborskii is a planktonic, Nostocalean cyanobacterium, which produces an alkaloid heptatoxin, cylindrospermopsin. PCR fingerprint analysis of short tandemly repeated repetitive (STRR) sequences, temperature tolerances and toxin analysis to characterize 24 strains of this toxic cyanobacterium isolated from Thailand and Japan are examined. All strains shared common morphological traits characteristic of C. raciborskii and showed high 16S rDNA sequence similarity, forming a defined cluster together with the reference strains from Australia. In particular, some of the Thai strains shared 99.9 per cent to 100 per cent similarity with the Australian strains. Various combinations of STRR primers revealed different and unique DNA band patterns among strains of *C. raciborskii*. The phylogenetic tree revealed two main clusters of *C. raciborskii* strains, the Thai/Japan-Shinobazugaike cluster (cluster I) and the Japan-Gonoike cluster (cluster II). Cluster I was further divided into two subclusters, A (only Thai strains) and B (one Thai strain and the Japan-Shinobazugaike strains). Thus, the results from 16S rDNA and STRR analyses showed no clear geographical distinction between Japanese and Thai strains and between Thai and Australian strains. Thai strains were separated into adaptive and non-adaptive groups to low temperature (15 and 17.5 degrees C) and Japanese strains were composed of only low-temperature-adaptive ones. The toxin cylindrospermopsin was detected in some strains of cluster I-A and in one strain of cluster II. The *C. raciborskii* is a species that has recently begun to invade, and a species with different physiological strains or ecotypes in temperature tolerance; the toxin is synthesized without any relation to phylogenetic or genetic clusters and to geography (Chonudomkul *et al.*, 2004).

ARDRA(Amplified 16S Ribosomal DNA Restriction Analysis)

The 16S-rDNA from 22 cyanobacteria isolated from biofilms on walls of modern and historic buildings in Brazil was partially sequenced (~350 bp) using specific primers. The cyanobacteria with the closest matching sequences were found using the BLAST tool. The sequences were combined with 52 other cyanobacterial sequences already deposited in public data banks and a dendrogram constructed, after deletion from each sequence of one of the variable 16S rDNA regions (VI). The newly sequenced organisms fitted well within their respective families, but their similarities to other members of the groups were generally low, less than 96 per cent. Close matches were found only with one other terrestrial (hot dry desert) cyanobacterium, *Microcoleus sociatus*, and with *Anabaena variabilis*. Phylogenetic analysis suggested that the deletion of the hypervariable regions in the RNA structure is essential for meaningful evolutionary studies. The results support the standard phylogenetic tree based on morphology, but suggest that these terrestrial cyanobacteria are distant relatives of their equivalent aquatic genera and are, indeed, a distinct population. (Peter *et al.*, 2005)

The bacterial community associated with black band disease (BBD) of the scleractinian corals Diploria strigosa, Montastrea annularis and Colpophyllia natans was examined using culture-independent techniques. Two complementary molecular screening techniques of 16S rDNA genes (amplified 16S ribosomal DNA restriction analysis (ARDRA) of clone libraries and denaturing gradient gel electrophoresis (DGGE)) were used to give a comprehensive characterization of the community. Findings support previous studies indicating low bacterial abundance and diversity associated with healthy corals. A single cyanobacterial ribotype was present in all the diseased samples, but this was not the same as that identified from *Phormidium corallyticum* culture isolated from BBD. The study confirms the presence of *Desulfovibrio* spp. and sulphate-reducing bacteria that have previously been associated with the BBD consortium. However, the species varied between diseased coral samples. No evidence of bacteria from terrestrial, freshwater or human sources in any of the samples was found. The presence of previously unrecognized potential pathogens (a Cytophaga sp. and an alpha–proteobacterium identified as the aetiological agent of juvenile oyster disease (JOD) that were consistently present in all the diseased coral samples was reported. The molecular biological approach described here gives an increasingly comprehensive and more precise picture of the bacterial population associated with BBD (Cooney *et al.*, 2002).

Genotypic diversity of several cyanobacterial strains mostly isolated from marine or brackish water, belonging to the genera *Geitlerinema* and *Spirulina* was investigated by amplified 16S ribosomal DNA restriction analysis and compared with morphological features and response to salinity. Cluster analysis was performed on amplified 16S rDNA restriction profiles of these strains along with profiles obtained from sequence data of five Spirulina-like strains, including three representatives of the new genus *Halospirulina*. Our strains with tightly coiled trichomes from hypersaline waters could be assigned to the *Halospirulina* genus. Among the uncoiled strains, the two strains of hypersaline origin clustered together and were found to be distant from their counterparts of marine and freshwater habitat. Moreover, another cluster, formed by alkali-tolerant strains with tightly coiled trichomes, was well delineated (Margheri *et al.*, 2003).

PC-IGS Region Sequencing (Phycocyanin Operon Intergenic Spacer)

The filamentous diazotrophic cyanobacterium *Nodularia* forms water blooms each year in the Baltic Sea. Filaments isolated from such water blooms vary in their trichome width, degree of coiling, and properties of their gas vesicles; previously, these characters have been used to classify individuals to species level. To test the validity of such a phenotypic classification, we determined the nucleotide sequences for a region of the phycocyanin locus that includes a noncoding intergenic spacer (PC-IGS), the IGS between two adjacent copies of the *gvpA* gene (which encodes the main structural gas vesicle protein) and the rDNA internal transcribed spacer (rDNA-ITS), for 13 clonal *Nodularia* isolates from the Baltic Sea during August 1994. The complete 16S-rDNA sequence was determined for three isolates and was found to be identical in each of them. Molecular sequences for noncoding regions of the genome were used to assign isolates to three groups on the basis of PC-IGS, two groups on the basis of *gvpA*-IGS, and three groups on the basis of rDNA-ITS. No consistent correlation was found between genotype and any of the phenotypic features examined, and no link was found between any of these features themselves, indicating that these characters are not useful for placing *Nodularia* isolates into meaningful taxonomic groups. The PC-IGS, *gvpA*-IGS, and rDNA-ITS genotypic groupings were not congruent. This might indicate that gene flow occurs between individuals in *Nodularia* populations (Gary *et al.*, 1999).

Some cyanobacteria have been shown to exchange genetic information under laboratory conditions, but it has not been clear whether such genetic exchange occurs in the natural environment. To address this, a population genetic study was carried out on the filamentous diazotrophic cyanobacterium *Nodularia* in the Baltic Sea. *Nodularia* filaments were collected from 20 widely distributed sampling stations in the Baltic Sea during June and July 1998. Allele-specific PCR (AS-PCR) was used to characterize over 2000 filaments at three loci: a non-coding spacer between adjacent copies of the main structural gas vesicle gene gvpA (gvpA-IGS), the phycocyanin intergenic spacer (PC-IGS) and the rDNA internal transcribed spacer (rDNA-ITS). The three loci were all found to be polymorphic in the 1998 population: two alternative alleles were distinguished at the gvpA-IGS and PC-IGS loci, and three at the rDNA-ITS locus. All 12 possible combinations of alleles were found in the filaments studied, but some were much more common than others. The index of association (IA) for all possible pairwise combinations of isolates was found to differ significantly from zero, which implies that there is some linkage disequilibrium between loci. The IA values for 16 out of 20 individual sampling stations also differed significantly from zero: this shows that the observed linkage disequilibrium is not due to pooling data from genetically distinct subpopulations. Monte-Carlo simulations with random subsets of the data confirmed that some combinations showed significantly more linkage disequilibrium than expected by chance alone. It is concluded that genetic exchange occurs in the natural *Nodularia* population, but the frequency is not high enough for the loci to be in linkage equilibrium. The distribution of the 12 genotypes across the Baltic Sea was found to be non-random, but did not correlate with temperature, salinity or major nutrient concentrations (Barker *et al.*, 2000).

Morphological and genetic diversity of cultured cyanobacterial strains of species of the genera *Arthrospira*, *Spirulina* and *Phormidium* from two geographically different regions and habitats (Kenyan saline-alkaline lakes and Indian freshwater bodies) were investigated. Light microscopy observations were used to determine morphological diversity of the cyanobacteria. Three independent molecular techniques, sequencing of 16S rRNA gene, internally transcribed spacer region between 16S and 23S rDNA (ITS) and the phycocyanin locus (PC-IGS) were conducted for the examination of phylogenetic relationship. Despite differences in morphology and habitats the Kenyan and Indian *Arthrospira* strains belong to the same cluster in phylogenetic trees of the 16S rDNA (AY575923-AY575932) or PC-IGS (AY575937-AY575946). The DNA similarity in both methods was 100 per cent. In the ITS tree, the two Indian *Arthrospira* strains PD1998/pus (AY575930) and PD2002/ana (AY575932) form their own sub-cluster. The *Phormidium* strain AB2002/07 (AY575933) from Lake Nakuru, Kenya is included in the *Arthrospira* cluster in the ITS tree and very closely related in the 16S and PC-IGS trees. Based on 16S rDNA and PC-IGS phylogeny the sequences of the *Spirulina* strains form a separate cluster distinct from the *Arthrospira* cluster. The Kenyan and Indian *Spirulina subsalsa* strains show a considerable genetic variability as similarities in 16S rRNA gene sequence is 91.5 per cent only. Molecular characterizations of cyanobacterial strains in the present study demonstrate that several distinct morphotypes may be genetically similar and vice versa (Ballot *et al.*, 2004).

Sequencing of 16S rRNA Gene

Toxic and non-toxic cyanobacterial strains from *Anabaena*, *Aphanizomenon*, *Calothrix*, *Cylindrospermum*, *Nostoc*, *Microcystis*, *Planktothrix* (*Oscillatoria agardhii*), *Oscillatoria* and *Synechococcus* genera were examined by RFLP of PCR-amplified 16S rRNA gene sequencing. With both methods, high 16S rRNA gene similarity was found among planktic, and a toxin producing *Anabaena* and non-toxic *Aphanizomenon*, microcystin-producing and non-toxic *Microcystis*, and microcystin-producing and non-toxic *Planktothrix* strains of different geographical origins. The respective sequence similarities were 99.9-100 per cent, 94.2-99.9 per cent and 99.3-100 per cent. Thus the morphological characteristics (*e.g.Anabaena* and *Aphanizomenon*), the physiological (toxicity) characteristics or the geographical origins did not reflect the level of 16S rRNA gene relatedness of the closely related strains studied. In addition, cyanobacterial strains were fingerprinted with repetitive extragenic palindromic (REP)–and enterobacterial repetitive intergenic consensus (ERIC)-PCR. All the strains except two identical pairs of *Microcystis* strains had different band profiles. The overall grouping of the trees from the 16S rRNA gene and the REP–and ERIC-PCR analyses was similar. Based on the 16S rRNA gene sequence analysis, four major clades were formed. (i) The clade containing filamentous heterocystous cyanobacteria was divided into three discrete groups of *Anabaena/Aphanizomenon, Anabaena/ Cylindrospermum/Nodularia/Nostoc* and *Calothrix* strains. The three other clades contained (ii) filamentous non-heterocystous *Planktothrix,* (iii) unicellular non-heterocystous *Microcystis* and (iv) *Synechococcus* strains (Lyra *et al.*, 2001).

The taxonomic coherence and phylogenetic relationships of 11 planktonic heterocystous cyanobacterial isolates were examined by investigating two areas of the rRNA operon, the 16S rRNA gene (rrnS) and the internal transcribed spacer (ITS) located between the 16S rRNA and 23S rRNA genes. The rrnS sequences were determined for five strains, including representatives of Anabaena flos-aquae, Aphanizomenon flos-aquae, Nodularia sp. and two alkaliphilic planktonic members of the genera Anabaenopsis and Cyanospira, whose phylogenetic position was previously unknown. Comparison of the data with those previously published for individual groups of planktonic heterocystous cyanobacteria showed that, with the exception of members assigned to the genus Cylindrospermopsis, all the planktonic strains form a distinct subclade within the monophyletic clade of heterocystous cyanobacteria. Within this subclade five different phylogenetic clusters were distinguished. The phylogenetic groupings of Anabaena and Aphanizomenon strains within three of these clusters were not always consistent with their generic or specific assignments based on classical morphological definitions, and the high degree of sequence similarity between strains of Anabaenopsis and Cyanospira suggests that they may be assignable to a single genus. Ribotyping and additional studies performed on PCR amplicons of the 16S rDNA or the ITS for the 11 planktonic heterocystous strains demonstrated that they all contain multiple rrn operons and ITS regions of variable size. Finally, evidence is provided for intra-genomic sequence heterogeneity of the 16S rRNA genes within most of the individual isolates (Isabelle *et al.*, 2002)

Modern stromatolites represent a significant resource for studying microbial ecology and evolution. A preliminary investigation was undertaken employing specific genetic probes to characterize the cyanobacteria responsible for stromatolite construction in a range of environments, including microbial mats found in Australia not previously examined with molecular methods. Isolates of cyanobacteria were collected from stromatolites in thermal springs, hypersaline lakes, and oceanic fringes on two continents. A polymerase chain reaction specific for DNA of cyanobacterial 16S rRNA was developed, the resulting products of the DNA amplification reaction were sequenced, and the data were used to infer relatedness between the isolates studied and other members of the cyanobacterial radiation. Complete sequence was generated for the region from position 27 to 408 for 13 strains of cyanobacteria associated with stromatolites. All stromatolite-derived sequences were most closely related to cyanobacteria, as indicated by local sequence alignment. It was possible to correlate genetic identity with morphological nomenclatures and to expand the phylogeny of benthic cyanobacteria. These inferences were also expanded to temporal variation in the dominant resident cyanobacterial species based on sampling of surface and core sinter laminations. Under the methods employed, only one cyanobacterial strain was detected in each sample, suggesting the possible dominance of a specific clonal population of cyanobacteria at any one time in the biota of the samples tested. The data indicate that internal core samples of a stromatolite at least 10 years old can be successfully analyzed by DNA-based methods to identify preserved cyanobacteria (Brett *et al.*, 2002).

Characterization of the 16S-23S ITS

The 16S-23S ITS region from single isolates of a variety of cyanobacterial taxa: *Calothrix parietina, Scytonema hyalinum, Coelodesmium wrangelii, Tolypothrix distorta*, and a putative new genus in the Microchaetaceae were amplified. All isolates were found to carry ITS regions containing the sequences coding for two tRNA molecules (tRNAIle and tRNAAla). Additional sequences without tRNA features from both C. parietina and S. hyalinum were retrieved. Furthermore, in S. hyalinum, two of these non-tRNA-encoding regions identical in length, but different in sequence were found. This is the first report of ITS regions from a single cyanobacterial isolate different not only in configuration but also, within one configuration, different in sequence. The potential of the ITS region as a tool for studying molecular systematics and population genetics is significant, but the presence of multiple non-identical rRNA operons poses problems. Multiple non-identical rRNA operons may impact both studies which depend on comparisons of phylogenetically homologous sequences, and studies which employ restriction digests of PCR products (Boyer *et al.*, 2002).

The genetic diversity of ten symbiotic *Nostoc* strains isolated from different *Gunnera* species was investigated. The strains were analyzed using molecular methods with different taxonomic resolutions, including restriction fragment length polymorphisms (RFLP) of the PCR-amplified 16S ribosomal gene and the 16S-23S internal transcribed spacer (ITS) region combined with computer-assisted analyses. The functional gene hetR, assigned to heterocyst differentiation, was used for denaturing gradient gel electrophoresis. A high genetic diversity was observed among the isolates even in the conserved gene coding for the small ribosomal unit. No correlation was observed between clustering of cyanobacteria and the host species of *Gunnera* (Rasmussen and Svenning, 2001).

The genetic diversity of *Trichodesmium* spp. from natural populations (off Bermuda in the Sargasso Sea and off North Australia in the Arafura and CoralSeas) and of culture isolates from two regions (Sargasso Sea and Indian Ocean) was investigated. Three independent techniques were used, including aDNA fingerprinting method based on a highly iterated palindrome (HIP1), denaturing gradient gel electrophoresis of a hetR fragment, and sequencing ofthe internal transcribed spacer (ITS) of the 16S-23S rDNA region. Low genetic diversity was observed in natural populations of *Trichodesmium* spp. From the two hemispheres. Culture isolates of *Trichodesmium thiebautii, Trichodesmium hildebrandtii, Trichodesmium tenue*, and *Katagnymene spiralis* displayed remarkable similarity when these techniques were used, suggesting that *K. spiralis* is very closely related to the genus *Trichodesmium*. The largestgenetic variation was found between *Trichodesmium erythraeum* and all otherspecies of *Trichodesmium*, including a species of *Katagnymene*. Our data obtained with all three techniques suggest that there are two major clades of *Trichodesmium* spp. The HIP1 fingerprinting and ITS sequence analyses allowed the closely related species to be distinguished. This is the first report of the presence of HIP1 in marine cyanobacteria (Orcutt *et al.*, 2002).

References

Avise J C. 1994. *Molecular Markers, Natural History and Evolution*. Chapman and Hall, New York.

Ballot A, Dadheech P and Krienitz L 2004. Phylogenetic relationship of *Arthrospira, Phormidium*, and *Spirulina* strains from Kenyan and Indian waterbodies. *Algological Studies* 113: 37-56.

Barker G L A, Handley B A, Vacharapiyasophon P, Stevens J R and Hayes P K 2000. Allele-specific PCR shows that genetic exchange occurs among genetically diverse *Nodularia* (cyanobacteria) filaments in the Baltic Sea. *General Microbiology* 146: 2865-2875.

Bogorad L, Gendel S M, Haury J H and Koller K P 1983. Photomorphogenesis and complementary chromatic adaptation in *Fremyella diplosiphon*, In: Papageorgiou, G. C. and Packer, L. (ed.), *Photosynthetic prokaryotes: Cell differentiation and function*. Elsevier, New York 119–126 pp.

Boyer S L, Flechtner V R and Johansen J R 2002. Is the 16S-23S rRNA internal transcribed spacer region a good tool for use in molecular systematics and population genetics? A case study in cyanobacteria. *Mol. Biol. Evol.* 18: 1057-1069.

Brett A, Brendan P and David A 2002. Molecular identification of cyanobacteria associated with stromatolites from distinct geographical locations. *Astrobiol.* 2: 271-280.

Chonudomkul D, Yongmanitchai W, Theeragool G, Kawachi M, Kasai F, Kaya K and Watanabe M M 2004. Morphology, genetic diversity, temperature tolerance and toxicity of *Cylindrospermopsis raciborskii* (Nostocales, Cyanobacteria) strains from Thailand and Japan. *FEMS-Microbiol. Ecol.* 48: 345-55.

Cohen Y, Jorgensen B B Revsbech N P and Poplawski R 1986. Adaptation to hydrogen sulfide of oxygenic and anoxygenic photosynthesis among cyanobacteria. *Applied and Environmental Microbiology* 51: 398–407.

Cooney R P, Pantos O, Tissier M D A, Barer M R, Donnell A G and Bythell J C 2002. Characterization of the bacterial consortium associated with black band disease in coral using molecular microbiological techniques. *Environmental-Microbiol.* 4: 401-413.

Gary L A, Barker P K, Sarah L, Mahony O, Panmuk V and Anthony E 1999. A molecular and phenotypic analysis of *Nodularia* (cyanobacteria) from the Baltic sea. *Journal of Phycol.* 35: 931-33.

Glazer A N 1983. Comparative biochemistry of photosynthetic light-harvesting systems. *Annual Reviews of Biochemistry* 52: 125–157.

Golden J W, Robinson S J and Haselkorn R 1985. Rearrangement of nitrogen fixation genes during heterocystis differentiation in the cyanobacterium *Anabaena*. *Nature* 314: 419–423.

Guevara R, Armesto J J and Caru M 2002. Genetic diversity of *Nostoc* microsymbionts from *Gunnera tinctoria* revealed by PCR-STRR fingerprinting. *Microbial-Ecology* 44: 127-36.

Hauman J H 1981. Is a plasmid(s) involved in the toxicity of *Microcystis aeruginosa*? In: W. W. Carmichael (ed.), The water environment: Algal toxins and health, vol. 20. Plenum, New York, 97–102 pp.

Herdman M, Janvier M, Rippka R, Stanier R Y 1979. Genome size of cyanobacteria. *Journal of General Microbiology* 111: 73–85.

Herrero A and Wolk C P 1986. Genetic mapping of the chromosome of the cyanobacterium *Anabaena variabilis*. *Journal of Biological Chemistry* 261: 7748–7754.

Hisbergues M, Christiansen G, Rouhiainen L, Sivonen K and Borner T 2003. PCR-based identification of microcystin-producing genotypes of different cyanobacterial genera. *Archives of Microbiol.* 180: 402-10.

Isabelle I, Rosmarie R, Nicole T M and Michael H 2002. rDNA analyses of planktonic heterocystous cyanobacteria, including members of the genera *Anabaenopsis* and *Cyanospi*. *Microbiol.* 148: 481-96.

Janssen P, Coopman R, Huys G, Swings J, Bleeker H, Vos P, Zabeau M and Kersters K 1996 Evaluation of the DNA fingerprinting methods: AFLP as a new tool in bacterial taxonomy. *Microbiol.* 142: 1881-93.

Lee-Hyun M, Vazquez-Bermudez M F and Marsac N T 1999. The global nitrogen regulator NtcA regulates transcription of the signal transducer PII (GlnB) and influences its phosphorylation level in response to nitrogen and carbon supplies in the cyanobacterium *Synechococcus* sp. strain PCC 7942. *Journal of Bacteriol.* 181: 2697-2702.

Lupski J R and Weinstock G M 1992. Short, interspersed repetitive DNA sequences in prokaryotic genomes. *J. Bacteriol.* 174: 4525-29.

Lyra C, Suomalainen S, Gugger M, Vezie C, Sundman P, Paulin L and Sivonen K 2001. Molecular characterization of planktic cyanobacteria of *Anabaena, Aphanizomenon, Microcystis* and *Planktothrix* genera. *International-Journal-of-Systematic-and-Evolutionary-Microbiol.* 51: 513-526.

Oberholster P J, Botha A M, Muller K and Cloete T E 2005. Assessment of the genetic diversity of geographically unrelated *Microcystis aeruginosa* strains using amplified fragment length polymorphisms (AFLPs). *African Journal of Biotechnology* 4: 389-99.

Orcutt K M, Rasmussen U, Webb E A, Waterbury J B, Gundersen K and Bergman B 2002. Characterization of *Trichodesmium* spp. by genetic techniques. *Applied and Environmental Microbiol.* 68: 2236-2245.

Oren A and Shilo M 1979. Anaerobic heterotrophic dark metabolism in the cyanobacterium *Oscillatoria limnetica*: sulfur respiration and lactate fermentation. *Archives of Microbiology* 122: 77–84.

Otsuka S, Suda S, Shibata S, Oyaizu H, Matsumoto S and Watanabe M M 2001. A proposal for the unification of five species of the cyanobacterial genus *Microcystis*. *Int. J. Syst. Evol. Microbiol.* 51: 873-879.

Margheri M C, Piccardi R, Ventura S, Viti C and Giovannetti L 2003. Genotypic diversity of oscillatoriacean strains belonging to the genera *Geitlerinema* and *Spirulina* determined by 16S rDNA restriction analysis. *Current Microbiol.* 46: 359–364.

Nishihara H, Miwa H, Watanabe M, Nagashima M, Yagi O and Takamura Y 1997. Random amplified polymorphic DNA (RAPD) analyses for discriminating genotypes of *Microcystis* cyanobacteria, *Bioscience, Biotechnology and Biochemistry* 61: 1067-72.

Padan E and Cohen Y 1982. Anoxygenic photosynthesis. In: Carr, N. C., and Whitton, B. A. (ed.), *The biology of cyanobacteria*. Blackwell Scientific Oxford, 215–235 pp.

Pan-Hui, LiRong S, ShouDong L, YongDing L 1999. Phylogenetic relationship among seven strains of *Microcystis* based on the randomly amplified polymorphic DNA PCR. *Acta-Hydrobiologica-Sinica.* 23: 469-74.

Peter M, Gaylar D, Porto A, Cesar A and Crispim A 2005. Cyanobacteria from Brazilian Building Walls Are Distant Relatives of Aquatic Genera. *Journal of Integrative Biol.* 9: 30-42.

Rasmussen U and Svenning MM 2001. Characterization by genotypic methods of symbiotic *Nostoc* strains isolated from five species of *Gunnera*. *Arch. Microbiol.* 176(3): 204-210.

Rasmussen U 1998. Fingerprinting of Cyanobacteria based on PCR with primers derived from short and long tandemly repeated repetitive sequences. *Appl-environ-microbiol.* 64: 265-72.

Reaston J R, Van den Hondel C A M J J, Vanden Ende A, Van Arkel G A, Stewart W D P, and Herdman M 1980. Comparison of plasmids from the cyanobacterium *Nostoc* PCC7524 with two mutant strains unable to form heterocysts. *FEMS Microbiological Letters* 9: 185–188.

Rebiere M C, Castets A M, Houmard J and Tandeau de Marsac N 1986. Plasmid distribution among unicellular and filamentous cyanobacteria: occurrence of large and megaplasmids. *FEMS Microbiological Letters* 37: 269–275.

Redfield E, Barns S M, Belnap J, Daane L L and Kuske C R 2002. Comparative diversity and composition of cyanobacteria in three predominant soil crusts of the Colorado Plateau. *FEMS-Microbiol. Ecol.* 40: 55-63.

Schwabe W, Weihe A, Borner T, Henning M and Kohl J G 1988. Plasmids in toxic and non-toxic strains of the cyanobacterium *Microcystis aeruginosa. Current Microbiology* 17: 133–137.

Stanier R Y and Cohen-Bazire G 1977. Phototrophic prokaryotes. The cyanobacteria. *Annual Reviews of Microbiology* 31: 225–274.

Subhashini R, Kumar K and Kannaiyan S 2002. Variation revealed by RAPD-PCR profiles of microsymbiont *Anabaena azollae* strains from four N_2 fixing *Azolla* cultures. *Acta-Biotechnologica* 20: 169-74.

Tandeau de Marsac N and Houmard J 1987. Advances in cyanobacterial molecular genetics In: Fay P., and Van Baalen, C. (ed.), *The cyanobacteria*. Elsevier Amsterdam, 251–302 pp.

Weising K, Nybom H, Wolff K and Meyer W 1995. DNA finger-printing in plants and fungi. CRC Press, Boca Raton, Florida

Welsh J and McClelland M 1990. Fingerprinting genome using cyanobacterial isolates. *Nucleic Acids Res.* 18: 7213-18.

Whitehead P R and Brown N L 1985. Three restriction endonucleases from *Anabaena* flos-aquae. *Journal of General Microbiology* 131: 951–958.

Williams J G K 1988. Construction of specific mutations in photosystem II photosynthetic reaction center by genetic engineering methods in *Synechocystis* 6803. *Methods in Enzymology* 167: 766–778.

Zheng W W, Nilsson M, Bergman B and Rasmussen U 1999. Genetic diversity and classification of cyanobacteria in different *Azolla* species by the use of PCR fingerprinting. *Theor-appl-genet.* 99: 1187-93.

© R.K. Gupta & V.D. Pandey (*Editors*)
Advances in Applied Phycology, 2007
Daya Publishing House, New Delhi

229–249

Chapter 18
Organic Farming and Environmental Biodiversity

Y.V. Singh*

Centre for Conservation and Utilization of Blue Green Algae, Indian Agricultural Research Institute, New Delhi – 110012

ABSTRACT

Organic farming systems are diverse and occur throughout the world. They are linked to common objective of economic, environmental and social sustainability. Organic farming production system aims at promoting and enhancing agro-ecosystem health, biodiversity, biological cycles and soil biological activities. Conventional farming has resulted in negative effect on the environment, availability of chemical residues in food and an overall reduced quality of food which have led to a marked increase in various diseases, use of fertilizers and pesticides. Organic farming has the capability to take care of these problems. In several investigations, a higher diversity and abundance of useful flora and fauna was found in organic farming compared to conventional farming. Because of its biodiversity benefits, organic farming offers an important agricultural management option. Organic farming has a high and possibly decisive potential for reversing the dramatic decline of biodiversity. With respect to energy consumption, organic farming performed better than conventional agriculture. Organically managed Orchards and crops have a lower incidence of insect pests than orchards/crops treated with pesticides, mainly because of an increased abundance and efficiency of predators and parasitoids. Besides, the obvious, immediate and positive effects organic farming has on the environment and quality of food, it also greatly helps a farmer to become self-sufficient in their requirements for agro-inputs and reduce production cost. Organic farming area and trade of organic produce is growing rapidly worldwide. Organic farming has the potential of developing huge export and local market and create employment opportunities in India.

* Tel: 011-25848431 (O); 09868416215 (M); Fax:011-25741648; E-mail: yvsingh63@yahoo.co.in; yvsingh_algal@iari.res.in

Keywords: *Organic farming, Conventional farming, Environment, Biodiversity, Micro-organisms.*

Introduction

Modern agricultural methods have brought spectacular increases in productivity of crops and animals (meat and milk). But most farmers in developing countries are poor and marginalized from input and product markets, and some 790 million people still go hungry. Thus an important question arises on the extent to which farmers can improve domestic food production with cheap, low-cost, locally available technologies and inputs. A supplementary issue relates to the extent to which they can do this with methods that do not add to existing environmental harm caused by conventional farming. In contrast to conventional farming systems, organic farming represents a deliberate attempt to make the best use of local natural resources. The aim of organic farming is to create integrated, humane, environmentally and economically viable agriculture systems in which maximum reliance is placed on locally or farm-derived renewable resources, and the management of ecological and biological processes.

Soil is a habitat for plants, animals and micro-organisms. As plants build up organic matter, soil animals feed on them and their debris, whilst microbes decompose the complex organic compounds to their mineral components and to carbon dioxide. A living soil is central to soil fertility because it is the activity of soil organisms that makes available the elements in plant residues and organic debris entering the soil. Part of this material, however, remains in the soil and contributes to its stabilization by humus build-up.

A Living and Healthy Soils for Agriculture

Soils contain enormous numbers of diverse living organisms assembled in complex and varied communities. Soil biodiversity reflects the variability among living organisms in the soil–ranging from the myriad of invisible microbes, bacteria and fungi to the more familiar macro-fauna such as earthworms and termites. Plant roots can also be considered as soil organisms in view of their symbiotic relationships and interactions with other soil components. These diverse organisms interact with one another and with the various plants and animals in the ecosystem, forming a complex web of biological activity. Environmental factors, such as temperature, moisture and acidity, as well as anthropogenic actions, in particular, agricultural and forestry management practices, affect to different extents soil biological communities and their functions.

Various research results show that the activity of micro-organisms is higher in organically than in conventionally-managed soil. As a consequence, in organically managed soils, nutrients are recycled faster and soil structure is improved. For example, Fliessbach *et al.* (2001) found in Switzerland up to 90 percent higher total mass of micro-organisms in organically managed soils. Elmholt (1996) found a higher number and abundance of saprophytic soil fungi with a higher potential of decomposition of organic material. Mycorrhizae, important representatives of soil fungi, live as symbionts with plant roots. The degree of mycorrhizal root colonization was found to be distinctly higher in organic plots as compared to conventional plots (Mäder *et al.*, 2000).

To capture the benefits of soil biological activity for optimum agricultural production we need to adhere to the following ecological principles:

Supply Organic Matter

Each type of soil organism occupies a different niche in the web of life and favours a different substrate and nutrient source. Most soil organisms rely on organic matter for food; thus a rich supply and varied source of organic matter will generally support a wider variety of organisms.

Increase Plant Varieties

Crops should be mixed and their spatial-temporal distribution varied, to create a greater diversity of niches and resources that stimulate soil biodiversity. It is possible by encouraging the presence of a wider variety of organisms, improve nutrient cycling and natural processes of pest and disease control.

Protect the Habitat of Soil Organisms

The activity of soil biodiversity can be stimulated by improving soil living conditions, such as aeration, temperature, moisture, and nutrient quantity and quality. In this regard, reduced soil tillage and minimized compaction–and refraining chemical use–are of particular note.

Organic Farming

Organic farming, through its systemic approach and avoidance of agro-chemicals, prevents natural resource degradation and the loss of land and productive potential. In organic agriculture, nature is both instrument and aim. As organic farmers cannot use synthetic substances (*e.g.* fertilizers, pesticides, pharmaceuticals) they need to restore the natural ecological balance because ecosystem functions are their main productive "input". For example: many unspecific pests like aphids, thrips, whiteflies or spider mites, economically damaging in many crops, can be kept below the economic threshold with naturally occurring or purposely released predators and parasitoids. The former are direct goods and services of hedges, botanically diverse field margins, intercropping or weedy undergrowth, the latter do better when released in botanically and ecologically enriched habitats. The other way to suppress soil-borne pests and diseases in organic agriculture are wide crop rotations. Adhering to such rotations is crucial to provide agro-ecosystem diversity. Soils with a high functional diversity of micro-organisms, which occur very often after decades of organic agriculture practice develop disease suppressive properties and can help to induce resistance in plants(Fliessbach *et al.*, 2001).

Principles of Organic Farming

In organic farming three basic principles are applied to ensure the success of organic farming. First, and perhaps most important, is the principle of interdependency. In organic farming the farm is viewed as an ecosystem, realizing that a change to one part of the system may disrupt the interrelationships on the farm. The second principle is diversity. In organic farming crop rotation is used to maintain habitat as natural as possible. Most organic farmers must keep livestock in order to keep a balance between crops and livestock. The diversity of crops and livestock gives the farmer both flexibility and a diverse income, and also creates an ecosystem that helps to prevent one insect, weed or disease from becoming a problem. The third principle is recycling. In organic farming plant and animal residues are recycled to develop self-sufficiency of nutrients at their own farm.

Need of Organic Farming

Organic farming is an age-old concept in India,which changed during the 1st green revolution. During the last few decades, approach and outlook towards agriculture and marketing of food has seen a quantum change worldwide. The focus is now more on quantity and outer appearance rather than intrinsic or nutritional quality. Pesticide and other chemical residues in food and an overall reduced quality of food have led to a marked increase in various diseases, mainly various forms of cancer and reduced bodily immunity.

This commercialization of farming has also had a very negative effect on the environment. The use of pesticides has led to enormous levels of chemical buildup in our environment, soil, water, air, animals and even in our own bodies. Fertilisers have a short-term effect on productivity but a longer-term negative effect on the environment where they remain for years after leaching and running off, contaminating ground water and water bodies. In the name of growing more to feed the earth, we have taken the wrong road of unsustainability. The effects already show–farmers committing suicide in growing numbers with every passing year; the horrendous effects of pesticide sprays (endosulphan) by a government-owned plantation in Kerala, India some years ago and other parts of country; the pesticide-contaminated bottled water and aerated beverages are only some instances. The bigger picture that rarely makes news however is that millions of people are still underfed and where they do get enough to eat, the food they eat has the capability to eventually kill them. Yet, the picture painted for the future by agro-chemical and seed companies and governments is rosy and bright. Another negative effect of this trend has been on the fortunes of the farming communities worldwide. Despite this so-called increased productivity, farmers in practically every country around the world have seen a downturn in their fortunes.

This is where organic farming comes in. Organic farming has the capability to take care of each of these problems. Besides, the obvious immediate and positive effects organic farming has on the environment and quality of food, it also greatly helps a farmer to become self-sufficient in his requirements for agro-inputs and reduce his costs. Organic farming aspires to a combine mixtures of agronomic, environmental, social and ethic objectives (Stockdale et. al.,2001).

Origin of Organic Farming

The roots of organic farming practices can be traced in European literature back to late nineteenth/ early twentieth century (Conford,1988). In 1924, the Austrian philosopher Dr Rudolf Steiner presented an alternative vision of agriculture arising from his spiritual science of Anthroposophy (Steiner,1924). Slightly later, Dr Hans Muller, working in Switzerland and later in Germany founded a movement for agricultural reforms on the principles of self-sufficiency and economic viability of farms in which soil fertility was to be maintained through crop rotation, careful management and use of animal manures (Zeichaus-Hartelt, 1991). Later Dr Hans Peter Rusch contributed important ideas relating to soil fertility and soil microbiology, which led to the further development of organic-biological agriculture in German speaking countries (Rusch, 1968). In the U.K and other countries the world scientists like Sir George Stapledon and Sir Albert Howard were influenced by the agricultural ideas of Rudolf Steiner, although they did not adopt them directly (Howord,1943; Conford,1988).Stapleton's work with alternate husbandry system and Howard 's work on the role of organic matter in soil and composting provided a powerful stimulus to Lady Eve Belfour (Belfour,1943). In 1939, she began the Haughley Experiment which continued till 1969, to investigate the link between the way food is produced, food quality and human health(Belfour,1976). Howard and Belfour's ideas emphasizing the role of a healthy, fertile soil in the production of healthy crops and livestock and the link with human health were persued in the U.S by the Rodale family, who founded the Soil and Health Foundation in 1947 (Harwood,1990).The early organic movement focused strongly on the issues of human health and promotion of soil fertility through the use of composts and organic fertilizers. Pesticides did not become a major issue in organic agriculture until the publication of *Silent Spring* generated widespread public concern (Carson, 1963).

International Federation of Organic Agriculture Movement(IFOAM) established in 1972 gave an international framework for the discussion and codification of internationally recognized principles of organic farming. FAO and WHO have officially declared that International guidelines on organically produced food products should be considered important for consumer protection and information as they facilitate trade (Yussefi, 2003). The Codex Alimentarius Commission, a joint FAO/WHO food standards programme, the body that sets International food standards, started to develop Guidelines for the production, processing, labeling and marketing of organically produced food in 1991.The Codex guidelines are important for equivalence judgements under the rule of World Trade Organisation(WTO). Modern organic farming represents a merging of a number of different stream of thought (Conford,1988;Harwood,1990 and Tate,1994). There are strong evidences that organic farming systems can deliver agronomic and environmental benefits both through structural changes and the tactical daily management of the farming system.

Benefits of Organic Farming on Soil Biological Activity (FiBL, 2000)

Abundant Arthropods and Earthworms

Organic management increases the abundance and species richness of beneficial arthropods living above ground and earthworms, and thus improves the growth conditions of crops. More abundant predators help to control harmful organisms (pests). In organic systems the density and abundance of arthropods, as compared to conventional systems, has up to 100 per cent more carabids, 60-70 per cent more staphylinids and 70-120 per cent more spiders(Pfiffiner and Niggli,1996). This difference is explained by prey deficiency due to pesticide influence as well as by a richer weed flora in the standing crop that is less dense than in conventional plots. In the presence of field margins and hedges, beneficial arthropods are further enhanced, as these habitats are essential for over-wintering and hibernation. The biomass of earthworms in organic systems is 30-40 per cent higher than in conventional systems, their density even 50-80 per cent higher. Compared to the mineral fertilizer system, this difference is even more pronounced.

High Occurrence of Symbionts

Organic crops profit from root symbioses and are better able to exploit the soil. On an average, mycorrhizal colonization of roots is highest in crops of unfertilized systems, followed by organic systems. Conventional crops have colonization levels that are 30 per cent lower. Even when all soils are inoculated with active micorrhizae, colonization is enhanced in organic soil. This indicates that, even at an inoculum in surplus, soil nutrients at elevated levels and plant protection suppress symbiosis. This underlines the importance of appropriate living conditions for specific organisms.

High Occurrence of Micro-organisms

Earthworms work hand in hand with fungi, bacteria, and numerous other microorganisms in soil. In organically managed soils, the activity of these organisms is higher. Micro-organisms in organic soils not only mineralize more actively, but also contribute to the build up of stable soil organic matter. Thus, nutrients are recycled faster and soil structure is improved.

Microbial Carbon

The total mass of micro-organisms in organic systems is 20-40 per cent higher than in the conventional system with manure and 60-85 per cent than in the conventional system without manure. The ratio of microbial carbon to total soil organic carbon is higher in organic system as compared to conventional systems. The difference is significant at 60 cm depth while at 80 cm depth, no difference is observed. Organic management promotes microbial carbon and soil carbon sequestration potential.

Enzymes

Microbes have different functions in the soil system and level of soil enzymes indicate these functions. The total activity of micro-organisms can be estimated by measuring the activity of a living cell-associated enzyme such as dehydrogenase. This enzyme plays a major role in the respiratory pathway. Proteases in soil, where most organic N is protein, cleave protein compounds. Phosphatases cleave organic phosphorus compounds and thus provide a link between the plant and the stock of organic phosphorus in the soil. Enzyme activity in organic soils is markedly higher than in conventional soils. Microbial biomass and enzyme activities are closely related to soil acidity and soil organic matter content.

Wild Flora

Large organic fields (over 15 ha) featured flora six times more abundant than conventional fields, including endangered varieties. In organic grassland, the average number of herb species was found to be 25 percent more than in conventional grassland, including some species in decline. Weeds (often sown in strips in organic orchards to reduce the incidence of aphids) influence the diversity and abundance of arthropods and flowering weeds are particularly beneficial to pollinators and parasitoids.

High-energy Efficiency

Organic land management allows the development of a relatively rich weed-flora as compared to conventional systems. The presence of versatile flora attracts beneficial herbivores and other air-borne or above-ground organisms. Their presence improves the nourishment of predatory arthropods. When comparing diversity and the demand of energy for microbial maintenance (as indicated by the metabolic quotient), it becomes evident that diverse populations need less energy per unit biomass. A diverse microbial population, as present in the organic field plots, may divert a greater part of the available carbon to microbial growth rather than maintenance. In agricultural practice this may be interpreted as an increased turnover of organic matter with a faster mineralization and delivery of plant nutrients. Finally, more organic matter is diverted to build-up stable soil humus.

Erosion Control

Organic soil management improves soil structure by increasing soil activity and thus, reduces erosion risk. Organic matter has a positive effect on the development and stability of soil structure. Organic matter is adsorbed to the charged surfaces of clay minerals. The negative charge decreases with increasing particle size. Silt is very susceptible to erosion since it is not charged, but organic matter layers on the silt surface favor aggregates with silt too.

Management Practices in Organic Farming

Plant Nutrition

In organic farming we constantly work to build the healthy soil that translates into healthy plants. Crop plants removes varying amounts of different nutrients from soil (Table 18.1) and to compensate the loss from soil use of organic materials is essential. However in conventional farming we give only a few nutrient as those are available in fertilizers. We need to plan well in advance of planting, how the plants will be nourished. We must ensure that soil is rich in organic matter and has all the nutrients that the plants will need. In organic farming we feed the soil micro and macro–organisms, which, are the external digestive system that possesses organic matter, delivering a smorgasbord of minerals, vitamins and other nutrients to the crop at a metered pace.

Table 18.1: Average Nutrient Uptake by Crops per Tonne of Economic Yield

Nutrient	*Unit*	*Uptake by Crops per Tonne of Economic Yield*				
		Paddy	*Wheat*	*Ground Nut*	*Tea*	*Tobacco*
N	kg	20.0	25	58.0	110.0	16.5
P_2O_5	kg	11.0	9	20.0	37.8	4.4
K_2O	kg	30.0	33	30.0	44.0	26.2
S	kg	3.0	4.7	5.7	10.1	2.8
Ca	kg	7.0	5.3	28	31.0	22.6
Mg	kg	3.0	4.7	7.3	10.7	4.8
B	g	15.0	48	133	200	96
Cu	g	18.0	24	15	240	11
Fe	g	153	624	1500	900	692
Mn	g	675	70	118	59	132
Mo	g	2.0	2	4	NA	0.6
Zn	g	40	56	28	240	21

Source: Tandon (2002).

The food that soil organisms need to do their job comes in the form of organic matter. There are many effective practices like green manuring (Mandal *et al.*, 1992), use of biofertilizers, *Azolla sp.* (Singh,1961; Singh and Bisoyi, 1989; Singh and Mandal, 1987, 2000) use of crop rotation, incorporation of crop residues etc (Singh and Mandal, 1987, 2000). The green manuring and biofertilizers not only add nutrients such as nitrogen to the soil, but also help prevent weeds and increase organic matter to feed soil microorganisms. In organic farming we add the natural minerals that plants need to grow and that help improve the soil's consistency *e.g.* liming is needed to adjust the soil's pH. Most of the manures used in organic farming are recycled by-products from other enterprizes that would otherwise go to waste. Farmers can make compost out of animal manures and mushroom compost. Vermi-compost is also becoming popular in organic farming. Before compost is applied to the fields, it should be properly decomposed to kill any unwanted bacteria and weed seeds.

Freyer (1997) found in Switzerland that only 14 percent of all organic farms have an N-surplus, and only 1.5 percent had a P-surplus. Most of the organic farms have a negative N–and P-balance. Results from different European countries comparing phosphorous and potassium balances of conventional and organic farms concluded that the phosphorous and potassium surpluses of organic farms are significantly lower than on conventional ones.

Weed Management

No chemical herbicides is used in organic farming. Here weeding is done by machine or by hand. When the field is fallow, a cover crop may be planted to suppress weeds and build soil quality. Drip irrigation may be adopted wherever possible, to restrict weed growth by distribution of water to the plant line only. Another tool that sometimes may be used is a "flame weeder," a propane device that attaches to the back of a tractor and directs flames toward the ground. In this method land is irrigated to germinate weeds, and then "flame" is used to kill weeds and weed seeds before planting.

Pest Management

In Organic farming we try to anticipate in advance where and when different pests will be present, and adjust their planting schedules and locations as much as possible to avoid serious pest problems.Here main strategy is to combat harmful pests and to build up a population of beneficial insects, whose larvae feed off the eggs of pests. The key to building a population of beneficial insects is to establish borders around fields planted with blends of flowering plants that the beneficial insects particularly like. These are called "host crops." Periodically beneficial insects are released into the fields, where the host crops serve as their home base and attract more beneficial insects over time. But, if there is a pest outbreak that cannot be handled by beneficial insects, we sometimes use natural or organically approved insecticides like neem products (1 per cent Nethrin/Nimbecidine) which have low toxicity to people and other animals and low persistence in the environment. We can use Trichograma parasite and Pheromone traps also as per requirement to control pest problems in different crops. In the field experiments conducted at IARI, New Delhi on organic farming there was no serious attack of any insect pest or disease in organically grown rice or wheat crop(Singh and Singh,2006).

Disease Management

One of the biggest rewards of organic farming is healthy soil that is alive with beneficial organisms. These healthy microbes, fungi and bacteria keep the harmful bacteria and fungi that cause disease in check(Singh and Singh,2006). Organic farmers, working with nature, build soil that protects their crops from disease. They also try to be diligent about crop rotation. They do not plant the same crop in the same location time after time because that encourages the build-up of diseases and pests that plague that particular crop.

Organic Farming and Abiotic Resources

Soil

Soil is one of the chief natural resource and central basis for all agricultural activities. Since organic farmers cannot compensate for a loss in soil fertility by inputs of synthetic nutrients, the building and maintenance of soil fertility is a central objective of organic farming(Lampkin, 1990; IFOAM, 2000).

Organic Matter Content

In Organically farmed plots soil organic carbon content is higher than on conventional ones. Fertilization in organic farming is done by organic substances such as farmyard manure, compost, green manure, plant residues and commercial organic N-fertilizers. Consequently, there is an extensive supply of organic matter passing through aerobic decomposition processes. Mineralization and decomposition processes are influenced by humidity, temperature and oxygen. Under humid tropical conditions these processes run faster and all year round, whereas under temperate conditions they are slower and come to a halt during the colder months. Soil type also plays a role. Sandy soils dry out quickly, slowing down the decomposition process, ferralitic soils on the other hand are generally not very fertile, but they encourage fast decomposition and the building-up of stable organic matter.

Soil Micro-organisms

In contrast to conventional agriculture, organic farmers depend more on a high and sustained supply of organic substances including crop rotations, catch crops, green and animal manure. In different experiments on organic farming higher soil microbial population(+30 to 40 percent) and their activity(+ 50 to 80 percent) as compared to conventional agriculture has been reported at different locations. At IARI, New Delhi also Microbial population (Actinomycetes, Bacteria, Fungi and BGA) in soil was found to be enhanced due to the application of organic amendments in comparison to control and recommended fertilizer application which accordingly resulted in a notable enhancement in dehydrogenase enzyme activity (Table 18.2).

Table 18.2: Effect of Different Organic Treatments on Microbial Population (CFU/gm of soil) and Enzymatic Activity at Mid Crop Stage of Rice Grains During Kharif, 2005

Tr.No.	*Treatments*	*Actinomycetes x 10^3*	*Bacteria x 10^3*	*Fungi x 10^3*	*BGA x 10^3*	*Activity* (mgm)*
1	Azolla [A]	332	369	31	59	131.11
2	BGA [B]	341	356	63	74	124.13
3	FYM [F]	261	322	51	61	110.04
4	Vermicompost [V]	276	365	43	48	108.76
5	A+B	287	380	32	23	121.20
6	A+F	279	364	33	42	134.74
7	A+V	195	321	32	35	113.38
8	B+F	267	386	34	55	113.45
9	B+V	243	364	37	68	127.62
10	F+V	267	368	34	57	112.25
11	A+B+F	256	376	41	78	122.14
12	A+F+V	380	402	65	98	124.74
13	B+F+V	376	378	75	86	132.52
14	A+B+F+V	301	334	61	87	125.99
15	$N_{80}P_{30}K_{30}$	164	332	69	23	101.23
16	$N_0P_0K_0$	160	312	29	12	101.46

*Dehydrogenase enzyme activity.

Ground and Surface Water

The detrimental effects of intensive agriculture on ground and surface water are largely due to erosion and to nitrate and pesticide pollution. The chief threats to water quality posed by agriculture are: high organic fertilization levels in combination with high stocking rates, the excessive application of mineral N-fertilizers; the lack of a protective soil cover; narrow crop rotations and frequent tillage; high levels of available nitrogen after harvest, and contamination of water with synthetic pesticides.

As organic farming uses no synthetic pesticides, there is no risk of synthetic pesticide pollution of ground and surface waters. As regards nitrate leaching, nitrate leaching rates per hectare were significantly lower in organic farming than in conventional agriculture systems.

Organic Farming and Biodiversity

For hundreds of years, agriculture has contributed substantially to the diversity of species and habitats, and agriculture has formed many of today's landscapes. Over the last century, however, modern intensive agriculture, with its high input of synthetic pesticides and fertilizers and mono crop specialization, has been detrimental to the diversity of genetic resources of crop varieties and livestock breeds, to the diversity of wild flora and fauna species and to the diversity of ecosystems.

Organic farming is dependent upon stabilizing agro-ecosystems, maintaining ecological balances, developing biological processes to their optimum, and linking agricultural activities with the conservation of biodiversity. Wild species perform a variety of ecological services within organic systems: pollination, pest control, maintenance of soil fertility. Thus, higher levels of biodiversity can strengthen functions essential for farming systems and therefore, agricultural performance. Enhancing functional biodiversity is a key ecological strategy to bring sustainability to production on organic farms. Organic systems also use substantially fewer external inputs and do not use synthetic chemical fertilizers, pesticides, genetic modified organisms and pharmaceuticals. Instead, systems are designed to work in harmony with nature in order to determine agricultural yields and disease resistance. By respecting the natural capacity of plants, animals and the landscape, organic agriculture aims to optimize quality in all aspects of agriculture and the environment.

Biological pest control on organic farms, for example, relies on maintaining healthy populations of pest predators and parasitoids. A study in California comparing conventional with organic tomato fields showed a higher abundance of natural enemies and greater richness of species in organic tomato fields. There was no significant difference for any type of damage to tomato foliage or fruit, showing that the organic system achieves the same levels of pest control without having to apply synthetic chemical pesticides (Letourneau and Goldstein, 2001).

Numerous scientific studies, mainly from Europe and North America, give evidence that on organic farms biodiversity is higher than that on conventional farms. Biodiversity is generally assessed at three distinct levels(European Environment Agency, 2002).

Genetic Diversity

The variation between individuals and between populations within a species;

Species Diversity

The different types of plants, animals and other life forms within a region or community;

Ecosystem Diversity

The variety of habitats found within an area (grassland, marsh, and woodland for instance).

Organic Farming Nurtures Soil Biodiversity

Scientific research has demonstrated that organic farming significantly increases the density and species of soil's life. Suitable conditions for soil fauna and flora as well as soil forming and conditioning and nutrient cycling are encouraged by organic practices such as: manipulation of crop rotations and strip-cropping; green manuring and organic fertilization (animal manure, compost, crop residues); minimum tillage; and of course, avoidance of pesticides and herbicides use.Twenty years of scientific research have demonstrated that organic agriculture significantly increases the density and species richness of indigenous invertebrates, specialized endangered soil species, beneficial arthropods, earthworms, symbionts and microbes(FiBL, 2000). The living soil generates following ecological services:

Soil Forming and Conditioning

Invertebrates (*e.g.*, earthworms and termites) decompose plant litter and create conditions that allow nutrients, oxygen and water to circulate;

Waste Disposal

Micro-organisms (*e.g.*, bacteria and fungi) reduce organic detritus to elemental nutrients and recycle nutrients and detoxify ecosystems;

Soil Stabilization

Invertebrates and micro-organisms influence the physical, chemical and biological characteristics of soils and thereby play a key role against erosion and floods;

Carbon Sequestration

The higher biomass and diversity of microbial population in organic systems contributes to the carbon retention potential of soils.

Organic Farming and Agroecosystems

Natural disease resistance and pest predation are strengthened in organic systems. Crop rotation is considered the cornerstone of organic management, functioning as a tool for pest management and soil fertility. This, together with inter-cropping, integrated crop-tree-animal systems, the use of traditional and under-utilized food and fodder species and the creation of habitats attracts pest enemies and pollinators and spreads the risk of crop failure across the agro-ecosystem. Agro-biodiversity is conserved and developed through the use of locally adapted landraces, improvement of genotypes of many plant varieties and animals and following functions(IFOAM, 2000).

Nitrogen Cycling

Atmospheric nitrogen is fixed by legumes and other nitrogen-fixing plants (*e.g.*, Azolla) which are used during rotations;

Symbiosis and Parasitism

Symbionts (*e.g.*, rhizobia and mycorrhiza) play a most important role in absorbing nutrients and reducing pathogen invasions. Parasitism is used in the biological control of insects;

Predation

Inter-specific competition between predator and prey populations keeps pest in check;

Pollination

Enhanced habitats and absence of chemical use on organic farms reverse the trend of pollinator population decline. One third of agricultural crops and the majority of flowering plants are pollinated by insects (*e.g.*, bees, butterflies, beetles) and other animals (*e.g.*, bats).

Organic Farming and Nature Conservation

The maintenance of natural areas of vegetation adjacent to crops and plant corridors is common in organic systems, providing alternative food and refuge for many insect predators, wild flora, birds and other wildlife. The absence of pesticide drifts and herbicides and on-farm integration of natural habitats (*e.g.*, productive perennial plants, hedgerows) and other structures (*e.g.*, stepping stones and corridors for migrating species) attract new or re-colonizing species to the area. Ultimately, the diversity of landscape and wildlife brings people in the form of eco-tourism, providing an important source of off-farm income (Mc Neely J.A. and S.J. Scherr, 2001).

Organic Habitats Conserve Wildlife

Studies have shown that organic farming conserve weed species at risk of extinction. On-farm diversity and biomass of arable flora was found to be higher in organic fields (*e.g.*, vineyards and olive groves in Greece).The abundance of food sources and habitats attracts micro and macro fauna to organic farms. Surveys have found that the quantity of organic land is very important for migratory birds. The abundance of birds in organic shade coffee is 90 per cent more than in sun-grown coffee plantations. Organic agriculture has been found to have positive effects on declining ground-breeding bird species (*e.g.*, skylark, and yellow wagtail). The use of organic farming is encouraged for wildlife conservation where agriculture is a dominant land-use in buffer zones.

Diversity of Genetic Resources in Organic Farming

Since organic farms are mostly mixed farms, integrating animal husbandry with crop production, using vast and diverse rotations, intercrops and green cover crops, and maintaining soil fertility by cultivating nitrogen fixing legumes, they display a higher diversity of domesticated species than conventional farms. Hausheer *et al.* (1998) evaluated crop rotations on 110 organic, integrated and conventional farms in a Swiss pilot farm project and found more diverse crop rotations (4.5 different crops in organic as opposed to 3.4 different crops in integrated farming) and a higher number of crops, including perennials, vegetables and herbs (10.2 in organic and integrated farms; 7.4 in conventional farms).

The Maintenance of Genetic Resources

Today, the adoption of high-yielding, uniform cultivars and varieties has led to a considerable reduction in the number of plants and animals used in agriculture. Only 120 cultivated plant species and 14 mammalian and avian species provide 90 percent of human food supply.

Evidence of the trend towards monoculture and uniformity is given by the fact that in India, under the Green Revolution, the number of cultivated rice varieties has decreased from more than 100 000 to 10; also, 50 percent of the goat breeds, 20 percent of the cattle breeds and 30 percent of the sheep breeds are in danger of extinction(Shiva,2001). In Mexico, only 20 percent of the maize varieties reported in 1930 are now known. In China, nearly 10 000 wheat varieties were used in production in 1949; by the 1970s, only about 1000 remained in use(FAO,1998). This trend is just as visible in animals: 740 animal breeds became extinct during the twentieth century. Currently, 1350 breeds face extinction, with two breeds being lost each week(FAO,2000).

Organic producers look for productive varieties suited to their local climatic and soil conditions and that are not susceptible to disease and pest attack. Organic farming standards recommend the cultivation of site adapted crop varieties(IFOAM,2000) characteristics often found in the older native cultivars. This, however, does not necessarily mean that organic agriculture sets narrow limits to the use of modern maximum yield varieties, which are often chosen for pest/disease resistance purposes. Still, the preservation of native varieties and breeds is an important initiative of the organic movement, but their actual use depends on individual farmers.

Genetic Engineering and Agricultural Biodiversity

Apart from the adoption of high-yielding, uniform cultivars, a further possible threat to genetic diversity, and biodiversity in general, are the side-effects of the release of genetically engineered organisms into the environment.

Genetically engineered plants designed to control pests can have negative side-effects on beneficial insects and further non-target organisms as well. Oilseed rape with genetically induced resistance to insects has been reported to damage beneficial insects such as honey bees(Crabb,1997). The use of herbicide resistant plants can result in greater use of herbicides, increasing the negative effects of intensive farming on natural biodiversity. Furthermore there is the danger that transgenic plants could become feral and thus suppress indigenous flora. Feral oilseed rape populations in Canada are resistant to three herbicides and have become one of the most troublesome weeds(Spears,2001).

Concerns have been raised about the potential effects of transgenic introductions on the genetic diversity of crop landraces and wild relatives in areas of crop origin and diversification, as this diversity is considered essential for global food security. As organic farming is dependent upon maintaining ecological balances and diverse agro-ecosystems, genetic engineering is a contradiction to the principal aims of organic agriculture. Organic agriculture does not allow genetic engineering in its standards as genetic engineering focuses on genetic makeup without taking into account the complete organism or farming system in which the organism functions.

Floral Diversity

Today, the diversity of the typical wild flora on arable fields, which is the main habitat for a wide range of species, is at risk. In conventional agriculture weeds are considered competitive to the crop and are eliminated by herbicides. In organic systems some of the accompanying plants are desired to a certain degree and considered useful, as they provide a wide range of ecological services. These services include protection from soil erosion, providing shelter and alternative food resource for beneficial organisms and pollinators.

In most cases, floral species are conserved better in organic farming than conventional agriculture. A survey comparing organic with conventional fields showed that in the organic system the share of the fields with endangered floral species after 27 years was 79 per cent as opposed to 81 per cent, showing that the share had not changed much. In the conventional fields the rate had dropped from 61 per cent to 29 per cent (Frieben,1997).The higher floral diversity and abundance in organic arable fields is generally due to the ban on synthetic N-fertilizers and herbicides. The limited availability and input of nitrogen, the application of mechanical and thermal weed control and more diverse crop rotations and a higher crop diversity lead to more favourable conditions for many wild plant species.

Effects on Faunal Diversity

Weeds influence the diversity and abundance of arthropods (*e.g.* beetles, ants and spiders), acting as natural food resource and shelter. Weeds like Umbelliferae, Leguminosae and Compositae play an especially important ecological role as they provide food and thus improve reproduction of many arthropod species(Altieri,1999). Research carried out on tomato plots on the effects of weed control on surface-dwelling arthropod species found the abundance of species is clearly influenced by weed biomass. Species numbers were lowest where mulching with rye straw was controlling the weeds. However, removing weeds within 20 cm of each plant reduced weed biomass but retained higher arthropod populations than in the plots treated with herbicide or mulch (Yardim and Edwards,2000).With regard to pollinators like butterflies, bees and wasps (Feber et al.1997) the organic system was favourable. Flowering plants are also important for many beneficial arthropods such as predators and parasitoids(Hald,1999).

In order to diversify the farming system and attract beneficial arthropods and pollinators, wild flower strips are sown in organic agriculture orchards. In a Swiss organic orchard, it was found that the strip management favoured beneficial insects and spiders, which reduced the density of aphids.

The density of aphids was reduced due to higher mortality caused by increased numbers of predators feeding on aphids(Wyss,1994;Wyss *et al.*, 1995). Measures aimed at managing appropriate habitats and thus increasing floral and structural diversity is a key strategy for improved natural pest control.

Earthworms and Organic Farming

Earthworms are highly suitable bio-indicators of soil fertility, and they are known for their sensitivity to synthetic pesticides and to many agricultural practices. Due to their biology, earthworm populations can indicate the structural, microclimatic, nutritive and toxic situation in soils. In conventional agriculture, earthworms are affected by the use of harmful pesticides and intensive soil cultivation.

Earthworms generally increase nutrient cycling rates. Their casts greatly help to improve soil structure and have high concentration of nutrients in an accessible form to plants. The burrowing activity of earthworms enhances aeration, porosity and drainage of the soil, all of which are important factors in the development of healthy and well-developed crop root system. Earthworms also play an important role in pest and disease control, including the reduction of leaf miner pupae and scab pathogens in orchards(Kennel,1990).

Many investigations in Europe and North America provide evidence that generally, organically managed soils exhibit a higher abundance and species number than conventionally-managed plots or farms(Pfiffiner,1997). The biomass of earthworms in the organic system in the Swiss DOC long-term trial was 30 to 40 percent higher than in the conventional system, the number of individuals even 50 to 80 percent higher(Pfiffiner and Mader,1997).

Organic matter constitutes an important food source for earthworms and can be maintained by an appropriate fertilizing and crop rotation system. Investigations have shown that earthworms also benefit greatly from green manuring and the planting of grass-clover in the crop rotation(Pfiffiner,1997).

Effects of Organic Farming on Environmental Quality

Overall, organic farming is much better for the environment than conventional farming. One of the greatest environmental problems today is energy consumption, and organic farming uses far less energy than does conventional farming. As a matter of fact, energy efficiency is around seven percent greater for the organic farming system. Other positive environmental aspects of organic farming include the use of much less fertilizer, and the complete avoidance of synthetic fertilizers, which are harmful to soil, water, animals, and people. Also, the nitrate content of organic fields is significantly lower than on conventional farms due to the absence of soluble fertilizers. Too much nitrogen can throw the soil community out of balance and lead to algal blooms in water that suffocate other aquatic organisms. In fact, algal blooms and dead zones are now a regular feature of coastal life in many places around the world because of the impact of conventional farming. Organic fields also promote biodiversity–a great variety of animal and plant species–which is essential to the future of all species on Earth. Furthermore, organic farmers focus on preserving the habitats of all species and their surrounding environments, including the air and water. Finally, organic farming releases much less carbon dioxide than does conventional farming. Carbon dioxide is the leading greenhouse gas that causes global warming.

International Guidelines on Organic Farming

The formation of the International Federation of Organic Agriculture Movement (IFOAM) in 1972 gave an international framework for the discussion and codification of internationally recognized principles of organic farming. FAO and WHO have officially declared that International guidelines on organically produced food products should be considered important for consumer protection and information as they facilitate trade. The Codex Alimentarius Commission, a joint FAO/WHO food standards programme, the body that sets International food standards, started to develop Guidelines for the production, processing, labeling and marketing of organically produced food in 1991.The Codex guidelines are important for equivalent judgements under the rule of World Trade Organisation(WTO).

Organic Certification

IFOAM is International umbrella organisation of Organic Agriculture organizations. It has 750 members in 200 countries. USDA–accredited and many pvt./govt. certification agencies are working for certification of organic produce.

In India, Ministry of Commerce launched National Programme of Organic Production (NPOP) in March 2000 to establish national standards for organic production (NSOP) which could be sold under " India organic" logo. To ensure implementation of NPOP, the National Accreditation Policy and Programme (NAPP) was formulated in May 2001. Agencies engaged in Inspection and certification should be accredited by NAPP. APEDA is having important role in promotion of organic products in India.

Quality of Organic Food

There is a growing demand for organic foods driven primarily by consumers' perceptions of the quality and safety of these foods and to the positive environmental impact of organic agriculture practices. This growth in demand is expected to continue in the foreseeable future. In view of consumer expectations, it is important that governments, industry and consumer groups carefully examine issues related to organic food quality and safety and make whatever interventions may be necessary to ensure an appropriate level of consumer protection.

The establishment of regional (EC) and international (Codex) guidelines for the production, processing, labelling and marketing of organic foods has been an important step in the international harmonization of requirements for organic foods. Harmonization is necessary to ensure that consumers receive what they expect, regardless of the origin of the organic food. This important work should be continued in order to broaden the scope of organic guidelines; ensure that the international guidelines reflect the existing differences in consumer perception in different regions; and, continue to adapt the guidelines according to changes and developments in organic production systems.

The "organic" label is not a health claim, it is a process claim. Pesticide residue persistence in agricultural produce and food commodities is a big threat to human health. In large no. of conventionally grown fruit and vegetable samples pesticide residue have been found above maximum residue limit (CCSHAU,2003). But in organic farming there is the commitment of no chemical/pesticide use in the organic food chain all the way to the consumer. Nevertheless, in view of the reduced use of chemically synthesized inputs in organic farming, many studies have been carried out to investigate safety and quality implications of the production system. It has been demonstrated that organically produced foods have lower levels of pesticide and veterinary drug residues and, in many cases, lower nitrate contents. Animal feeding practices followed in organic livestock production, also lead to a reduction in contamination of food products of animal origin. In addition, the "organic" label provides assurance to consumers that no food ingredient has been subject to irradiation and that GMOs have been excluded. No clear trends have been established in terms of organoleptic quality differences between organically and conventionally grown foods. Besides, novel methods and approaches may be required to guarantee freshness and safety during storage and transport chains in the marketing of organic foods. This challenge requires intensive research and appropriate support and incentives, but will eventually benefit conventional as well as organic production.

Where to Buy Organic Food?

More and more businesses, including restaurants, co-operatives, and even supermarkets, are making organic foods available to consumers. Organic farming systems operate at an financial disadvantage relative to other producers, because they seek to deliver environmental and sustainability benefits and internalize the externalities within the farming system(Pretty,1998). Many people are surprised to find out how much more expensive organic foods are than non-organic foods. But organically produced foods have to meet stricter regulations governing from production to harvesting than do conventional foods(Tate,1994). So farmers must get premium price of their produce as the end result is-healthier and more environmentally friendly food-and it is well worth the higher price tag.

Present Status of Organic farming

India was having 1,426 certified organic farms producing approximately 14,000 tons of organic food/produce annually in mid 2003 (SOEL Survey,2004),. There are a large number of farms in India which have either never been chemically–managed/cultivated or have converted back to organic farming because of their farmers' beliefs or purely for reason of economics. These thousands of farmers cultivating hundreds of thousands of acres of land are not classified as organic though they are. Their produce either sells in the open market along with conventionally grown produce at the same price or sells purely on goodwill and trust as organic through select outlets and regular specialist bazaars. These farmers may never opt for certification because of the costs involved as well as the extensive documentation that is required by certifiers.

Organic trade has become a growing reality and the growth is naturally heading towards what can be called a boom. According to SOEL Survey (February,2003), almost 23 million hectares(m ha) are managed organically worldwide. Currently, the major part of this area is located in Australia (10.5 m ha), Argentina (3.2m ha) and Italy (1.2m ha) (Hussein, 2003). Recent years have seen very rapid growth in organic farming, particularly in Europe and United States but also in many other regions of the world including China, Latin America, Australia and Africa. In European Union, area under policy supported organic production increased to 2.1 per cent of total utilizable agricultural area (UAA) in 1998 from 0.1 per cent UAA in 1985 which is a 30-fold increase in 13-years (Limpkin, 1999). Several countries have now achieved 3-10 per cent of their agricultural area managed organically, and in some cases more than 30 per cent on regional basis. But many countries are still at or below 1 per cent level and India is in this category where only 0.03 area is under organic agriculture.

A large, but very competitive western European market exists for organically produced food, so there is a need to create an environment conducive organic production sector. It is clearly foreseeable that in order to maintain reliable organic certification and verification and ensure adherence to all safety standards and special regulations, both the private and public sectors have to make tremendous efforts in research, training/education and marketing.

Scope of Organic Farming in India

In India fertilizer use on dry lands covering 65 per cent area is always low as chemical fertilizers require sufficient water to respond. Pesticide use also in these lands is low as the economics of these hardy or "not-so profitable" crops will not permit expensive inputs. These areas are at least "relatively organic" or perhaps even "organic by default". While neither of these terms necessarily denotes a healthy farm or a recommended agriculture system, it would at least imply a non-chemical farm that can be converted very easily to an organic one providing excellent yields and without the necessity and effort of a conversion period. Uttaranchal government has already declared Uttaranchal as "Organic" state and created special Export Zones like Basmati Export Zone is created by comprising Dehradun, Haridwar, Udham Singh Nagar and Nainital districts. A large area of North-Eastern states and many other areas of other states may be developed as commodity based "organic" production areas. The development of organic farming systems and markets illustrates that this is a valid alternative approach to intensification. With great political will and investment in research, more of this potential could be realized(Stockdale et.al,2001).

References

Acquistucci R (1999) Technological and Nutritional Quality Indexes of Wheat Produced with Organic Agricultural Techniques. National Institute of Nutrition, Laboratory of Cereal Studies.

Andersson A and Bergh T (1995) Pesticide residues in some vegetables and berries. Vor foda/Our Food, *Journal of Swedish National Food Administration*, 47 (8) : 22-24).

Altieri MA (1995) The ecological role of biodiversity in agroecosystem. *Agric. Ecosystem Environ.* 74: 19-31.

Belfour EB (1943) *The living soil.* Faber and Faber. London.

Belfour EB (1976) In: *"The living soiland the Haughley Experiment"*. Universe Books, New York.

Braae L, Nohr H and Petersen BS (1988) Bird faunas of conventionally and ecologically farmed land. *Miljoprojeckt* 120

Bognale UD and Bharathi SG (1990) Algal flora of cultivated soils of Karnataka state, India, *Phykos* 19: 95-113.

CCSHAU (2003) CCS Haryana Agricultural University's Monogram on Emerging challenges to Haryana's agricultural economy vis-à-vis *Diversification in Agriculture.* 1-52.

Carson R (1963) *"Silent Spring"*. Humish Hamilton, London. 304 pp.

Chakraborty PK, Mandal LN and Majumdar A (1988) Organic and chemical sources of nitrogen: its effect on nitrogen transformation and rice productivity under submerged conditions. *J. of Agricultural Sciences*, U. K. 11 (1) : 91-94.

Christensen KD, Jacobsen EM and Nohr H (1996) Acomerative study of bird faunas in conventionally and organically farmed areas. *Dansk Ornitol. Forenings Tidsskrift*. 90: 21-28

Conford P. E. (1988) In: "The Organic Tradition: An Anthology of Writings on organic Farming 1900-1950". Green Classics, Green books, Bideford, Devon, UK.

Conklin N and Thompson G 9 1993) Product quality in organic and conventional produce: is there a difference? *Agribusiness*: 9; 3; 295-307

Couzin J (1998) Cattle Diet Linked to Bacterial Growth. *Science* Vol. 281, 1578-1579

Dahlsted and Dlouhy (1995) Vor föda/Our Food, *Journal of Swedish National Food Administration*, 47 (8): 39-41

Darrington H (1999) Going Organic. *Food Manufacture*: 74; 4; 38-39.

FAO (1997) *Biological Farming in Europe*. REU Technical Series 54. Rome.

FAO (1999) Organic Agriculture. 15th Session of the Committee on Agriculture, Rome 25-29 January 1999.

FAO/WHO (1999) Codex Alimentarius Commission Guidelines for the Production, Processing, Labelling and Marketing of organically produced foods *cac/gl 32–1999*.

Feber RE, Johnson PJ, Firbank LG, Hopkins A and Macdonald DW (1997) The effect of organic farming on pest and non-pest butterfly abundence. *Agric. Ecosystems Environ*. 64: 133-139.

Fischler F (1999) Organic Farming and the new Common Agricultural Policy. Proceedings of Organic Farming in the European Union: Perspectives for the 21st Century. 27-28 May 1999, Baden, Austria.

Foster C and Lampkin N (1999) In: *Organic farming in Europe: Economics and Policy*. University of Hohenheim, Hohenheim.

FiBL (2000) *Extracted from Organic Farming Enhances Soil Fertility and Biodiversity*: Results from a 21 Year Old Field Trial, Research Institute of Organic Farming (FiBL), Frick, Switzerland.

Gupta AB and Shukla AC (1967) Studies on nature of algal growth promoting substance and their influence on growth, yield and protein content of rice plants. *J. Sci. Tech. Kanpur* 5: 162-163.

Haglund and Johansson L (1995) Sensorisk undersökning av morötter och tomater/Sensory testing of carrots and tomatoes. Vr föda/Our food, *Journal of Swedish National Food Administration*, 47 (8) : 52-55.

Harwood, RR (1990) A history of sustainable agriculture. In: "*Sustainable agriculture systems*". (C. A. Edwards, R. Lal, P. Maddee, R. H. Miller and G. House, Eds.), pp. 3-9.

Howord A (1943) *An Agricultural Testement*. Oxford University Press, London.

IFOAM (1998) In "*IFOAM basic standards for organic production for organic production and processing*" IFOAM Publications, Germany

IFOAM (2000) *Statistics Central and Eastern Europe*. IFOAM Publications, Germany

IFOAM (2000) The Relationship between Nature Conservation, Biodiversity and Organic Agriculture. *Proceedings of an international workshop held in Vignola*, Italy, by IUCN, IFOAM, WWF and AIAB. Stolton S., B. Geier and J. A. McNeely (eds.).

ITC (1999) *Product and Market Development. Organic Food and Beverages*. World Supply and Major European Markets. Geneva.

Jörgensen K, Rasmussen G and Thorup I (1995) Ochratoxin A in Danish cereals 1986-92 and daily intake by the Danish population. *Food Additives and Contaminants*.

Kuiper-GoodmanT (1998) Food safety: mycotoxins and phycotoxins in perspective, in *Mycotoxins and phycotoxins–Developments in chemistry, toxicology and food safety*, eds M. Miraglia, H. van Egmond, C. Brera and J. Gilbert. IUPAC.

Kuhn F (1999). Bestimmung von Trichothecenen im Weizen aus verschiedenen Anbausystemen mittels HPLC-MS. In *Organische Analytische Chemie*, Universität Basel. Basel, pp. 60.

Lampkin NH (1990) In "*Organic farming*" 715pp. Farming press, Ipswich, UK.

Lampkin NH (1999) Organic farming in the European Union-overview, policies and perspective. In: *Organic farming in the European Union-Perspective for the 21st century*. "Proceedings EU commission/ Austrian Government conference. Baden/Vienna. May 1999.

Langlais C. 1999 Des pesticides dans les produits bio, Biofil no. 3.

Lecerf JM (1995) L'agriculture biologique: intér?t en nutrition humaine? *Cah. Nutr. Diet.*, 30, 349-357.

LockeretzW (1990) Major issues confronting sustainable agriculture. In: *"Sustainable agriculture in temperate zone"*. (C. A. Francis, C. B. Flora, and L. D. King Eds.) pp. 423-438. John Wiley and Sons, New York.

Lyon (2000) Proceedings of "L'agriculture biologique face ? son developpement–Les Enjeux futurs", 6-8 Dec. 1999, Lyon, France.

Mader P, Pfiffiner L, Niggli U, Balzer F, Plochberger A, Velimirov A and Besson JM (1993) Effect of three farming systems (bio-dynamic, bio-organic, conventional) on yield and quality of beetroot *(Beta vulgaris L.)* in a seven year crop rotation. *Acta Hort* 339: 11-31

Mc Neely JA and Scherr SJ (2001) Common Ground, Common Future. How Ecoagriculture can Help Feed the World and Save Wild Biodiversity. *IUCN and Future Harvest*, May 2001.

Mandal BK, Subudhi UK and SinghYV (1992) Response of rice *(Oryza Sativa)* to green leaf manure and prilled urea. *Indian J. of Agricultural Sciences*, 62 (7) : 432-35.

Marx H, Gedek B and Kollarczik B (1995) Comparative investigations of mycotoxological status of alternatively and conventionally grown crops. *Z Lebensm Unters Forsch*, 201 (1), 83-6.

Offerman F and Nieberg H (1999) Economic performance of organic farms in Europe. Organic farming in Europe: Economic and policy 5. University of Hohenheim.

Olsen M and Möller T Mögeloch (1995) mykotoxiner I spannm?l/Mould and mycotoxines in grain. V?r föda/Our food, *Journal of Swedish National Food Administration*, ; 47 (8) : 30-33.

Pearce J, Leach CK and Carr NG (1969) The Incomplete tricarboxylic acid cycle in the blue green alga Anabaena variabilis. *J. Gen. Microbid.*, 55: 371-378.

Pfiffiner L and Mader P (1998) Effects of biodynamic, organic and conventional farming systems on earthworm populations. *Biol. Agric. Hort.* 15: 181-196.

Pfiffiner L and Niggli U (1996) Effects of biodynamic, organic and conventional farming systems on ground beetles *(Col. Carbidae)* and other arthopods in winter wheat. *Biol. Agric. Hort.* 12: 353-364.

Pretty, J (1998). In *"The living land"*. Earthscan Publications, London.

Raupp J (1996) Quality of plant products grown with manure fertilization. Fertilization systems in organic farming (concerted action AIR3-CT4-1940) Proceedings of the 4th meeting in Juva, Finland, July 6-9. Publications of the Institute for Biodynamic Research, Vol. 9, Darmstadt.

Reavel H (1999) The wholesome quest. *World of Ingredients*: 58, 60-61.

Rusch HP (1968) In: *"Bodenfruchtbarkeit: eine studie Biologischen Denkens*. Verlag. K. F. Haug, Heidelberg.

Schulze PL (1994) Extensivierungsprogramn und okologischer Landbau. Fur wen rechnet sich die Umstellung?*Forsch. Rep. Ernahrung Landwirtchaft Forsten*. 9: 11-15

Singh RN (1950) Reclamation of usar lands in India through blue green algae. *Nature* 165: 325-326.

Singh PK and Bisoyi RN (1989) Blue green algae in rice fields. *Phykos* 28: 181-195.

Singh AL and Singh PK (1987) Nitrogen fixation and balance studies of rice soils. *Biology and fertility of Soil*. 4 (1-2) : 15-19

Singh PK (1976) Algal inoculation and its growth in water logged rice fields. *Phykos* 15: 5-10.

Singh RN (1939) An investigation in to the algal flora of paddy field soils of the united provinces. *Indian J. Agric. Sci.* 9: 55–77.

Singh RN (1961) *Role of blue green algae in nitrogen economy of Indian agriculture.* Indian Council of Agricultural Research, New Delhi.

Singh YV and Mandal BK (1997) Nutrition of rice (Oryza sativa) through Azolla, organic materials and urea. *Indian J. Agronomy,* 42 (4) : 626-633.

Singh YV and MandalBK (2000) Rate of mineralization of *Azolla,* other organic materials and urea in water logged soils. Tropical Agriculture (Trinidad). 77 (1) : 119-122.

Singh YV and Singh BV (2006). Role of cyanobacteria in scented rice farming. *Glimpses of cyanobacteria.* Daya Publishing House, New Delhi, 48-66

Sloan (1998.) Organics: grown by the book. *Food Technology*: 52(5) : 32.

Steiner R (1924). "In: *Agriculture a course of Eight lectures*". Rudolph Steiner Press/Biodynamic agriculture Association. London.

StockdaleEA, Lampkin N. H, Hovi M, Keatinge R, Lennartsson EKM, Macdonald DW, Padel S, Tattersall FH, Wolfe MS and Watson CA (2001) Agronomic and environmental implications of organic farming systems. *Advances in Agronomy*. 70: 261-327.

Tandon, H. L. S. (2002). Fertilizer guide, 21.

Tate WB (1994) The development of the organic industry and market: an international perspective. In: *The economics of organic farming: An international perspective*". (N. H. Lampkin and S. Padel, Eds.). pp. 11-25. CABInternational, Wallingford. Oxon. U. K.

Thomas J, Wolk CP, Shaffer PW, Austin SM and Galonsky A (1975) The initial organic products of fixation of 15N–labeled nitrogen gas by the blue green alga Anabaena cylindrical. *Biochem. Biophys. Res. Commun.*, 67: 501–507.

UNDP (2000) Changing consumption and production patterns: Organic agriculture. Commission on Sustainable Development: 8th Session, 24 April–5 May 2000, New York.

WHO (1999) Surveillance programme for control of foodborne infections and intoxications in Europe. 5th Report 1993-1997. Federal Institute for Health Protection of consumers and veterinary medicine. In preparation.

Wallace HA (1998) *Beyond the Big Three. A Comprehensive Analysis of the Proposed National Organic Program*. Institute for Alternative Agriculture, Docket TMD-94-00-2, April 30, 1998.

WatsonCA (2001) Agronomic and environmental implications of organic farming systems. *Advances in Agronomy* 70: 261-311.

Weibel FP, Bickel R, Leuthold S, Alfoeldi T and Niggli U (1999) Are organically grown apples tastier and healthier? A comparative field study using conventional and alternative methods to measure fruit quality.

Woese K, Lange D, Boess C and Bögl KW (1995) Ökologisch und K onventionell erzeugte Lebensmittel im Vergleich–Eine Litteraturstudie. Bundeinstitut für gesundheitlichen Verbraucherschutz und Veterinärmedizin (BgVV-Hefte 4-5).

Woese K., Lange D, Boess C and Bögl K W (1997) A comparison of organically and conventionally grown foods–results of a review of the relevant literature. *J. Sci. Food Agric.* : 74; 281-293.

Wyatt JT and Silvey JKG (1969) Nitrogen fixation by Gloeocapsa. *Science*, 165: 908-909.

Yussefi M (2003) The world of organic future agriculture 2003–Statistics and prospects.

Zeichaus-Hartelt C (1991) Bioland-ein Verbandentwickelt sich. *Bioland*. 2, 13-14.

© R.K. Gupta & V.D. Pandey (*Editors*)
Advances in Applied Phycology, 2007
Daya Publishing House, New Delhi

250–260

Chapter 19

Cyanobacteria as a Source of Pharmaceutical Compounds

V.D. Pandey[1], R.K. Gupta[1*] and S.K. Singh[2]

[1]*Department of Botany, Pt. L.M.S. Govt. P.G. College, Rishikesh – 249201 (Dehradun) Uttaranchal,*
[2]*Department of Experimental Medicine and Surgery, Institute of Medical Sciences, Banaras Hindu University, Varanasi – 221 005, UP*

ABSTRACT

Cyanobacteria are a large, diverse and widely distributed group of O_2-evolving photoautotrophic prokaryotes endowed with metabolic flexibility and biochemical diversity. Many species produce a wide variety of pharmacologically important novel bioactive compounds, which include anticancerous agents, antimicrobial agents, antiviral agents, enzyme inhibitors, anti-inflammatory agents and muscle-relaxants. Because of their rich pharmacological potential, cyanobacteria have been recognized as suitable candidates for drug–discovery. Drug–discovery from cyanobacteria is a multistep process, which involves isolation and culturing of cyanobacteria, preparation and pharmacological screening of test samples, bioactivity–guided fractionation for the isolation of active compounds, structure elucidation and chemical modification of isolated compounds.

Keywords: *Cyanobacteria, Bioactive compounds, Pharmaceutical compounds, Chemotherapeutic agents, Antineoplastic agents.*

Introduction

Cyanobacteria (Blue-green algae) constitute an ancient, large and morphologically diverse group of oxygen-evolving photoautotrophic prokaryotes which resemble green plants in oxygenic

* Corresponding Author: E-mail: rajankgupta1@rediffmail.com.

photosynthesis. They are classified by bacteriologists as the class Oxyphotobacteria in the eubacterial division Gracilicutes (Castenholz and Waterbury, 1989) and by phycologists as the class Cyanophyceae or Myxophyceae in the algal division Cyanophyta. Endowed with effective protective mechanisms against different abiotic stress and remarkable adaptability to varying environmental conditions, they successfully colonize almost all kinds of terrestrial and aquatic habitats, including extreme ones (Tandeau de Marsac and Houmard, 1993; Pandey *et al.*, 1995; Ward and Castenholz, 2000; Oren, 2000). They grow as free-living organisms, in symbiotic associations with a wide range of plants, and in microbial mats (Adams, 2000; Stal, 1995). Morphologically, they range from simple unicellular and coccoid to complex filamentous forms with or without branching. Many filamentous forms differentiate specialized cells, such as heterocysts for nitrogen fixation, and akinetes for perennation and survival under adverse environmental conditions.

The biochemical diversity of cyanobacteria is reflected by the presence of a wide variety of photosynthetic pigments, storage products, and primary and secondary metabolites. The profound ability of many cyanobacteria to perform phtoheterotrophy, chemoheterotrophy, diazotrophy and bacterial-like anoxygenic photosynthesis under certain conditions (Van Baalen *et al.*, 1971; Rippka, 1972; Adams and Duggan, 1999; Padan and Cohen, 1982), and their ability to biodegrade and metabolise aromatic compounds (Wurster *et al.*, 2003; Cerniglia *et al.*, 1980) suggest that these are one of the metabolically versatile organisms on this planet. Besides, the dual ability of many cyanobacteria to perform both photosynthesis and nitrogen fixation together with their adaptation to low light intensity and efficient nutrient uptake at low ambient concentrations make them highly productive and efficient biological system.

The potential and actual applications of cyanobacteria in diverse fields, such as agriculture, aquaculture, human nutrition, energy production and bioremediation are well known. More recently, they have received considerable attention as a potential source of a wide variety of pharmacologically important novel bioactive compounds.

Cyanobacteria-Derived Pharmaceutical Compounds

Over the past few years, pharmacological screening and phytochemical investigations of cyanobacteria from diverse habitats have revealed the presence of novel bioactive compounds in many species. A large number of cyanobacteria–derived bioactive compounds have been proved to be pharmacologically important as potential chemotherapeutic agents. Bioactive or biologically active compounds refer to the compounds that at low concentrations affect the life processes in beneficial or detrimental manner (Skulberg, 2000). Cyanobacterial bioactive compounds represent a herterogenous group of secondary metabolites belonging to phenolics (Papendorf *et al.*, 1998; Cano *et al.*, 1990), polychlorinated aromatic compounds (Falch *et al.*, 1993), alkaloids (Moore *et al.*, 19987a, 1987b, Carmeli *et al.*, 1990a), cyclic peptides and depsipeptides (Moore, 1996), lipopeptides (Hooper *et al.*, 1998; Jimenez and Scheuer, 2001), glyco–and sulfolipids (Gustafson *et al.*, 1989; Shirahashi *et al.*, 1993; Reshef *et al.*, 1997), fatty acids (Murakami *et al.*, 1992; Mundt *et al.*, 2003), amides (Gerwick *et al.*, 1987; Orsini *et al.*, 2001), macrolides (Ishibashi *et al.*, 1986), isonitriles (Carmeli *et al.*, 1990b; Park *et al.*, 1992), lactones (Stierle *et al.*, 1998; Singh *et al.*, 1999) and nucleosides (Barchi *et al.*, 1983; Stewart *et al.*, 1988). They may be either constitutively produced throughout exponential growth phase or synthesized during the pre-stationary or stationary phase (Armstrong *et al.*, 1991). They may be released extracellularly either actively or passively or by autolysis (Metting and Pyne, 1986).

Anticancerous Agents

Large–scale screening programs, conducted by many workers with the aim of discovering new anticancerous drugs from prokaryotes, have revealed the *in vitro* anticancerous (antineoplastic) activity of several cyanobacteria against variety of cancerous (tumour) cell lines. The active principles in many cases have been isolated and chemically characterized. Following the screening of extracts of about 1000 cyanophytes from diverse habitats, Patterson *et al.* (1991) identified the blue-green algal families Scytonemataceae and Stigonemataceae as prolific producers of new cytotoxic compounds with *in vitro* activity against human epidermoid carcinoma (HEP-2), human colorectal adenocarcinoma (LoVo), human nasopharynx carcinoma (KB) and lymphocytic leukemia (P-388) cells. An alkyl phenolic compound debromoaplysiatoxin, isolated from the filamentous marine cyanobacterium *Lyngbya gracilis* was found to diplay cytotoxicity against lymphocytic mouse leukemia cell line (Mynderse *et al.*, 1977). Scytophycins, the potent cytotoxic and antifungal gaents, are congeneric macrolides (macrocyclic lactones) which are produced mainly by certain species of the genera *Scytonema* and *Tolypothrix* of the family Scytonemataceae (Ishibashi *et al.*, 1986; Carmeli *et al.*, 1990c). Tolytoxin (6-hydroxy-7-O-methyl-scytophycin b), a well-characterized scytophycin, was originally isolated from *Tolypothrix conglutinata* var. colorata in 1977 (Moore, 1981). The cyanophytes which are now known to produce scytophycins and tolytoxin include *Scytonema pseudohofmanni, S. mirabile, S. burmanicum* and *S. ocellatum* (Ishibashi *et al.*, 1986; Carmeli *et al.*, 1990c). Scytophycins, including tolytoxin, exert cytostatic effect in eukaryotic cells by disrupting actin microfilaments and by inhibition of actin polymerization (Patterson *et al.*, 1993; Smith *et al.*, 1993). These compounds, in addition to cytotoxicity towards cancerous cell lines, exhibit potent antifungal activity (Ishibashi *et al.*, 1986; Patterson and Carmeli, 1992).

Calothrixin A and B, which are chemically pentacyclic indolophenanthridine, are produced by the cyanobacterium *Calothrix* (Chen *et al.*, 2003; Rickards *et al.*, 1999). Both compounds posses potent activity against cancerous cells, HeLa and Jurkat cells. They induce apoptotic killing of cancerous cells, and cause cell cycle arrest in the G_2/M phase (Chen *et al.*, 2003; Rickards *et al.*, 1999). The filamentous cyanobacterium *Nostoc sphaericum* produces indolocarbazoles, which exhibit cytotoxicity against KB and LoVo human carcinoma cell lines and antiviral activity against herpes simplex virus, HSV-II (Knübel *et al.*, 1990). *Nostoc* sp. GSV224 produces cyclic depsipeptides, named as cryptophycins, which possess strong antitumor activity against cancerous cell lines (Trimurtulu *et al.*, 1994). The cyanobacteria *Tolypothrix byssoidea, T. tenuis* and *Plectonema radiosum* are known to produce a pyrimidine nucleoside, called tubercidin with activity against a variety of cancerous cell lines. Additionally, the compound shows antifungal activity against several fungi (Barchi *et al.*, 1983; Stewart *et al.*, 1988). The marine cyanobacterium *Lyngbya majuscula* has been recognized as a rich source of thiazoline-containing lipids, called curacins, which show strong antimitotic or antiproliferative activity (Gerwick *et al.*, 1994; Yoo and Gerwick, 1995). The cytotoxin westiellamide, a cyclic peptide, has been isolated from the terrestrial cyanobacterium *Westiellopsis prolifica* possessing activity against KB and LoVo cell lines (Prinsep *et al.*, 1992).

Antimicrobial Agents

The development of resistance in pathogenic microorganisms against commonly used antibiotics has necessitated the search for new antimicrobial compounds from sources other than the traditional microorganisms. Over past two decades, systematic screening of cyanobacteria isolated from fresh water, marine water and terrestrial habitats have confirmed the presence of varying degree of antimicrobial activities in many species against a wide range of bacteria and fungi, including human pathogens, such as *Pseudomonas aeruginosa, Staphylococcus aureus, S. epidermidis, Enterobacter aerogenes, Salmonella typhi, Mycobacterium tuberculosis, Candida albicans* and *Tricophyton mentagrophytes*. A number of potent antibacterial and antifungal compounds have been isolated with chemical characterization and structure elucidation. Hormothamnins, which display antibacterial and antifungal activities, are cyclic undecapeptides isolated from the marine cyanobacterium *Hormothamnion enteromorphoides* (Gerwick *et al.*, 1989; Gerwick *et al.*, 1992) The cyclic undecapeptide schizotrin A, which is active against bacteria and fungi, is produced by the terrestrial cyanobacterium *Schizothrix* sp. (Pergament and Carmeli, 1994). The marine filamentous blue-green alga *Lyngbya majuscula* produces an antifungal cyclic depsipeptide, called majusculamide C (Carter *et al.*, 1984). A group of antifungal compounds, named as laxaphycins, were isolated from *Anabaena laxa*. Chemically, they belong to cyclic undeca- and dodecapeptides (Frankmölle *et al.*, 1992a; Frankmölle *et al.*, 1992b). The epilithic cyanobacterium *Nostoc spongiaeforme* var. *tenue* has been found as a source of a group of cyclic hexapeptides, designated as tenuecyclamides, which exhibit antibacterial activity (Banker and Carmeli, 1998). Muscoride A, an antibacterial oxazole peptide alkaloid, is produced by the freshwater cyanobacterium *Nostoc muscorum* (Nagatsu *et al.*, 1995). Malyngamides, amides of the fatty acid (-)–7 (S)–methoxytetradec-4 (E)–enoate, are antibacterial principles of the marine cyanobacterium *Lyngbya majuscula* (Gerwick *et al.*, 1987). Tanikolide, an antifungal lactone, is another antimicrobial compound isolated from *Lyngbya majuscula* (Singh *et al.*, 1999). The freshwater cyanophyte *Oscillatoria redekei* produces unsaturated hydroxy fatty acids α–dimorphecolic acid and coriolic acid, which exhibit potent antibacterial activity (Mundt *et al.*, 2003). Ambigols, isolated from the terrestrial cyanobacterium *Fischerella ambigua*, are polychloroinated aromatic compounds possessing antibacterial and antifungal activities (Falch *et al.*, 1993). The same species have been reported to produce another antibacterial and antifungal compound, parsiguine (Ghasemi *et al.*, 2004). Hapalindoles, which are indole alkaloids, are antibacterial and antifungal agents isolated from an edaphic cyanobacterium *Hapalosiphon fontinalis* (Moore *et al.*, 1987a). Ambiguine isonitriles are fungicidal hapalindole-type alkaloids known to be produced by blue-green algae *Fischerella ambigua, Hapalosiphon hibernicus* and *Westiellopsis prolifica* (Smitka *et al.*, 1992). *Fischerella muscicola*, a terrestrial cyanophyte, has been reported to produce an antifungal compound, fischerindole L, which is related to hapalindoles in chemical structure (Park *et al.*, 1992). γ-linolenic acid isolated from Neem tree bark inhabiting cyanobacterium *Fischerella* sp. possess antibacterial activity (Asthana *et al.*, 2006)

Antiviral Agents

Many cyanobacterial species have been reported to produce novel compounds showing activities against viral pathogens, including human immunodeficiency virus (HIV). Indolocarbazoles isolated from *Nostoc sphaericum* exhibit antiviral activity against herpes simplex virus, HSV-II in addition to cytotoxicity towards carcinoma cell lines (Knübel *et al.*, 1990). The cyanobacterium *Spirulina platensis* produces a sulphated polysaccharide, calcium spirulan, which shows anti-herpes simplex virus and anti-HIV activity (Hayashi *et al.*, 1996). Carbolines, which are active against herpes simplex virus (HSV-II), are antiviral compounds in blue-green alga *Dichothrix baueriana* (Larsen *et al.*, 1994). A novel proteinaceous compound, cyanovirin N, produced by *Nostoc ellipsosporum* have been reported to show anti-HIV activity (Boyd *et al.*, 1997). Other anti-HIV compounds derived from cyanobacteria include glycolipids from *Oscillatoria limnetica* (Reshef *et al.*, 1997) and *Oscillatoria trichoides* (Loya *et al.*, 1998), and sulfolipids from *Lyngbya lagerheimii* and *Phormidium tenue* (Gustafson *et al.*, 1989). The anti–HIV activity of these compounds is mainly due to the inhibition of reverse transcriptase enzyme of HIV (Reshef *et al.*, 1997; Loya *et al.*, 1998).

Enzyme Inhibitors

Several cyanobacteria have been proved as a rich source of enzyme inhibitors of pharmacological importance. The toxigenic cyanobacterium *Microcystis aeruginosa* produces cyclic depsipeptides, micropeptins, which inhibit the enzyme plasmin (Ishida *et al.*, 1997, Ishida *et al.*, 1995). Plasmin is a serine protease that regulates blood coagulation, and is related to cardiovascular diseases, such as stroke and coronary artery blockade. Plasmin inhibitors are potential chemotherapeutic agents for such diseases. Cyclic depsipeptides, nostopeptins, and cyclic peptides, microviridins are elastase (a serine protease) inhibitors produced by the cyanobacterium *Nostoc minutum* (Okino *et al.*, 1997) and *Microcystis aeruginosa* (Okino *et al.*, 1995), respectively. The enzyme elastase is suggested to be involved in pulmonary emphysema, rheumatoid arthritis and respiratory distress syndrome. Its inhibitors might be useful in the treatment of these ailments. Another enzyme inhibitor isolated from *Microcystis aeruginosa* includes microginin, a linear peptide, which inhibits angiotensin-converting enzyme (ACE) (Okino *et al.*, 1992). ACE is involved in catalyzing the release of angiotensin from its precursor angiotensinogen. Excessive release of angiotensin leads to hypertension and associated cardiovascular disorders. ACE inhibitors appears to be suitable targets as antihypertensive drugs.

Anti-inflammatory Agents and Muscle Relaxants

Inflammation is a non-specific response to tissue injury that protects the host from further damage. It stimulates immune reactivity and blocks the spread of an infectious agent. Prinsep *et al.* (1996) isolated an anti-inflammatory diterpinoid, called tolypodiol, from the cyanobacteroum *Tolypothrix nodosa*. The capsular polysaccharide of the thermal cyanobacterium *Mastigocladus laminosus* have been repoted to possess anti-inflammatory properties (Gloaguen *et al.*, 2003). The toxic cyanobacterium *Anabaena flos-aquae* produces cyclic peptides, named as anabaenopeptins, which have been reported to cause concentration-dependent relaxations in rat aortic preparations (Harada *et al.*, 1995).

Drug Discovery and Development from Cyanobacteria

The process of drug discovery from natural sources is driven largely by the desire to identify structurally novel compounds that possess potentially useful pharmacological activities. Drug discovery from cyanobacteria, as from other plant and microbial sources, is a multistep process which involves isolation of cyanobacteria from different habitats, establishment of clonal and axenic culture, preparation of test samples using solvents of different polarities, pharmacological screening of test samples using sensitive and reliable bioassays, high-density culture of active species, isolation of active compounds employing bioactivity-guided fractionation of active crude extracts, chemical characterization and structure elucidation of active compounds, and clinical evaluation, pharmacological profile analysis and toxicological testing of active compounds. In many cases, the cyanobacteria - derived natural products may not be effective drugs, but they may nevertheless provide templates (pharmacophores) for future drug development. In such cases, chemical modification of the structure of isolated natural products, either by direct modification (semisynthesis) or by total chemical synthesis, can yield effective drugs.

Conclusion

Pharmacological screening and phytochemical investigations aimed at discovering novel pharmaceuticals from natural sources has largely been carried out with higher plants and non-cyanobacterial microorganisms. Algae, in general, and cyanobacteria, in particular, have not received much attention. The production of pharmacologically important novel bioactive compounds by

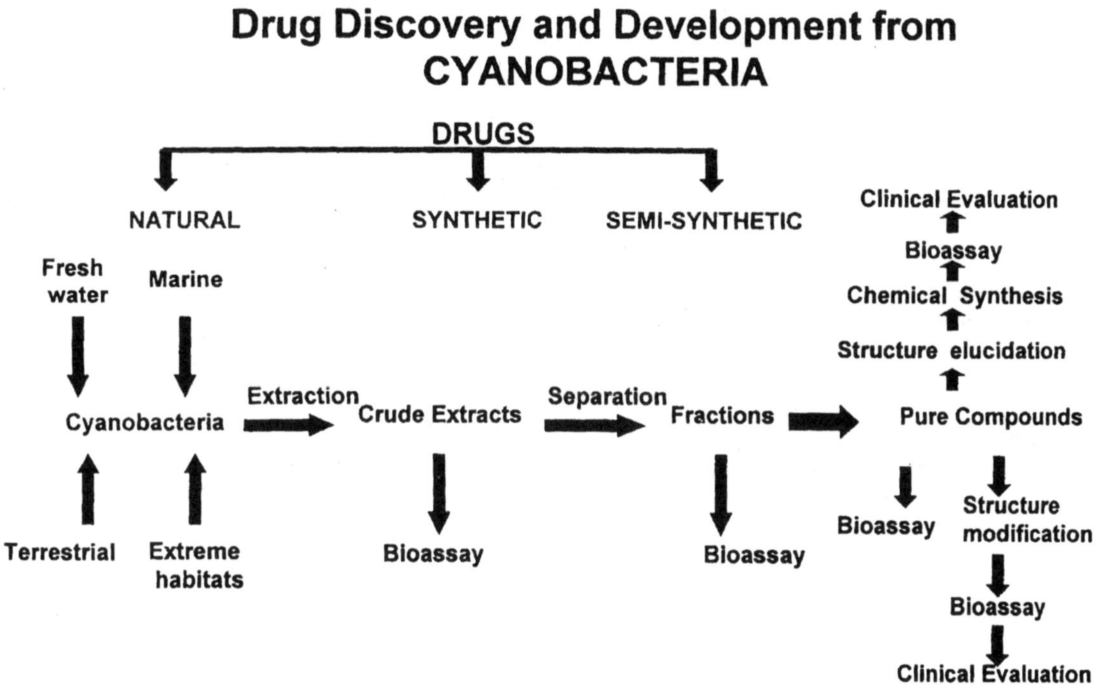

cyanobacteria open up new possibilities for their exploitation as potential source of different pharmaceuticals. Till now, our knowledge of pharmacologically active compounds of cyanobacterial origin is based on few species. The chance of discovering new pharmaceuticals from these underexplored organisms seems to be high. Ubiquitous distribution, rapid growth rate, simple growth requirements and amenability to controlled laboratory culture and genetic modification make them suitable candidates for drug discovery programs. In order to realize the full potential of cyanobacteria, increased research effort needs to be directed towards extensive screening, purification and characterization of active compounds, strain improvement, optimization of growth media and culture conditions, and high–density culture for mass production of bioactive species.

References

Adams, D.G. and Duggan P.S. (1999). Heterocyst and akinete differentiation in cyanobacteria. *New Phytol.* 144: 3-33.

Adams, D.G.(2000). Symbiotic interactions. In: Whitton, B.A. and Potts, M. (eds.) *The Ecology of Cyanobacteria: their diversity in time and space.* Kluwer Academic Publishers, Dordrecht, The Netherlands, pp. 523 –561.

Armstrong, J.E., Janda, K.E., Alvarado, B. and Wright, A.E. (1991). Cytotoxin production by a marine *Lyngbya* strain (cyanobacterium) in a large-scale laboratory bioreactor. *J. Appl. Phycol.* 3: 277–282.

Asthana, R.K., Srivastava, A., Kayastha, A.M., Nath, G. and Singh, S.P. (2006). Antibacterial potential of γ-linolenic acid from *Fischerella* sp. colonizing Neem tree bark. *World J. Microbiol. Biotechnol.* 22: 443-448.

Banker, R. and Carmeli, S. (1998).Tenuecyclamides A–D, cyclic hexapeptides from the cyanobacterium *Nostoc spongiaeforme* var. *tenue*. *J. Nat. Prod*. 61: 1248–1251.

Barchi, J.J.Jr., Norton, T.R., Furusawa, E., Patterson, G.M.L. and Moore, R.E. (1983). Identification of a cytotoxin from *Tolypothrix byssoidea* as tubercidin. *Phytochemistry* 22: 2851–2852.

Boyd, M.R., Gustafson, K.R., Mc Mohan, J.B., Shoemaker, R.H., O'Keefe, B.R., Mori, T., Gulakowski, R.J., Wu, L., Rivera, M.I., Laurencot, C.M., Currens, M.J., Cardellina II, J.H., Buckheit Jr, R.W., Nara, P.L., Pannell, L.K., Sowder II, R.C. and Henderson, L.E. (1997). Discovery of cyanovirin-N, a novel human immunodeficiency virus-inactivating protein that binds viral surface envelope glycoprotein gp120: potential applications to microbiocide development. *Antimicrob. Agents Chemother.* 41: 1521–1530.

Cano, M.M.S.de., Mule, M.C.Z.de., Caire, G.Z.de. and Halperin, D.R.de. (1990). Inhibition of *Candida albicans* and *Staphylococcus aureus* by phenolic compounds from the terrestrial cyanobacterium *Nostoc muscorum*. *J. Appl. Phycol.* 2: 79–81.

Carmeli, S., Moore, R.E. and Patterson, G.M.L. (1990c). Tolytoxin and new scytophycins from three species of *Scytonema*. *J. Nat. Prod.* 53: 1533–1542.

Carmeli, S., Moore, R.E., Patterson, G.M.L., Corbett, T.H. and Valeriote, F.A. (1990a). Tantazoles: unusual cytotoxic alkaloids from the blue-green alga *Scytonema mirabile*. *J. Am. Chem. Soc.* 112: 8195–8197.

Carmeli, S., Moore, R.E., Patterson, G.M.L., Mori, Y. and Suzuki, M. (1990b).Isonitriles from the blue-green alga *Scytonema mirabile*. *J. Org. Chem.* 55: 4431–4438.

Carter, D.C., Moore, R.E., Mynderse, J.S., Niemczura, W.P. and Todd, J.S. (1984). Structure of majusculamide C, a cyclic depsipeptide from *Lyngbya majuscula*. *J. Org. Chem.* 49: 236–241.

Castenholz, R.W. and Waterbury, J.B. (1989). Oxygenic Photosynthetic Bacteria. Group I. Cyanobacteria. In: Staley, J.T., Bryant, M.P. and Pfennig, N. (eds.), *Bergey's Manual of Systematic Bacteriology*, Vol. 3. Williams and Wilkins, Baltimore, Maryland, pp. 1710–1798.

Cerniglia, C.E., Gibson, D.T. and Van Baalen, C. (1980). Oxidation of naphthalene by cyanobacteria and microalgae. *J. Gen. Microbiol.* 116: 495-500.

Chen, X. Smith, G.D. and Waring, P. (2003). Human cancer cell (Jurkat) killing by the cyanobacterial metabolite calothrixin A. *J. Appl. Phycol.* 15: 269–277.

Falch, B.S., König, G.M., Wright, A.D., Sticher, O., Rüegger, H. and Bernardinelli, G. (1993). Ambigol A and B: new biologically active polychlorinated aromatic compounds from the terrestrial blue-green alga *Fischerella ambigua*. *J. Org. Chem.* 58: 6570–6575.

Frankmölle, W.P., Knübel, G., Moore, R.E. and Patterson, G.M.L. (1992b). Antifungal cyclic peptides from the terrestrial blue-green alga *Anabaena laxa*. II. Structures of laxaphycins A, B, D and E. *J. Antibiotics* 45: 1458–1466.

Frankmölle, W.P., Larsen, L.K., Caplan, F.R., Patterson G.M.L., Knübel, G., Levine, I.A. and Moore, R.E. (1992a). Antifungal cyclic peptides from the terrestrial blue-green alga *Anabaena laxa*. I. Isolation and biological properties. *J. Antibiotics* 45: 1451–1457.

Gerwick, W.H., Jiang, Z.D., Agarwal, S.K. and Farmer, B.T. (1992). Total structure of hormothamnin A, a toxic cyclic undecapeptide from the tropical marine cyanobacterium *Hormothamnion enteromorphoides*. *Tetrahedron* 48: 2313-2324.

Gerwick, W.H., Mrozek, C., Moghaddam, M.F. and Agarwal, S.K. (1989). Novel cytotoxic peptides from the tropical marine cyanobacterium *Hormothamnion enteromorphoides*. 1. Discovery, isolation and initial chemical and biological characterization of the hormothamnins from wild and cultured material. *Experientia* 45: 115–121.

Gerwick, W.H., Proteau, P.J., Nagle, D.G., Hamel, E. Blokhin, A and Slate, D.L. (1994). Structure of curacin A, a novel antimitotic, antiproliferative, and brine shrimp toxic natural product from the marine cyanobacterium *Lyngbya majuscula*. *J. Org. Chem.* 59:1243-1245.

Gerwick, W.H., Reyes, S. and Alvarado, B. (1987). Two malyngamides from the Caribbean cyanobacterium *Lyngbya majuscula*. *Phytochemistry* 26: 1701–1704.

Ghasemi, Y., Yazdi, M.T., Shafiee, A., Amini, M., Shokravi, S. and Zarrini, G. (2004). Parsiguine, a novel antimicrobial substance from *Fischerella ambigua*. *Pharmaceut. Biol.* 42: 318-322.

Gloaguen, V., Garbacki, N. Petit, D., Morvan, H. and Hoffmann,L. (2003). Bioactive capsular polysaccharide from the thermophilic *Mastigocladus laminosus* (cyanophyceae/cyanobacteria): demonstration of anti-inflammatory properties. *Algological Studies* 108: 63-73.

Gustafson, K.R., Cardellina, J.H., Fuller, R.W., Weislow, O.S., Kiser, R.F., Snader, K.M., Patterson, G.M.L. and Boyd, M.R. (1989). AIDS-antiviral sulfolipids from cyanobacteria (blue-green algae). *J. Natl. Canc. Inst.* 81: 1254–1258.

Harada, K.I., Fujii, K., Shimada, T., Suzuki, M., Sano. H., Adachi, K. and Carmichael, W.W. (1995). Two cyclic peptides, anabaenopeptins, a third group of bioactive compounds from the cyanobacterium *Anabaena flos-aquae* NRC 525–17. *Tetrahedron Lett.* 36: 1511–1514.

Hayashi, K., Hayashi, T., and Kojima, I. (1996). A natural sulfated polysaccharide, calciums spirulan, isolated from *Spirulina platensis: in vitro* and *ex vivo* evaluation of anti-herpes simplex virus and anti-human immuno deficiency virus activities. *AIDS Res. Hum. Retroviruses*. 12: 1463-1471.

Hooper, G.J., Orjala, J., Schatzman, R.C. and Gerwick, W.H. (1998). Carmabins A and B, new lipopeptides from the Carribean cyanobacterium *Lyngbya majuscula*. *J. Nat. Prod.* 61: 529–533.

Ishibashi, M., Moore, R.E., Patterson G.M.L., Xu, C. and Clardy, J.(1986). Scytophycins, cytotoxic and antimycotic agents from the cyanophyte *Scytonema pseudohofmanni*. *J. Org. Chem.* 51: 5300–5306.

Ishida, K., Matsuda, H., Murakami, M. and Yamaguchi, K. (1997). Micropeptins 478-A and–B, plasmin inhibitors from the cyanobacterium *Microcystis aeruginosa*. *J. Nat. Prod.* 60: 184–187.

Ishida, K., Murakami, M., Matsuda, H., and Yamaguchi, K. (1995). Micropeptin 90, a plasmin and trypsin inhibitor from the blue-green alga *Microcystis aeruginosa* (NIES–90). *Tetrahedron Lett.* 36: 3535–3538.

Jimenez, J.I. and Scheuer, P.J. (2001). New lipopeptides from the Carribean cyanobacterium *Lyngbya majuscula*. *J. Nat. Prod.* 64: 200–203.

Knübel, G., Larson, L.K., Moore, R.E., Levine, I.A. and Patterson, G.M.L. (1990). Cytotoxic, antiviral indolocarbazoles from a blue-green alga belonging to the Nostocaceae. *J. Antibiotics* 43: 1236–1239.

Larsen, L.K., Moore, R.E. and Patterson, G.M.L. (1994). Carbolines from the blue-green alga *Dichothrix baueriana*. *J. Nat. Prod.* 57:419-421.

Loya, S., Reshef, V., Mizrrachi, E., Silberstein, C., Rachamim, Y., Carmeli, S. and Hizi, A. (1998). The inhibition of the reverse transcriptase of HIV–1 by the natural sulfoglycolipids from cyanobacteria: contribution of different moieties to their high potency. *J. Nat. Prod.* 61: 891–895.

Metting, B. and Pyne, J.W. (1986). Biologically active compounds from microalgae. *Enzyme Microb. Technol.* 8: 386–394.

Moore, R.E. (1981). Constituents of blue-green algae. In: Scheuer, P.J.(ed.), *Marine Natural Products,* vol. 4, Academic Press, New York, pp. 1–52.

Moore, R.E. (1996). Cyclic peptides and depsipeptides from cyanobacteria: A review. *J. Indust. Microbiol.* 16: 134–143.

Moore, R.E., Cheuk, C., Yang, X-Q.G., Patterson, G.M.L., Bonjouklian, R., Smitka, T.A., Mynderse, J.S., Foster, R.S., Jones, N.D., Swartzendruber, J.K, and Deeter. J.B. (1987a). Hapalindoles, antibacterial and antimycotic alkaloids from cyanophyte *Hapalosiphon fontinalis. J.Org. Chem.* 52: 1036–1043.

Moore, R.E., Yang, X-Q.G. and Patterson, G.M.L. (1987b). Fontonamide and anhydrohapaloxinodole A, two new alkaloids from the blue-green alga *Hapalosiphon fontinalis. J.Org. Chem.* 52: 3773–3777.

Mundt, S., Kreitlow, S. and Jansen, R.(2003). Fatty acids with antibacterial activity from the cyanobacterium *Oscillatoria redekei* HUB 051. *J. Appl. Phycol.* 15: 263–267.

Murakami, N., Shirahashi, H., Nagatsu, A. and Sakakibara, J. (1992). Two unsaturated 9R–hydroxy fatty acids from the cyanobacterium *Anabaena flos-aquae. Lipids* 27: 776– 778.

Mynderse, J.S., Moore, R.E., Kashiwagi, M. and Norton, T.R. (1977). Antileukemia activity in the Oscillatoriaceae: Isolation of debromoaplysiatoxin from *Lyngbya. Science* 196: 538–540.

Nagatsu, A., Kajitani, H. and Sakakibara, J. (1995). Muscoride A: a new oxazole peptide alkaloid from freshwater cyanobacterium *Nostoc muscorum. Tetrahedron Lett.* 36: 4097–4100.

Okino, T., Matsuda, H., Murakami, M. and Yamaguchi, K. (1992). Microginin, an angiotensin–converting enzyme inhibitor from the blue green alga *Microcystis aeruginosa. Tetrahedron Lett.* 34: 501–504.

Okino, T., Matsuda, H., Murakami, M. and Yamaguchi, K. (1995). New microviridins, elastase inhibitors from the blue–green alga *Microcystis aeruginosa. Tetrahedron* 51: 10679–10686.

Okino, T., Qi, S., Matsuda, H., Murakami, M. and Yamaguchi, K. (1997). Nostopeptins A and B, elastase inhibitors from the cyanobacterium *Nostoc minutum. J. Nat. Prod.* 60: 158-161.

Oren, A. (2000) Salts and brines. In: Whitton, B.A. and Potts, M. (eds.). *The Ecology of Cyanobacteria: Their diversity in time and space.* Kluwer Academic Publishers Dordrecht, The Netherlands, pp. 281 –306.

Orsini, M.A., Pannell, L.K. and Erickson, K.L. (2001). Polychlorinated acetamides from the cyanobacterium *Microcoleus lyngbyaceus. J.Nat. Prod.* 64: 572–577.

Padan, E. and Cohen, Y.C. (1982). Anoxygenic photosynthesis. In: Carr, N.G. and Whitton, B.A. (eds.), *The Biology of Cyanobacteria.* Blackwell Scientific Publication Ltd. Oxford, pp. 215–235.

Pandey, K.D. Kashyap, A.K. and Gupta, R.K. (1995). Nutrient status, algal and cyanobacterial flora of six freshwater streams of Schirmacher Oasis, Antarctica. *Hydrobiologia* 299: 83-91.

Papendorf, O., König, G.M. and Wright, A.D. (1998). Hierridin B and 2,4-dimethoxy-6-heptadecyl-phenol, secondary metabolities from the cyanobacterium *Phormidium ectocarpi* with antiplasmodial activity. *Phytochemistry* 49: 2383–2386.

Park, A., Moore, R.E. and Patterson, G.M.L. (1992). Fischerindole L, a new isonitrile from the terrestrial blue-green alga *Fischerella muscicola*. *Tetrahedron Lett*. 33: 3257– 3260.

Patterson, G.M.L., and Carmeli, S. (1992). Biological effects of tolytoxin (6-hydroxy–7-O-methyl-scytophycin b), a potent bioactive metabolite from cyanobacteria. *Arch. Microbiol*. 157: 406–410.

Patterson, G.M.L., Smith, C.D., Kimura, L.H., Britton, B.A., Carmeli, S. (1993). Action of tolytoxin on cell morphology, cytoskeletal organization and actin polymerization. *Cell Motil Cytoskeleton* 24: 39-48.

Patterson, G.M.L., Baldwin, C.L., Bolis, C.M., Caplan, F.R., Karuso, H., Larsen, L.K., Levine, I.A., Moore, R.E., Nelson, C.S., Tschappat, K.D., Tuang, G.D., Furusawa, E., Furusawa, S., Norton, T.R. and Raybourne, R.B. (1991). Antineoplastic activity of cultured blue-green algae (cyanophyta). *J. Phycol*. 27: 530–536.

Pergament, I. and Carmeli, S. (1994). Schizotrin A; a novel antimicrobial cyclic peptide from a cyanobacterium. *Tetrahedron Lett*. 35: 8473–8476.

Prinsep, M.R., Moore, R.E., Levine, I.A. and Patterson, G.M.L. (1992). Westiellamide, a bistratamide-related cyclic peptide from the blue-green alga *Westiellopsis prolifica*. *J. Nat. Prod* 55: 140-142.

Prinsep, M.R., Thompson, R.A., West, M.L. and Wylie, B.L. (1996). Tolypodiol, an anti –inflammatory diterpenoid from the cyanobacterium *Tolypothrix nodosa*. *J. Nat. Prod*. 59: 786-788.

Reshef, V., Mizrachi, E., Maretzki, T., Silberstein, C., Loya, S., Hizi, A. and Carmeli, S. (1997). New acylated sulfoglycolipids and digalactolipids and related known glycolipids from cyanobacteria with potential to inhibit reverse transcriptase of HIV-1. *J. Nat. Prod*. 60: 1251–1260.

Rickards, R.W., Rothschild, J.M., Willis, A.C., de Chazal, N.M., J., Kirk, J., Kirk, K., Saliba, K.J. and Smith, G.D. (1999). Calothrixins A and B, novel pentacyclic metabolites from *Calothrix* cyanobacteria with potent activity against malaria parasites and human cancer cells. *Tetrahedron* 55: 13513–13520.

Rippka, R. (1972). Photoheterotrophy and chemoheterotrophy among unicellular blue-green algae. *Archiv für Mikrobiologie* 87: 93–98.

Shirahashi, H., Murakami, N., Watanabe, M., Nagatsu, A., Sakakibara, J., Tokuda, H., Nishino, H. and Iwashima, A. (1993). Isolation and identification of anti-tumor-promoting principles from the freshwater cyanobacterium *Phormidium tenue*. *Chem. Pharm. Bull*. 41: 1664–1666.

Singh, I.P., Milligan, K.E. and Gerwick, W.H. (1999). Tanikolide, a toxic and antifungal lactone from the marine cyanobacterium *Lyngbya majuscula*. *J. Nat. Prod*. 62: 1333–1335.

Skulberg, O.M. (2000). Microalgae as a source of bioactive molecules–experience from cyanophyte research. *J. Appl. Phycol*. 12: 341–348.

Smith, C.D., Carmeli, S., Moore, R.E. and Patterson, G.M.L. (1993). Scytophycins: novel microfilament-depolymerizing agents which circumvent p-glycoprotein-mediated multidrug resistance. *Cancer Res*. 53: 1343-1347.

Smitka, T.A., Bonjouklian, R., Doolin, L., Jones, N.D., Deeter, J.B., Yoshida, W.Y., Prinsep, M.R., Moore. R.E., and Patterson, G.M.L. (1992). Ambiguine isonitriles, fungicidal hapalindole–type alkaloids from three genera of blue-green algae belonging to the Stigonemataceae. *J.Org. Chem*. 57: 857–861.

Stal, L.J. (1995). Physiological ecology of cyanobacteria in microbial mats and other communities. *New Phytol*. 131: 1–32.

Stewart, J.B., Bornemann, V., Chen, J.L., Moore, R.E., Caplan, F.R., Karuso, H., Larsen, L.K. and Patterson, G.M.L. (1988). Cytotoxic, fungicidal nucleosides from blue-green algae belonging to the Scytonemataceae. *J. Antibiotics* 41: 1048–1056.

Stierle, D.B., Stierle, A.A., Bugni, T. and Loewen, G. (1998). Gloeolactone, a new epoxy lactone from a blue-green alga. *J.Nat. Prod.* 61: 251–252.

Tandeau de Marsac, N. and Houmard, J. (1993). Adaptation of cyanobacteria to environmental stimuli: new steps towards molecular mechanisms. *FEMS Microbiol. Rev.* 104: 119–190.

Trimurtulu, G., Ohtani, I., Patterson, G.M.L., Moore, R.E., Corbett, T.H., Valeriote, F.A. and Demchik, L. (1994). Total structures of cryptophycins, potent antitumor depsipeptides from the blue-green alga *Nostoc* sp. Strain GSV 224. *J. Am. Chem. Soc.* 116: 4729–4737.

Van Baalen, C., Hoare, D.S. and Brandt, E. (1971). Heterotrophic growth of blue-green algae in dim light. *J. Bacteriol.* 105: 685–689.

Ward, D.M. and Castenholz, R.W. (2000). Cyanobacteria in geothermal habitats. In: Whitton, B.A. and Potts, M. (eds.), *The Ecology of Cyanobacteria: Their diversity in time and space.* Kluwer Academic Publishers, Dordrecht, The Netherlands, pp. 37–59.

Wurster, M., Mundt, S., Hammer, E., Schauer, F. and Lindequist, U. (2003). Extracellular degradation of phenol by the cyanobacterium *Synechococcus* PCC 7002. *J. Appl. Phycol* 15: 171-176.

Yoo, H.D. and Gerwick, W.H. (1995). Curacins B and C, new antimitotic natural products from the marine cyanobacterium *Lyngbya majuscula. J. Nat. Prod.* 58: 1961–1965.

© R.K. Gupta & V.D. Pandey (*Editors*)
Advances in Applied Phycology, 2007 261–274
Daya Publishing House, New Delhi

Chapter 20

Cellular Water in Desiccation-Tolerant Cyanobacteria

B.S. Tiwari*[1] and S.N. Tripathi[2]

[1]*Department of Plant Biology and Forest Genetics, Swedish Agricultural University, Uppsala Genetic Centre, PO Box 7080, SE 75 007, Sweden*
[2]*Center of Advanced Study in Botany, Banaras Hindu university, Varanasi – 221 005*

ABSTRACT

Desiccation tolerance of an organism is positively correlated with cellular bound water present in it. Analyzing cellular water can provide a useful clue towards extent of desiccation tolerance of the organism. We have characterized state of cellular water of sub aerial and desiccation tolerant cyanbacteria , *Scytonema geitleri and Lyngbya arboricola*,. For characterizing cellular water we opted thermodynamic as well as kinetic probe. Our finding suggests that water present in dry cyanobacterial mats can be divided in to three different types *viz.* very strongly bound on to the ionic sites, weakly bound through hydrogen binding and water that are loosely associated with polar surfaces through van der waal interaction. Grown mats of both the cyanobacteria showed low value of water binding onto strong binding indicating a loss in their desiccation tolerance.

Introduction

Water as an indispensable component of the cell is required for the maintenance of biological integrity and function. There are numerous organism which survive the removal of all or almost all their cellular water without irreversible damage mainly by suspending their metabolism in dry state. Such organisms are referred as desiccation tolerant (Bewely *et al.*, 1979). Desiccation-tolerance is wide spread phenomenon occurring over wide range of taxa including bacteria plants and animals (Crow

* Corresponding Author: E-mail: budhi@rediffmail.com; budhi.tiwari@vbsg.slu.se.

and Clegg 1978). Many cyanobacteria express marked tolerance to desiccation (Potts and Friedmann 1981, Brock 1975, Potts et al., 1983, Whitton et al., 1979, Whitton and Potts 1982, Tripathi 1980a&b, Tripathi et al., 1991, Tiwari and Tripathi 1998, Tripathi and Srivastava 2002).

Desiccation tolerant cyanobacteria have been reported in extremely diverse range of habitats like surfaces of bare rocks, building and building materials bark of trees and soil receiving low and uncertain amount of water mainly from rainfall. Noticeably most of the desiccation tolerant cyanobacteria growing on the surfaces of bark of the trees, buildings, soil and rocks do not possess specialized structure like spore (Akinate) but survive extremes of environmental perturbations like temperature, light and desiccation.

Defining Cellular Water

Water plays an important role in the folding and reactivity of native biomolecules and hence become an inevitable requirement for the maintenance of integrity of biological system by forming hydrogen bonds with polar groups like hydroxyl, carboxyl and amine and by spatial association with non polar groups like aliphatic chains (Multon 1989). Conformational changes due to water in proteins can lead to their structural folding and defolding: and 30-40 per cent of side chains of proteins have no affinity for water, indicating hydrophobic interactions to play an important role in structural status of proteins (Yon 1969, Multan 1989). On the other hand, cyanobacteria contain a large number of hydroxyl groups through which water forms hydrogen bonds and provides spatial conformational ordering of the structure. Beyond hydrophilic interactions and specific affinity between water molecules and different polar groups, steriochemical aspect appears to be the determining factor for hydration properties of polyalcohol and carbohydrates (Suggett 1975). Water also affects structure of nucleic acids by changing hydrogen bond and hydrophobic interactions between stacked bases (Parsegian et al., 1986). In addition, water is important substrate as well as solvent for many physiological and biochemical reactions.

Generally cellular water exists in two forms (*i*) water, which is in direct contact with sub-cellular surfaces or matrix ie. bound water and (*ii*) water molecules which are not in direct with cellular surfaces, but are associated with water molecules ie. bulk water. However, these terms are being abandoned these days in order to avoid ambiguity and terms like solvent water, freezable and unfreezable water are being preferred for defining structural water.

Characterizing Cellular Water

Clegg (1978a) has categorized all the techniques for defining cellular water in to two broad probes: *Thermodynamic probe*, which includes water sorption isotherm, differential scanning calorimetry dealing with energy state and type of water binding with biomolecular surfaces and *Kinetic probe*, which includes NMR. dielectric, IR and X-ray diffraction techniques related to motional behaviour of cellular water.

In dry organisms, defining structural water is essential for understanding the role of water in the integrity of the cells. It is believed that there are at least three types of interstitial water, very strongly bound, weakly bound and water that are loosely associated with polar surfaces which probably display different structural characteristics. Hence, the application of thermodynamic probes, which are capable to reveal the strength of water association with surfaces, are supposed to be appropriate in defining cellular water in dry systems.

Water sorption isotherm has successfully been applied for characterizing water binding in several biological systems especially in seeds and sub aerial cyanobacteria (Vertucci and Leopold 1984,

Tiwari and Tripathi 1998). It is a composite curve representing absorption of water by a dried material kept in the atmosphere of known humidity at constant temperature, and comprises of component isotherm arising from three separate processes;

1. Monolayer adsorption at strong binding sites,
2. Monolayer adsorption at weak binding sites and
3. Formation of multilayer, the extent of which is limited by properties of substrates.

The component A and B have the form of Langmuir isotherm for monolayer adsorption at the surfaces, while component C describing formation of multilayer adsorption yields a modified Langmuir type equation. Generally, the component B can be approximated by adsorption which is linear with vapour. Sorption isotherm is analyzed:

1. By using a mathematical model developed by D'Arcy and Watt (1970) as well as
2. By comparing sorption at two temperatures via Clausius Clapeyron equation (Bull 1944, Luscher-Mattli and Ruegg 1982, Schneider and Schneider 1972 and Vertucci and Leopold 1987a&b).

Dielectric method has been used for characterization of motional properties of water in biopolymers and sub aerial cyanobacteria by several workers (Grigera and Mascarenhas 1979, Clegg 1978, and Luscher-Mattli and Ruegg 1982, Tiwari and Tripathi 1998). Application of this technique is based on the fact that polar molecules have electric dipoles oriented randomly when there is no electric field. In the presence of electric field, these dipoles are aligned in the direction of applied field and intermediate +ve and –ve charges neutralize each other except the charge on the extreme faces of dielectric, also known as free charge in contrast to bound charge and are responsible for the conduction, thus increasing the capacitance of the conductor. Being polar molecule, water also behaves as dielectric and measurement of the capacitance at various hydration reflects the idea of mobility of water in the system.

Interaction between electromagnetic radiation like IR and water has been used to get information regarding water binding with different polymers and intact plant tissues and sub aerial cyanobacteria (Ruegg and Hani 1975, Ressler et al., 1976, Vandermeulen, and Ressler N 1980, Luescher-Mattli and Ruegg 1982, Tiwari and Tripathi 1998). In this method, mainly vibrational motion of water molecules is undertaken to define the state of cellular water. A diatomic molecule is treated as simple harmonic oscillator and its motion is such that the force acting on it is proportional to it's displacement from the position of equilibrium. As the distance of displacement from position of equilibrium is increased, the bonds become weaker and eventually dissociate resulting in appearance of free group. For example, in case of water, there is generation of free OH^-group. This free OH^-group gives a sharp peak around frequency range of 3300-3600 cm^{-1} referred as OH stretching band. A measurement in the OH stretching region has successfully been used to characterize kinetic behaviour of cellular water.

Cellular Water in Cyanobacteria

State of cellular water in the terms of desiccation has been well characterized in seeds (Vertucci and Leopold 1987) and sub-aerial cyanobacterium *Scytonema geitleri* (Tiwari and Tripathi 1998). We have used two cyanobacterial system *Scytonema geitleri* and *Lyngbya arboricola* growing in almost unialgal form on the roof surfaces of buildings and barks of trees. Both of these forms have been reported to survive extremes of desiccation in combination with high temperature upto 68°C on the roof surfaces. (Tripathi and Talpasayi 1980, Tripathi *et al.*, 1991). For characterizing cellular water at different levels of hydration and dehydration we applied thermodynamic as well as kinetic probe. For the experiments,

1 cm^2 dried cyanobacterial mats were equilibrated in the atmosphere of various relative vapor pressure developed by sodium chloride solution of different concentrations and saturated solutions of different salts (Table 20.1) at different temperatures 5, 15, 25 and 35°C for four days, a time sufficient to achieve no further increase in the weight.

Table 20.1: Applications of D'Arcy/Watt Equation to Sorption Isotherms of Dry and Grown Mats of *Scytonema geitleri* and *Lyngbya arboricola* at Different Tempeatures

Organism	*Growth Conditions**	*Temperature*	*Sorption Region*				
			Strong		*Weak*	*Multimolecular*	
			K	*K'*	*c*	*k*	*k'*
Scytonema geitlri	Dry	5	58.33	0.03365	0.1188	1.069	0.0185
		15	50.75	0.0251	0.1056	1.0276	0.0109
		25	32.66	0.0195	0.0908	1.0091	0.0857
		35	58.33	0.02634	0.08558	1.0230	0.00635
S. geitleri	Grown	5	9800	004791	0.11796	1.07852	0.01837
		15	58.33	0.03963	0.11931	1.07389	0.01817
		25	43.75	0.03457	0.11931	1.06927	0.01691
		5	51.33	0.02561	0.11841	1.0184	0.01521
Lyngbya arboricola	Dry	15	35.66	0.01944	0.09383	1.00915	0.00548
		25	25.75	0.01676	0.071667	1.00452	1.00312
		35	75.25	0.0306	0.08358	1.03227	0.00748
L. arboricola	Grown	5	17383	0.06434	0.11845	1.1064	0.01822
		15	128.91	0.06143	0.11931	1.09702	0.02026
		25	63.00	0.0503	0.11886	1.0877	0.01931
		35	58.22	0.0482	0.11841	1.0785	0.01912

*: The mats of *S. geitleri* and *L. arboricola* equilibrated at 0 bar were grown under light intensity of 72 μM photon m^{2}s^{-1} at 25°C for 24 and 72 hr respectively whereas dry mats of the cyanobacteria were obtained after drying at 85°C for 120 hr

The change in the weight *i.e.* Water content of the samples after four days of equilibration in the atmosphere of different relative vapour pressure solutions at different temperatures was used to prepare isotherm by least square fitting with mathematical model proposed by D'Arcy and Watt (1970) and analyzed by comparing the isotherm data at two temperatures via Clausius-Clapeyron equation in order to get the parameters such as differential enthalpy, entropy and free energy, strength and number of binding sites of water at strong, weak and multi-molecular region. To analyzed isotherm data for the calculation of parameters of D'Arcy/Watt model, a computer program base on simplex method has been used. The experimentally obtained values of relative vapour pressure and corresponding water content is fed in the above said program which carries out a function minimization via particular subroutine determining the values for parameters K,K',k,k' and c which when substituted to D'Arcy Watt equation, gives a minimum sum of square of differences between experimentally determined points on the isotherm and values calculated from the isotherm. In order to optimized fit of predicted curve to the experimental points, the deviation between observed and calculated values are weighed by taking the error expressed either as percentage of observed value or as absolute magnitude of deviation so that in all the cases maximum value is assigned to the magnitude of the error. For dielectric studies, small pellets prepared from dried algal samples were equilibrated for hydration and capacitance of pellets were measured by Impedance Analyzer (Hewlett-Packyard HP 4192A-LF) in the frequency range of 10^2 to 10^6 Hz. For infra-red spectroscopic analyses, thin pellets of nearly 100 mg of four days dried experimental samples, powdered and mixed with dry KBr powder in the ratio of 1:2000 (w/w) (Ruegg and Hani 1975). Spectra were obtained at frequency range of 4000-200 cm^{-1} in Perkin Elmer IR spectrophotometer and spectral variation in OH stretching region (approximately in the range of 3400-3200 cm^{-1}) was studied.

Analyses of Water Sorption Isotherms in Cyanobacteria

Sorption isotherms of *S geitleri* and *L. arboricola* measured at the temperature between 5-35°C followed typical reverse sigmoid shape consisting of three regions–a knee at low relative vapour pressure (0.00–0.2), a linear region at intermediate relative vapour pressure (0.2-0.75) and a sharp upswing at high relative vapour pressure (0.75-0.95). The amount of water sorbed by green mats was much higher than their respective dry mats. Higher amount of water was sorbed at low temperature (Figures 20.1–20.4).

Table 20.2: Water Binding Characteristics Determined from D'Arcy/Watt Equation Applied to the Isotherm of Dry and Grown Mats of *Scytonema geitleri* and *Lyngbya arboricola* at Different Temperatures

Organism	*Growth Conditions*	*Temp (°C)*		*Number of Sorption Sites*[b]		
			$\Delta H_{int(s)}{}^{a}$	*Strong*	*Weak*	*Multimolecular*
Scytonema geitleri	Dry	5	75.3129	1.125	3.985	0.6187
15	218.39	0.8394	3.5338	3.3645		
25	324.21	.0.6521	3.039	0.2866		
35	527.19	0.8795	2.866	0.2122		
S. geitleri	5	84.68	1.6023	3.956	0.6143	
15	225.93	1.325	3.998	0.6076		
25	35.60	1.156	3.994	0.5655		
Lyngbya arboricola	Dry	5	73.00	0.8565	3.972	0.5086
15	199.28	0.6501	3.114	0.1832		
25	302.75	0.5605	2.566	0.1043		
35	559.37	1.0234	12792	0.2501		
L. arboricola	Grown	5	95.02	2.151	3.973	0.6093
15	268.88	2.054	3.998	0.6775		
25	383.51	1.682	3.979	0.6458		
35	527.19	1.612	3.956	0.6394		

*: The mats of *S. geitleri* and *L. arboricola* equilibrated at 0 bar were grown under light intensity of 72 μM photon m^2s^{-1} for 24 and 72 hr respectively whereas dry mats of the cyanobacteria were obtained after drying at 85°C for 120 hr.

[a]: Integrated heat or sorption (-kJ/mol); [b]: Number of sites g/dry wt × 10^{21}.

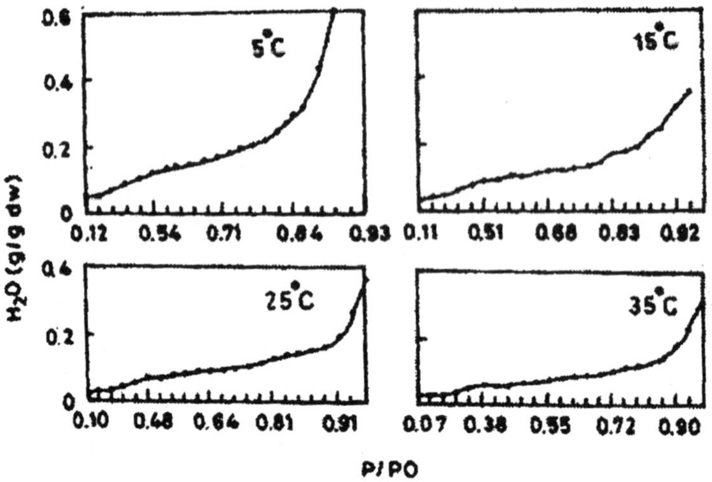

Figure 20.1: Sorption Isotherms of *Scytonema geitleri* Dry Mats (The mats dried al 90°C for five days and preweighed dry cyanobacterial mats were equilibrated in the atmosphere of relative vapour pressures ranging from 0.1–1.0 P/P_o at different temperatures (5, 15, 25 and 35°C) for 7 days, a time sufficient to sorb maximum water Amount of water sorb/g dry wt was calculated and analysed for other parameters in D'Arcy/Watt model)

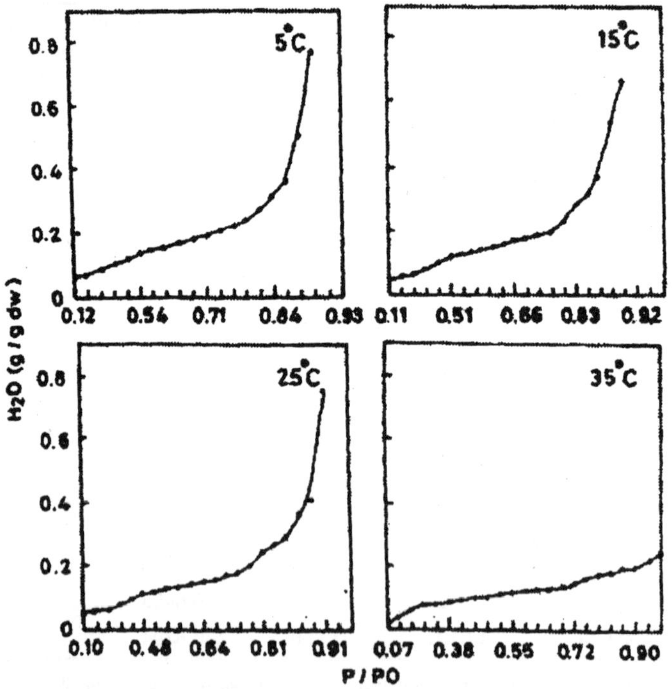

Figure 20.2: Sorption Isotherms of *Scytonema geitleri* Grown Mats (The grown mats were obtained by equilibrating the algal mats at bar under growth conditions for 72 hr. The other conditions were as described in the legend of Figure 20.1

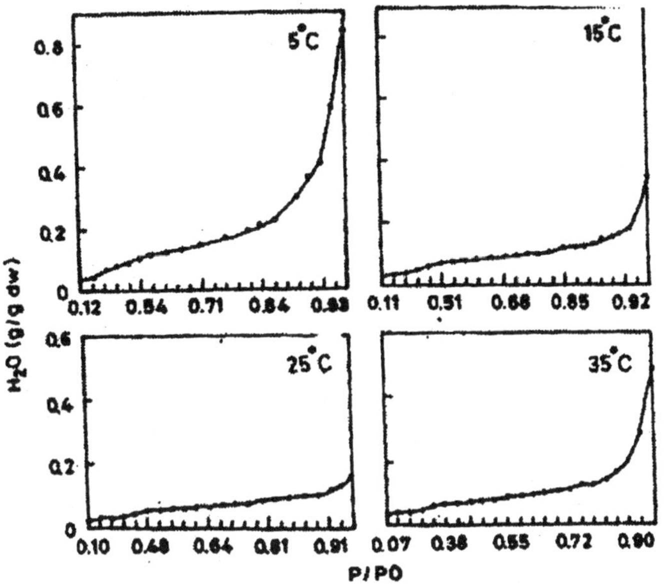

Figure 20.3: Sorption Isotherms of *Lyngbya arboricola* Dry Mats (The experimental conditions and other details were as described in the legend of Figure 20.1)

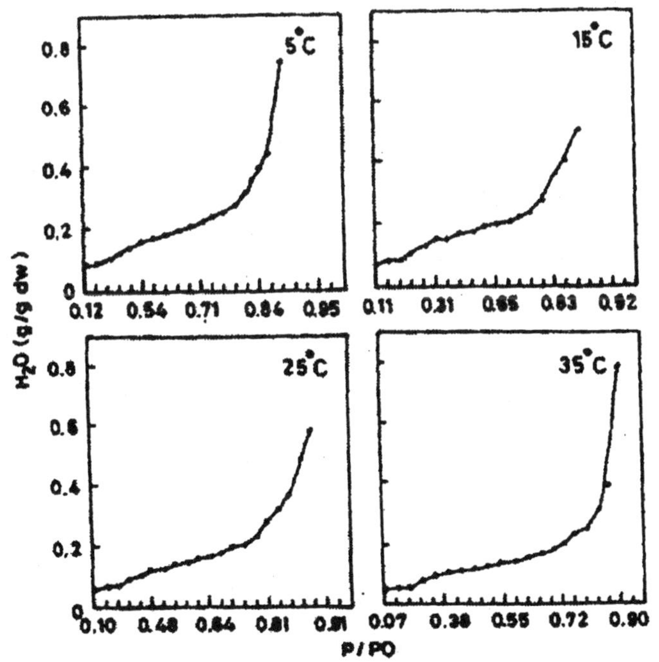

Figure 20.4: Sorption Isotherm of *Lyngbya arboricola* Grown Mats (The grown mats were obtained by equilibrating the algal mats at 0 bar under growth conditions for 120 hr. The other conditions were as described in the legend of Figure 20.1)

Sorption characteristics of both the cyanobacteria characterized (Table 20.2) showed that in dry mats, value of k ranged between 25.75 and 75.25. The strength of water onto strong binding site showed a decrease on increasing the temperature 5-25°C. At 35°C there was an increase which went upto 75.24 in the dry mats of *L. arboricola*. In the grown mats of both the cyanobacteria, the value of K decreased on increasing temperature from 5-25°C, though grown mats of *L. arboricola* did not follow the pattern of dry mats and they show a decrease in the value of K. The parameter K′ also followed the same pattern as the K. the coefficient c which is related to strength and number of water sorption at weak binding of isotherm was recorded in the range of 0.07667 to 0.11981g H_2O/g dw in both the cyanobacteria. In almost all the cases value of c decreased on increasing the temperature from 5-35°C. At the multi-molecular region of the isotherm, strength of water sorption (k) and number of sorption sites (k′) did not show much variation in their values.

Integrated values of water binding and number of strong weak and multi-molecular binding sites determined from the parameters of D′Arcy/Watt and applied to isotherm of grown and dry mats of cyanobacteria at different temperatures (Table 20.3) showed that integral enthalpy of water binding for first binding region was determined to be between–10.6 to–8.89 kJ/mol and–11.33 to–8.3 kJ/mol in dry mats of *S. geitleri* and *L arboricola* respectively. The number of sorption sites consistently decreased with increasing temperature from 5-35°C in all the regions in dry as well as grown mats of both the cyanobacteria. The sorption isotherm data of dry mats of both the cyanobacteria were also used to calculate differential enthalpy (ΔH), free energy (ΔG) and differential entropy (ΔS) in relation to water

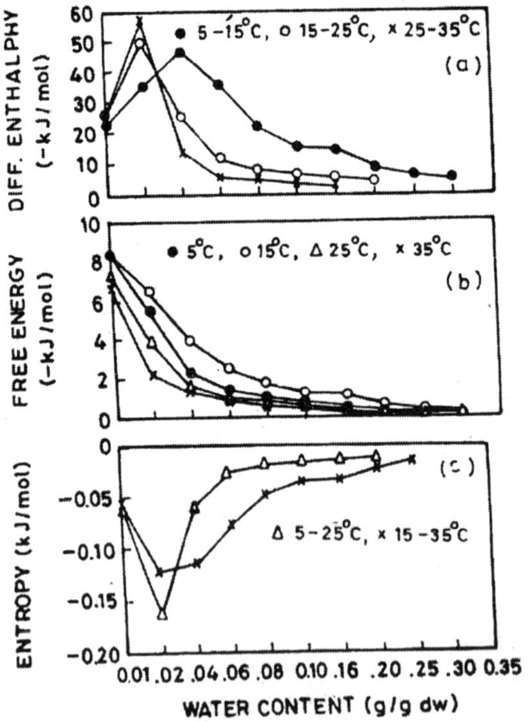

Figure 20.5: Thermodynamic Parameters Δ*H* (a) Δ*G* (b) and Δ*S* (c) Calculated for Three Sorption Regiosn in the Dry Mats of *L. arboricola* Equilibrated at Different Levels of Hydration (Calculation was made using isotherm data and Clausius-Clapeyron equations used by Vertucci and Leopold)

Table 20.3: Thermodynamic Properties of Water Sorption Calculated for Three Regions of Water Binding in Dry and Grown Mats of *Scytonema geitleri* at Different Temperature

Growth Conditions[a]	Temp (°C)	Sorption Regions: I				II				III			
		Water Content (g/g dw)	ΔH* (-kJ/mol)	ΔG (-kJ/mol)	ΔS (-kJ/mol)	Water Content (g/g dw)	ΔH* (-kJ/mol)	ΔG (-kJ/mol)	ΔS (-kJ/mol)	Water Content (g/g dw)	ΔH* (-kJ/mol)	ΔG (-kJ/mol)	ΔS (-kJ/mol)
Dry	5	≥0.08		5.460		0.1–0.25		0.788	–	0.25–0.39		0.07	–
			46.25				6.695				1.77		
	15	≥0.06		4.654	0.137	0.08–0.1		1.235	0.0315	0.15–0.3		0.3544	0.0136
			42.70				6.842				1.77		
	25	≥0.04		4.141	0.119	0.04–0.08		1.010	0.0210	0.1–0.3		0.4524	0.0140
			36.06				6.390				4.75		
	35	≥0.02	–	4.486	–	0.02–0.06		1.516	–	0.08–0.25	–	0.4832	–
Grown	5	≥0.08		8.396	–	0.09–0.17		1.802	–	0.18–0.40		0.473	
			13.866				8.472				1.155		
	15	≥0.07		7.318	0.0149	0.08–0.16		2.256	0.0169	0.17–0.40		0.449	0.00194
			20.202				5.572				1.102		
	25	≥0.06		6.871	0.0416	0.07–0.15		1.993	0.0125	0.16–0.4		0.434	0.0027
			18.259				4.062				0.871		
	35	≥0.07		6.489	–	0.08.0.14		1.869	–	0.15–0.4		0.396	

*: Differential enthalpy (Mf) has been calculated at temperature range of 5–15, 15–25, 25–35°C.

[a]: The mats of *S. geitleri* and *L. arboricola* equilibrated at 0 bar were grown under light intensity of 72 μM pholon m^2s^{-1} at 25°C for 24 and 72 hr respectively whereas dry mats of the cyanobacteria were obtained after drying at 85.C for 120 hr.

content at two temperatures by applying Clausius-Clapeyron equation as used by Clegg (1978a&b) (Figure 20.5). The differential enthalpy at strong water region was found to be very low. The values of ΔH increased on increasing temperature from 5-25°C and 25-35°C. The other cyanobacteria L. arboricola showed less negative values of ΔH. Similarly, grown mats of both the cyanobacteria reflected less negative values of ΔH at almost all water content and temperature compared to their respective dry mats. The value of free energy showed an increasing trend on the increment in hydration and temperature. At 25°C, on the average, *S geitleri* posed–4.33,–1.18 and–0.332 kJ/mol free energy at water binding region of I, II and III respectively whereas, at 35°C the average values went upto–3.19,–1.01 and–0.219 kJ/mol. Under such conditions, value of free energy increased in grown mats of both the organism. Similar trend of "S were observed in dry and grown mats of both the organisms.

Motional Behaviour of Sorbed Water in Cyanobacteria

The motional behaviour of sorbed water was characterized by dielectric and infra red spectroscopy, Capacitance and dielectric constant (ε') did not show apparent variation at hydration level below 0.07 g H_2O/g dw. At lower frequencies an upshift in the curves could be observed above the water content of 0.07 g H_2O/g dw. On increasing frequency, there was decrease in the value of both the parameters (Figures 20.6–20.7). Infrared spectroscopic analyses of *S geitleri* mats equilibrated at 0.15, 0.2, 0.3, 0.4 and 0.95 relative vapour pressure and recorded in the wave number range of 4000-200 cm^{-1} showed

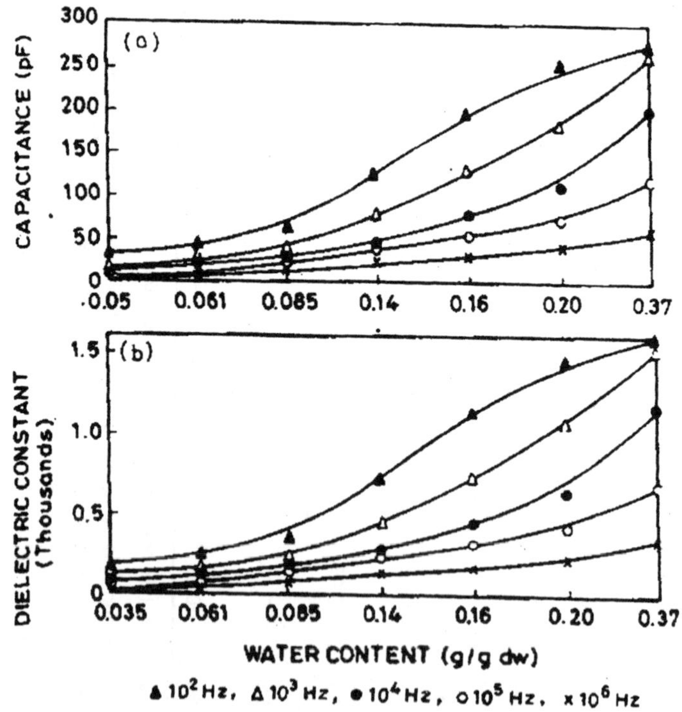

Figure 20.6: Capacitance (a) and Dielectric Constant (b) for Dry Pellets of *S. geitleri* Equilibrated at Different Levels of Hydration (For the measurement, dried mats were powdered and transformed to pellets. Algal pellets were equilibrated at different levels of hydration. Capacitance was measured in impedance Analyzer in the frequency range of 102–106 Hz. Dielectric constant was calculated using eq. 6)

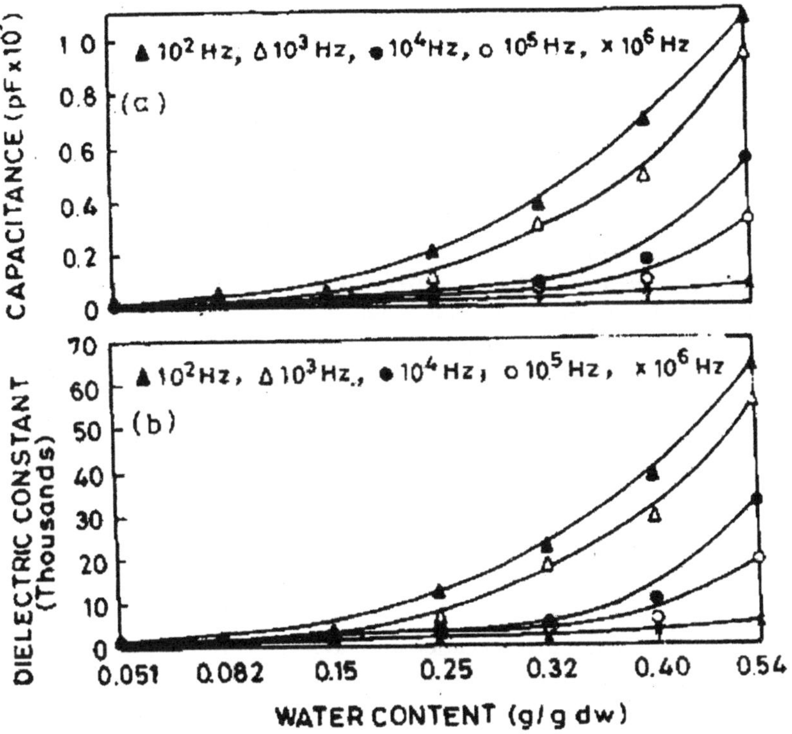

Figure 20.7: Capacitance (a) and Dielectric Constant (b) for Dry Pellets of *L. arboricola* Equilibrated at Different Levels of Hydration (Details of measurement and calculations were as given in caption for Figure 20.6)

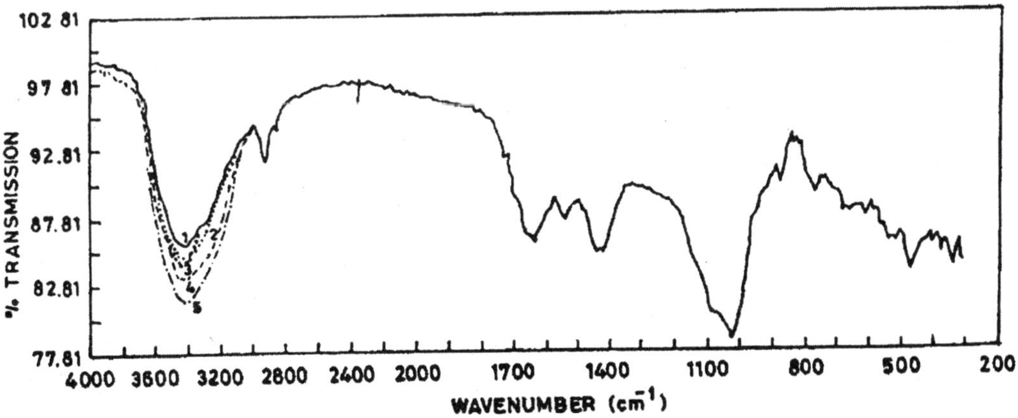

Figure 20.8: Infra-red Spectrum of *S. geitleri* Dry Mats Equilibrated at Dry State (1), 0.16 (2), 0.22 (3), 0.33 (4) and 0.96 (5) Relative Water Vapour Pressure (Cyanobacterial mats were mixed with KBr in the ratio of 1 : 200 (W/W) and transformed to pellet. Spectra were obtained at frequency range of 4000–200 cm^{-1}. Variation in OH stretching region between wave number of 3200–3600 cm^{-1} has been redrawn).

no significant variations in the spectra except in the wave number range of 3400 to 3200 cm^{-1} (Figure 20.8).

Conclusion

The water sorption characteristics of two cyanobacterial forms *S geitleri* and *L. arboricola* compared at different temperatures have been found useful in defining the state of cellular water at different levels of hydration. The pattern of sorption isotherms of these two forms under dry and growing conditions followed the usual sigmoid shape involving three regions in consistence with D'Arcy/Watt model. Like most of heterogeneous systems like legume seeds (Vertucci and Leopold 1987a&b), peeled wheat and sunflower seeds (Multan 1989) where water sorption at strong binding region is higher, the three regions of isotherms of two cyanobacteria can be divided in the ra nge of 0.00-20, 21-75 and 76-100 per cent RH respectively. How ever the values of water sorbed up to 2-5 per cent, 6-55 per cent and 56-100 per cent in strong, weak and multimolecular region in dry cyanobacteria are low as compared to legume and sunflower seeds. Matters like starch and chemotropism have been reported to sorb maximum water in weak and multimolecular region (Multan 1989 and Luscher-mattli and Ruegg 1982). It seems that dry mat of cyanobacteria possessing high amount of starchy substance or thick sheath made of polysaccharides might be responsible for showing starchy type of isotherm.

It has been proposed that very tightly sorbed water is necessary for the maintenance of structural integrity of biomolecules and desiccation tolerant organisms (Clegg 1978 a&b, Crow and Crow 1986). This doesn't seem to be relevant for two cyanobacteria taken for the study. It is relevant to mention that these two forms are highly tolerant to desiccation and high temperature (up to 100°C) under dry state. The unavailability of high strength of binding sites in strong binding region of isotherm in such desiccation tolerant system may be due to gradual replacement of water with certain substances such as trehalose, glycine betain, glycerol, sucrose etc. these substance are reported in many cyanobacteria inhabiting in water stress environments.

The motional behavior of water has been taken in to the consideration as another way to characterize the water in the system employing dielectric and IR spectroscopy. The observations made favored the concept of utilization of these techniques for the characterization cellular water in such type of desiccation-tolerant organisms. The pattern of changes in the value of capacitance and dielectric constant (ε') obtained at frequencies 10^2–10^6 Hz are consistent with findings made on lysozyme and ovalbumin. In S geitleri, curve of capacitance and dielectric constant plotted against hydration showed two critical hydration levels at 0.061 and 0.185 g H_2O/g dw reflecting the presence of relatively immobile water and sharp increase in the mobility of water respectively. IR studies followed the same pattern as observed in β casein. These observations reflect that water sorption isotherm of cyanobacteria fits well with D'Arcy/Watt model showing three sorption regions. Values calculated from D'Arcy/Watt equation fall within range of the values reported for other biomolecules. The observations made with kinetic probe supported the findings made with thermodynamic probe.

Acknowledgement

This work was supported by extra mural funding from Council of Scientific and Industrial Research, New Delhi, India.

References

Bull HB (1944) Adsorption of water vapour by proteins. *J. Am. Chem Soc.* 66, 1499-1507.

Brock TD (1975) Effect of water potential on *Microcoleus* (cyanophyceae) from desert crust. *J. Phycol.* 11, 316-320.

Bewely JD (1979) Physiological aspects of desiccation tolerance. *Ann. Rev. Plant Physiol.* 30, 195-238.

Crow JH and Clegg JS (1978) *Dry biological syatems*. Acad. Press. NY.

Crow JH and Crow LM (1986) Stabilization of membranes in anhydrobiotic organisms. In: *Membranes metabolism and dry organism*. (Ed. Leopold AC) Cornell University Press Ithaca, NY. Pp 188-209.

Clegg JS (1978) Hydration dependent metabolism and the state of cellular water in Artemia. In: *Dry biological systems* (eds. JH Crow and JS Clegg) Academic Press. NY 117-154.

D'Arcy RL and Watt JC (1970) Analysis of sorption isotherms of non homogeneous sorbents. *Trans Faradey Soc.* 66, 1236-1245.

Grigera JR and Mascarenhas S (1978) *Studies in Biophysics* 73, 19-24

Luscher-Mattli, M and Ruegg M (1982) Thermodynamic functions of biopolymer hydration. I Their determination by vapour pressure studies discussed in an analysis of primary hydration process. *Biopolymers.* 21, 403-418.

Multon JL (1989) Interaction between water and constituents of grains, seeds and by products. In: *Preservation and storage of grains, seeds and their by products cereals, oil seeds pulses and animal feed.* (ed. Multan, JL) CBS publishers. pp 89-159.

Potts M and Friedman EI (1981) Effect of water stress on cryptoendolythic cyanobacteria from hot desert rocks. *Arch. Microbiol.* 130, 267-271.

Potts M, Ocampo-Friedman R, Bowman MA, and Tozun B (1983) *Chroococcus* S24 and *Chroococcus* N41 (cyanobacteria):morphological, biochemical and genetic characterization and effect of water stress on ultrastructure. *Arch. Microbiol.* 135, 81-90.

Parsegian A, Rau D, Zimmerberg J (1986) Structural transitions induced by osmotic stress. In: *Membrane metabolism and dry organisms*. (ed. Leopold AC) Cornell University Press, Ithaca, NY pp.306-317.

Ruegg M and Hani H (1975) Infra-red spectroscopy of water vapour sorption process of caseins. *Biochem. Biophys. Acta.* 400, 17-23.

Ressler N and Ziauddin, Vygantas C, Tanzen W and Karachorlu K (1976) Improved technique for near infra red study of water binding by globular proteins and intact tissues. *Appl. Spectroscopy.* 30, 295-302.

Schneider MJT and Schneider AS (1972) Water in biological membranes: adsorption isotherms and circular dichroism as a function of hydration. *J. Membrane Biol.* 9, 127-140.

Suggett A (1975) Water carbohydrate interactions. In: *Water relation of foods.* (ed Duckworth R) Academic Press, London. Pp 23-36.

Tripathi SN (1980) Physiology and biochemistry of subaerial blue green algae. *Ph.D. thesis,* Banaras Hindu University.

Tripathi SN and Talpasayi ERS (1980) Sulphydryls and survival of subaerial blue green algae. *Curr. Sci.* 49, 31-32.

Tripathi SN, Tiwari BS and Talpasayi ERS (1991) Growth of cyanobacteria (blue green algae) on urban buildings. *Energy and Buildings.* 15-16, 449-505.

Tiwari BS and Tripathi SN (1998) Water binding in subaerial cyanobacteria. *Ind. J. Biochem and Biophys.* 35, 52-61.

Vandermeulen DL and Ressler N (1980) A near infra red analysis of water macromolecules interactions: hydration and the spectra of aqueous solutions of intact proteins. *Arch Biochem and Biophys.* 199, 197-205.

Vertucci CW and Leopold AC (1987) water binding in legume seeds. *Plant Physiol.* 85, 224-231.

Vertucci CW and Leopold AC (1987) The relationship between water binding and desiccation tolerance in tissues. *Plant Physiol* 85, 232-238.

Whittan BA, Donaldson A, and Potts M (1979) Nitrogen fixation by Nostoc colonies in terrestrial environments of Aldebra toll, Indian Ocean. *Phycologia.* 18, 278-287.

Whittan and Potts M (1982) Marine littoral. In: *The biology of cyanobacteria* (eds Carr NG and Whittan BA) Blackwell Scientific Publications, Oxford, pp. 515-542.

Yon J (1969) *Structure et dynamique conformationelle des proteins.* Ser. Methods. Herman press. Paris.

© R.K. Gupta & V.D. Pandey (*Editors*)
Advances in Applied Phycology, 2007
Daya Publishing House, New Delhi

275–293

Chapter 21

Nutraceutical, Pharmaceutical and Bioactive Potential of Cyanobacteria

N. K. Rathore, R. Raghuvanshi and S. Singh*

School of Studies in Microbiology,
Jiwaji University, Gwalior – 474011, India

ABSTRACT

Cyanobacteria are ubiquitous in nature and considered to be next to bacteria as for as their distribution is concerned. This large and diverse group of Gram negative prokaryote having traditional applications, is now being utilized and further exploited for biotechnological development. These organisms also possess immense potentials for application towards human welfare. However, except for the use of *Spirulina* as a health food, there are no other major products from cyanobacteria in India and elsewhere. This is due to the limitations in the economic production of cyanobacterial products in low quantity. Furthermore, mass cultivation technology has not yet been evolved for such potential organisms and in many cases the methods of industrial extraction are not even optimized. Intensive research is thus warranted to understand many of the basic aspects pertaining to the production of a metabolite with the concurrent evolution of applied research towards the large scale production of the product.

In recent years people throughout the world are focusing their attention towards cyanobacteria for their possible use for the photobiological production of biofuel, ammonia, amino acids, various metabolites, vitamins, toxins, therapeutic substances, aqua or animal feed. Various chemicals including restriction enzymes, pharmacological probes and labeled compounds for research, as

* Corresponding Author: Phone: 0751-2345338 (R), 0751-5016737 (O); Fax: 0751-2341450;
E.mail: surendrasinghg@indiatimes.com; surendrasinghg@yahoo.com

well as fluorescent probes for clinical diagnostics are now commercially available. Potentially useful compounds, including pharmaceuticals and industrial chemicals, have been identified and are currently being developed. This chapter summarizes the progress made so far in the field of biotechnological potentials of cyanobacteria, and highlights various areas of applications mainly nutraceutical, pharmaceutical and bioactive potential.

Introduction

Cyanobacteria are oxygenic photosynthetic prokaryotes, which have long been recognized as having potential in biotechnology. These organisms have been evolved nearly about three billion years ago and have been responsible for the initial oxygenation of the Earth's atmosphere. They are the large and diverse group of prokaryotic organism occupying unique taxonomic position. They combine an autotrophic mode of nutrition that is common to eukaryotic algae and plant cells with a metabolic system that is generally regarded as bacterial rather than the plant like. These prokaryotic organisms thus offer the growth potential of microbial cells growth with the light harvesting capabilities of plant cells, making them ideal candidates for biosolar energy conversion programs. Another advantage of cyanobacteria over eukaryotic organisms is the capacity of several strains to fix atmospheric nitrogen (N_2). Cyanobacteria perform various metabolic functions like other prokaryotic bacteria. Similarly they are also exposed to diverse conditions in their natural habitats and hence have evolved alternate metabolic pathways and elegant regulatory mechanisms for adaptation.

Cyanobacteria are now increasingly attracting the attention of the biologists in view of their importance in agriculture, industry, environmental pollution abatement, conservation, biodiversity and pharmaceutical markets. Cyanobacteria in general are one of the richest sources of known and novel bioactive compounds including toxins with wide pharmaceutical applications (Becker, 1994; Patterson, 1996). This has led to the systematic screening of cyanobacterial strains both to test their potential pharmaceutical applications and also to know their general effects on animal systems. In recent years people throughout the world are focusing their attention towards cyanobacteria for their possible use for the photobiological production of biofuels, ammonia, amino acids, various metabolites, vitamins, toxins, therapeutic substances and aqua or animal feed (Singh, 1992, 1993; Mohanty, 1996). Although, the screening of cyanobacterial strains having the ability to produce novel compounds with antimicrobial and antineoplastic activities were started as early as 1970 (Moore, 1981; Rinehart *et al.*, 1981), the systematic screening for therapeutic substances from micro-algae particularly cyanobacteria has, however, received greater attention recently.

Pharmaceuticals

Microorganisms have a significant role in Medical Sciences as they not only cause infections, but also produce organic substances that can cure infections. The discovery of penicillin in 1929 heralded the era of antibiotics and led to the realization that microorganisms can be considered as a rich source of clinically useful natural products. Since then several natural products with bioactive potentials have been discovered from microorganisms. The tremendous rate of discovery is a testament to the inherent ability of microorganisms to produce bioactive metabolites.

However, despite many discoveries, the pharmaceutical industries have focused their attention on the same group of microorganisms, mainly the actinomycetes as their primary source for new drugs. Although actinomycetes continue to be studied extensively, it is clear that the rate of discovery of novel metabolites from this group is decreasing and hence new sources of bioactive natural products must be explored. One such resource is cyanobacteria, which are recently being exploited for their potential as producers of biomedically efficient metabolites.

Cyanobacterial bioactive compounds will contribute significantly to drug development either directly or as models for developing analogs for increased activity and as research tools for studying physiological processes and diseases. Pharmacological evalutions of cyanobacterial products have been revolutionized over the past two decades, beginning with the early investigation of toxins, followed by studies of cytotoxic and antitumor activity, to the present where a myriad of activities based on whole animal models and receptor binding assays are being pursued.

Antimicrobials

Any compound, natural, synthetic or semi synthetic (*i.e.*, chemically derived from natural substances) that is clinically used for the treatment of microbial infections is generally known as antimicrobials. These antimicrobials include antivirals, antibacterials, antifungals and antiparasitic.

Antivirals

To discover a new antiviral agent from cyanobacteria, extracts from cyanobacteria were screened for anti HIV properties (Boyd, 1988). Cellular extracts of *Lyngbya langerhimii* and *Phormidium tenue* protected human lymphoblastoid T cells from the cytopathic effect of HIV–1 infection. With the use of an *in vitro* anti HIV-1 assay, a new class of HIV-1 inhibitory compound, the sulfonic acid containing glycolipids was isolated (Gustafson *et al.*, 1989). Recently a tetrazolium based microculture assay was developed to screen the extracts of cultured cyanobacteria for inhibition of the cytopathic effects of HIV (Gustafson *et al.*, 1989). At non-cytotoxic concentration, pure cyanobacterial compounds were strikingly active against HIV-1 in cultured human lymphoblastoid T cell-lines (Gustafson *et al.*, 1989).

An aqueous extract of *Spirulina platensis* inhibits HIV-1 replication in human derived T–cell lines and in human peripheral blood mononuclear cells. The extracts also blocked Raucher Murine Leukemia Virus (RLV) induced plaque (Ayehunie *et al.*, 1996, 1998). The aqueous extract of *Spirulina* also exhibited a dose–dependent inhibition of the replication of *Herpes simplex* virus type 1 (HSV-1) in Hela cells within the concentration range of 0.08-50 mg ml^{-1}. However, this extract does not show any virucidal activity and could not even interfere the adsorption of the virus to the host cells (Hayashi *et al.* 1993). Bioactivity-directed fractionation of a hot water extract from *S. platensis* led to the isolation of a novel sulfated polysaccharide, calcium spirulan (Ca-SP), as an antiviral principle. This polysaccharide was composed of rhamnose, ribose, mannose, fructose, galactose, xylose, glucose, glucuronic acid, galacturoinic acid, sulfate and calcium. Ca-SP was found to inhibit the replication of several enveloped viruses, including *Herpes simplex* virus type 1 (HSV 1), Human Cytomegalo Virus (CMV), Measles Virus, Mumps Virus, Influenza A Virus and HIV 1. Ca-SP also selectively inhibits the penetration of viruses into the host cells. Retention of the molecular conformation by chelation of calcium ion with sulfate groups was suggested to be indispensable to its antiviral effect (Hayashi *et al.* 1996).

A large number of cyanobacterial extracts have exhibited antiviral activity against *Herpes simplex* virus II (HSV-2) (Rinehart *et al.*, 1981). Whereas 10 per cent of the cyanobacterial extracts showed anti HSV-2 activity, 2.5 per cent of them, however, exhibited activity against respiratory syncytial virus (Patterson *et al.*, 1993). When antiviral activity of lipophilic and hydrophilic extracts of cyanobacteria against three pathogenic viruses were examined, a significant reduction in cytopathic effect normally associated with viral infection was recorded (Patterson *et al.*, 1993). Sulfoglycolipids isolated from cyanobacteria also exhibit strong antiviral properties. In *in vitro* studies, Helper T cells exposed to cyanobacterial sulfoglycolipids were protected from HIV-1 infection (Gustafson *et al.*, 1989). *Phormidium valderianum*, *Plectonema terebrans* and *Pseudanabena schimidelei* also exhibited HbsAg-clearing activity (Sundararaman, 1992), indicating their potential use against Hepatitis B virus.

Cyanovirin derived from the cultures of the fresh water cyanobacterium *Nostoc ellipsosporum* binds to GP-120 protien of HIV and thus inhibits its replication. *N. ellipsosporum* producing cyanovirin-N and cyanovirin-V is also capable of inhibiting the growth of HIV-I and HIV-II (Chauhan *et al.*, 2004). Recently, allophycocyanin, a red fluorescent protein isolated from *S. platensis.* is found to inhibit enterovirus 71-induced apoptosis, which may have a significant impact on protecting hosts from severe consequences associated with infection with enterovirus 71 (Shih *et al.*, 2003).

Antibacterial

The antibacterial activity of cyanobacteria in general has been poorly studied (Reichelt *et al.*, 1984; Moore *et al.*, 1988).The cell-extract fractions from *Spirulina subsalsa* resulted in the complete inhibition of the growth of a clinical isolate of *Pseudomonas aeruginosa*, which is considered to be one of the difficult pathogens to treat (Sundararaman, 1992). *Nostoc muscorum* also produces an antibacterial agent having broad-spectrum activity (Bloor and England, 1991).

The chloroform-methanol extracts obtained from several *Oscillatoria* species inhibit the Gram-positive bacterial pathogen, *Staphylococcus aureus*. Whereas, the compounds obtained from *Phormidium angustissimum* and *Lyngbya sp.* were found to be effective against Gram negative and Gram positive pathogenic bacteria, the compounds obtained from *S. subsalsa* were, however, found to possess antibacterial activity against Gram negative organisms only (Sundararaman, 1992). Similarly, the crude extracts of *P. angustissimum* and *Lyngbya sp.* were found to be potential sources of new broad-spectrum antibacterial compounds (Sundararaman, 1992).

Antifungal

A large number of cyanobacteria are reported to possess the ability to produce the compounds which exhibit the antifungal activity. The cyanobacterial extracts that inhibit the proliferation of tissue cultured mammalian cells also inhibit the fungal growth, indicating that the mechanism of inhibition involves a site or function similar to eukaryotic cells. Scytophycin, a potent bioactive metabolite from cyanobacteria also exhibited antifungal activity (Ishibashi *et al.*, 1986; Patterson and Carmeli 1992). Tjipanazoles, N-glycosides of indole-carbazoles isolated from the cyanobacterium *Tolypothrix tjipanasensis* exhibited significant fungicidal activity against phytopathogenic fungi (Bonjouklian *et al.*, 1991). The laxaphycins are a large family of cyclic undeca and dodecapeptides responsible for antifungal activity caused by crude extracts of *Anabaena laxa*. The laxaphycins structurally and biologically resemble to a group of cyclic peptides known as hormothamnins, which have recently been isolated from the *Hormothamnion enteromorphoides* (Gerwick *et al.*, 1989). Extracts of *Synechococcus elongatous*, *P. corium* and *Dichothrix baueriana* were found to be active against *Candida* species (Sundararaman, 1992). A compound, presumably a methoxylated disaccharide, purified from *Oscillatoria latevirens* exhibited anti-*Candida*, anti–*Mucor* and antibacterial activity (Deth, 1999).

Antialgal Activity

One hundred and ninety-eight cyanobacterial strains, isolated mainly from diverse habitats in Australia, Indonesia, Nepal, Thialand and Vietnam, were screened for their antibiotic activity against green algal species of the genera *Coelastrum*, *Scenedesmus* and *Monoraphidium* (Schlegel *et al.*, 1999). Out of these, ten strains of *Fischerella*, seven strains of *Nostoc* and three strains of *Calothrix* produce antialgal compounds with a broad spectrum activity. Some of these bioactive cyanobacteria are reported to inhibit the growth of *Coelastrum, Scenedesmus* and *Monoraphidium*, whereas others inhibited the growth of either two or one algae (Schlegel *et al.*, 1999). *P. valderianum* completely eliminated a dense population of marine diatoms inducing the extracellular production of antidiatom compound.

Other Potential Drugs

The hypoglycemic effect of *Spirulina* in non-insulin dependent diabetes mellitus (NIDDM) patients has been reported (Venkatraman, 1983). A significant reduction in the fasting and postprandial blood and urine sugar level has been reported following the uptake of about 2g of *Spirulina* (Venkatraman, 1983). This has also been confirmed in experiments using albino rats with allaxan-induced diabetes. The reduction in the sugar level following *Spirulina* uptake has been attributed to the possible stimulation of prostaglandins. However, more studies on the similar lines are necessarily required not merely to establish the levels of *Spirulina* tablets, but also to understand the mechanisms of its action (Venkatraman, 1983). β–carotene as a dietary supplement has been reported to inhibit the development of 7, 12 dimethyl benzanthrazene (DMBA)–induced salivary gland carcinomas in rats.

Nutraceuticals

There has always been a lack of harmony between the rapidly growing world population and an adequate supply of protein in human diets. The gap between the human population and the food supply has widened steadily in recent years and the balance is greatly in the favour of population growth. According to World Health Organization about more than two million people are suffering from malnutrition in different parts of the globe and majority of them belong to Third World Countries like India. Though efforts have been made to improve land and animal based conventional sources of protein (meat, fish, poultry, eggs, milk, pulses, vegetables etc.) production, frequent increase in food scarcity and rising prices of food commodities may not satisfy the demand of the ever increasing global population. Furthermore, in view of the appearance of new viruses, drug-resistant bacteria and ineffective antibiotics, scientists are discovering the possibility of certain foods and/or food additives to increase the immune system and confronts problems associated with early aging.

Cyanobacteria have found many tradition applications. However, its biotechnological applications have recently been recognized.

Spirulina, a photosynthetic, filamentous, spiral-shaped, multicellular blue-green alga, that has been existing on earth surface for more than 3-6 billion years, has been used as food and food supplement and may be one of the disease preventing and anti-aging wonderful nature gift to mankind (Puniya *et al.*, 1995). The high contents of protein, vitamin B_{12}, iron, β–carotene, iodine and g–linolenic acid make the *Spirulina* protein as " naturally made" promising dietary substitute that could provide significant nutritional input to the Third World (Puniya *et al.*, 1995). In India, the *Spirulina* technology has been extended even to the villages by training the rural people especially women to grow this alga in their backyards for nutrition and income generation. Department of Biotechnology (Govt. of India) has assisted in rehabilitating the earthquake-stricken villagers of Latur, Maharashtra by introducing the *Spirulina* technology as an income generating activity. In addition to that a number of different cyanobacteria, mostly *Nostoc* strains have traditionally been used as food or food supplements, however, these strains are not extensively exploited (Jassby *et al.*, 1988).

Spirulina is a wholesome food supplement rich in proteins, vitamins, amino acids, beta-carotene, linolenic acid, minerals, and other nutrients. *Spirulina* biomass is available mainly as health food in the form of powder, soup, noodles, candies, appetizers, vegetable pate, low calorie bread, bread spread, health drinks and multivitamin tablets. *Spirulina* is claimed as a non-toxic, nutritious food, with some corrective properties against viral attacks, anemia, cancer, hepatotoxicity, cardiovascular diseases, hyperglycemia, hyperlipidemia, immunodeficiency, and inflammatory processes and as a source of the yellow coloration of egg yolk when consumed by hens. Several of these activities are attributed to either *Spirulina* itself or to some of its components including fatty acids omega-3 or omega-6, beta-carotene, alpha-tocopherol, phycocyanin, phenol compounds and sulphated polysachharide, Ca-spirulan (Ca-SP). *Spirulina* is a food that has extremely long shelf life. It also contains 26 times the calcium of milk and has a good supply of niacin and phosphorus. Protien Efficiency Ratio (PER) of *Spirulina* protein has been reported to be higher than vegetables, cereals and soya proteins. It has been proved that 1 kg of *Spirulina* is equivalent to 1,000 kg of assorted vegetables. *Spirulina* also has six times more protein than eggs and 20 times more than milk. However, the digestibility coefficient and biological values of *Spirulina* protein are only marginally lower than the milk protein, casein.

Supplementation of *Spirulina* with cereals like rice and wheat on isoproteinic levels improved the protein quality (Venkataraman, 1983). The bioavailability of total carotenes and in particular β–carotene from *Spirulina* was found to be comparable to the values reported for carotenes from other plant sources like leafy vegetables and carrots (Annapurna *et al.*, 1991). *Spirulina* is also a good source of vitamin A and hence it can be used as a source of vitamin A in the diet (Annapurna *et al.*, 1991).

In a clinical study, feeding of *Spirulina* (1g day^{-1} for 150 days) to the children suffering from vitamin A deficiency (Bitot's spot) has resulted in the decrease in symptoms of vitamin A deficiency from 80 per cent to 10 per cent (Seshadri, 1991). It is also the richest source of iron and vitamin B12 and has all the essential amino acid besides enzymes and minerals.

Phycobiliproteins

Cyanobacteria in general possess all the known phycobiliproteins: phycocyanin, phycoerythrin, phycoerythrocyanin and allo-phycocyanin. Among them, phycocyanin and phycoerythrin are commercially valuable. 'Linablue' a phycocyanin product from Dainippan ink and Chemicals Inc, Japan is an odourless, non-toxic blue powder and used for colouring candy, ice cream, dairy products and soft drinks (Cohen, 1986). Phycocyanin is also obtained in a water insoluble form from *Spirulina* and used in eye shadow, eyeliner and lipstick preparations (Dainippon Patent, 1980). The blue or red chromophores are isolated by enzymatic or acid hydrolysis of the protein to yield more concentrated pigment which is used in cosmetics (Dainippon Patent, 1981). Phycoerythrin from *Spirulina* and other cyanobacteria is used as a food colour for products like icecream, yoghurt and it could also be used in cosmetics.

C-phycocyanin is a major biliprotein of *S. platensis*, the cyanobacterium having antioxidant (Upasani, *et al.*, 2001), anti-arthritic, and anti-inflammatory properties (Ramirez, *et al.*, 2002). Phycocyanin acts as a hepatoprotective agent and as a hydroxyl radical scavenger (Madhava *et al.*, 2000). This pigment also inhibits oxidative damage in DNA and hence may be used as a therapeutic agent (Bhat and Madyastha, 2001). However, the mechanism of action of phycocyanin is not yet clearly understood.

C-phycocyanin, which has been shown to be a selective COX-2 inhibitor, induces dose-dependent apoptosis in BC–8 cells with a concomitant decrease in cell viability. However, BC-8 cells transfected with Bcl-2 gene were found to be resistant to phycocyanin mediated apoptosis. These observations were further confirmed by staining of these cells with Annexin-V. Annexin-V staining is observed in BC-8 cells after treatment with phycocyanin, which was, however, not observed in Bcl-2-transfected cells. These observations clearly demonstrate the phycocyanin-induced apoptotic death in BC-8 cells and its inhibition by Bcl–2 (Bobbili *et al.*, 2003).

β-carotene

The β–carotene level in *Spirulina* at over 300 mg per cent is perhaps the highest content available from any other natural sources (Venkatraman, 1983). The β–carotene content of *Spirulina* is more than 18 times greater than carrot, which is the normally known source of β–carotene. Furthermore, the natural β–carotene of *Spirulina* is different from the synthetic β–carotene, since it contains a higher percentage of 9-cis-isomer as compared to over 47 per cent of all trans in the synthetic form. β–carotene as a dietary supplement has been reported to inhibit the development of DMBA–induced salivary gland carcinomas in rats and UV-induced skin cancer in hairless mice. Based on epidemiological studies, it has been suggested that β–carotene can also reduce the chances of occurrence of cancer in human (Normal and Temple, 1988). *Spirulina* protein is also rich in iron, magnesium and trace minerals and the iron of *Spirulina* can be easily absorbed as compared to other iron supplements (Henrikson, 1997).

Amino Acids

Overproduction of amino acids is common feature of microbial resistance to analogs of amino acids. In general, the overproducing organisms are heterotrophic and production of metabolites is supported by the organic sources supplied with the medium. To our knowledge, such a phenomenon has been reported rarely for obligate photoautotrophs, a category which includes most cyanobacteria. Cyanobacteria liberate small quantities of amino acids into their medium (Fogg 1971, Fogg *et al.*, 1973). The most abundant amino acids liberated by cyanobacteria were phenylalanine, threonine, glutamate and glycine, irrespective of the nitrogen source on which the cells were grown. Amino acid production was obtained with the mutant strains of *Anabaena variabilis*, which liberated alanine as the major abundant amino acid along with phenylalanine and tyrosine (Kerby *et al.*, 1987). Anolog-resistant mutants of *S. platensis* also produce the corresponding amino acids (Riccardi *et al.*1981). Continuous photobiological production of amino acids was achieved following immobilization of the mutant cells in calcium alginate gel (Kerby *et al.*, 1988). However, the effective isolation of strains for commercial production of amino acids is dependent on our knowledge of transport and metabolism of amino acids. Characterization of such strains can increase our basic knowledge thus permitting further strain improvement. Advances are being made in both fundamental and applied aspects by use of mutant strains of cyanobacteria. Recent advances in cyanobacterial molecular genetics can be expected to have a major impact in this field.

Ammonia

One of the biotechnological aspects receiving considerable interest, especially by using cyanobacteria, is the photobiological production of ammonia. Light-driven synthesis of ammonia at the expense of N_2 and water via photosynthesis is a photobiological process for the conversion of solar energy into chemical energy, namely ammonia which can be used as fertilizer. Since, cyanobacteria in natural conditions release limited or no fixed-N (ammonia), attempts are being made to generate strains which can fix N_2 at higher rate and release most of the fixed-N in the form of ammonia. In symbiosis about 50-90 per cent of the N_2-derived ammonia is liberated by the cyanobiont (Rai, 1990). Attempts have also been made for photobiological production of ammonia at the expense of N_2 and nitrate by using L-Methionine–DL-sulphoximine (MSX), a glutamate analogue and an irreversible inhibitor of GS (Vincenzini *et al.*, 1986; Singh and Rai, 1989; Musgrave *et al.*, 1982; Guerrero *et al.*, 1982;

Ramos *et al.*, 1987; Singh 1992, 1993). However, this approach seems to be impractical because MSX rapidly leads to deficiency of glutamine and other vital nitrogenous compounds with cessation of ammonia production and finally lysis of the cells. An alternative approach is to select mutant strains particularly deficient in GS activity which would liberate ammonia resulting from N_2 fixation into the medium. Mutant strains of *Anabaena variabilis* and *Nostoc* ANTH fixed N_2 and liberated most of the fixed N_2 in the form of ammonia into the external medium (Singh, 1992, 1993). Following immobilization in calcium alginate gel, sustained photoproduction of ammonia was obtained in reactors. The rate of ammonia production was comparable to the rate of its MSX-treated freely suspended wild type counterpart.

Fatty Acids

Fatty acids, in general, are of commercial value and many are pharmaceutical agents. Fatty acid of *Spirulina* possesses palmitic acid, g-linolenic acid (GLA), linolenic acid and oleic acid predominantly (Cohen *et al.*, 1987). The concentration of GLA ranged between 8 to 31.7 per cent of the total fatty acids. Since, *Spirulina* is one of the best sources of GLA, a rare essential fatty acid and the key to good health. (Henrikson, 1997), efforts are being made to produce GLA from other cyanobacterial strains.

Furthermore, the presence of essential fatty acids (EFAs) in *Spirulina* is very important. EFAs protect the cell membranes from destruction by short wave length electromagnetic radiations, carcinogens, bacteria and viruses (Horrobin, 1990). The EFAs decrease the aggregation of blood platelets and coronary thrombosis and also inhibit the salt-induced hypertension (Horrobin, 1990). Essential fatty acids are also recommended for the prevention and treatment of multiple sclerosis (Horrobin, 1990).

In diabetes mellitus, EFAs reduce the diabetic retinopathy. Diabetes induces the capillary permeability and fragility whereas EFAs increase the membrane intregrity. Because of their fortifying action in cellular membranes, EFAs are important in the control of gastrointestinal, connective tissue and skin disorders (Horrobin, 1990).

Vitamins

Although, a number of cyanobacteria are known to be rich sources of vitamins and many are known even to excrete them into their milieu (Borowitzka and Borowitzka, 1988), no report is, however, available on the commercial exploitation of these organisms for vitamin production. Some of the marine cyanobacteria appear to be the potential sources of vitamin B complex group and vitamin E (Sundararaman, 1992). *Spirulina* is considered to be the potential source of vitamin B_{12}, which is essential for healthy nerves and tissues especially for vegetarians (Henrikson, 1997). Studies among preschool children in India have demonstrated *Spirulina fusiformis* to be an effective source of dietary vitamin A (Annapurna *et al.*, 1991).

Bioactive Compounds

The discovery of new sources of bioactive natural products are of paramount importance to assure the supply of new drugs needed to treat those infectious diseases that have become resistant to currently available drugs and also to combat newly emergent diseases. Considerable efforts have been made in the past to identify interesting bioactive compounds from different sources including cyanobacteria (Table 21.1). The search for bioactive compounds in cyanobacteria has been reviewed extensively (Patterson, 1996). The potential impact of this field on all of cyanobacterial biotechnology is immense. The wide array of cyanobacterial strains and the range of habitats that they occupy guarantee statistically to predict success in finding a valuable compound.

Table 21.1: Bioactive Compounds from Cyanobacteria (Burja *et al.*, 2001)

Organisms	*Compounds*	*Activity*
Microcystis aeruginosa	Aeruginosin, Kawaguchipeptin, microcystilide, microcystin micropeptin, microviridin	Enzyme inhibitor, cytotoxic, cell-differentiation, hepatotoxin, promoter, endotoxic, tumor antibiotic, anticancer
Microcystis viridis	Cyanoviridin RR	Toxic
Synechocystis trididemni	Didemnin	Anticancer, antiviral, Immunosuppressive
Lyngbya langerheimii	Sulfolipid	Anti –HIV activity
Lyngbya majuscula	Antillatoxin, aplysiatoxin, Aparamide, barbamide, Sulfolipid, lyngbyatoxin	Anticancer, anti-inflammatory, antiviral, anti-HIV, antimicrobial, cytotoxic, toxin
Oscillatoria acutissima	Acutiphycin	Anticancer activity
Oscillatoria agardhii	Agardhipeptin, microcystin	Enzyme inhibitor, hepatotoxin
Oscillatoria nigroviridis	Oscillatoxin	anticancer
Phormidium tenue	Sulfolipid	anti-HIV activity
Plectonema radiosum	Radiosumin, tubercidin	Enzyme inhibitor
Spirulina platensis	Calcium spirulan, poly-β-hydroxybutyrate, phycocyanin	Anticancer, anti-HIV activity, free-radical scavenger
Anabaena basta	Bastadin, bastadin O-sulfate esters	Antibiotic, anti-inflammatory, cytotoxicity
Anabaena circinalis	Circinamide, microcystin	Enzyme inhibitor, toxic
Aphanizomenon flos-aquae	Aphanorphine, siatoxin	Antibiotic, anticancer toxic
Nostoc commune	Nostodione, microsporine, diterpenoid	Antifungal, antibiotic, antimitotic, cytotoxic, suncreen pigment
Nostoc ellipsosporum	Cyanovirin	Anti-HIV, antiviral
Nostoc muscorum	Muscoride	Antibiotic
Nostoc sphaericum	Staurosporine, indolecarbazole	Antiviral, cytotoxic
Tolypothrix nodosa	Tolyporphin	Antibiotic
Tolypothrix tenuis	Toyocamycin, tubercidin	Antifungal, cytotoxic

Screening for a wide variety of potentially useful bioactivities, including cytotoxic, multi-drug resistance reversal, anti-fungal and antiviral effects, has led to the discovery and identification of numerous novel bioactive compounds or secondary metabolites possessing unprecedented structures. Structural novelty can be combined with exceptionally high incidence of biological activity.

Enzymes

Microbial enzymes have been used in various industries for many centuries. Recently, with the advent of biotechnology, there has been a growing interest and demand for enzymes with novel properties. Cell bound and extracellular phosphatase activities of cyanobacterial isolates have also been reported (Whitton *et al.*, 1991). *Microcystis aeruginosa* presents a large capacity of mineralizing organic phosphorous per unit of biomass due to its high alkaline phophatase activity (Giraudet *et al.*, 1997). A number of studies have been conducted on the detection of enzymes including phosphatase, arylsulfatase, chitinase, L-asparginase, L-glutaminase, amylase, protease, lipase, cellulase, urease and lactamase produced by cyanobacteria (Chandrasekaran, 1997). Freely-suspended and immobilized cells of *Anacystis nidulans* contain amino acid oxidase activity, which is further increased by illumination with red light (Wikstrom *et al.*, 1982). Superoxide dismutase exists in all aerobic organisms and has been purified from *Spirulina*.

Restriction Enzymes

Although more than 500 restriction enzymes with at least 100 different site specificities have been reported (Roberts, 1985), enzymes with new specificities are continuously sought to increase the possibilities of DNA manipulation. Many species of cyanobacteria, both filamentous and unicellular, appear to be a promising source of such enzymes (Table 21.2).

Table 21.2: Important Restriction Enzymes Identified in Cyanobacteria (Ciferri *et al.*, 1989)

Organisms	Enzymes	Isoshizomer of	Recognition Sequence
Anacystis nidulans	Ani I	—	CCNNGG
Anabaena variabilis	Ava I	—	C ↓ PyCGPuG
	Ava II	—	G ↓ G(A/T)CC
	AvaIII	—	ATGCAT
Anabaena cylindrica	Acy I	—	Gpu ↓ CGPyC
Nostoc sp. 6705	Nsp BI	AsuII	TTCGAA
	Nsp BII	—	C(G/C)G ↓ C(T/G)G
Nostoc sp. 8009	NspMACI	BglII	A ↓ GATCT
Nostoc sp. 7413	NspHI	Nsp(7524)I	PuCATG ↓ Py
	NspHII	AvaII	GG(A/T)CC
Spirulina platensis	SplI	—	C ↓ GTACG
	SplII	TthIIII	GACNNNGTC
	SplIII	HaeIII	GGCC
Tolypothrix tenuis	TtnI	HaeIII	GGCC

Cyanobacteria contain different endonucleases in different combinations, some of which are isoschizomers even in unrelated strains. For example, the isoschizomers AquI and AvaI, that recognize and cleave the sequence CPyCGPuG are produced by a unicellular cyanobacterium *Synechococcus* and a filamentous cyanobacterium *Anabaena*. Similarly, besides other restriction endonucleases, *Fremyella* and *Nostoc* contain enzymes with the same specificity as Ava II (Whitehead and Brown, 1985), the enzyme which was first identified in *Anabaena*. This is not unusual as the same restriction enzyme has been found in different bacterial genera (Roberts, 1985). Such findings may indicate that some of the activities are plasmid or phage-borne, although, at the moment, there is no evidence correlating restriction activity and presence of plasmids or phages.

Immunomodulators

Spirulina is commercially produced for human consumption as well as for feed ingredients. Recently, whole *Spirulina* and cell extracts were found to enhance immunity in animals by increasing phagocytic activity. The whole cells of cyanobacteria and lipopolysaccharides (LPS) isolated from cyanobacterial cells were shown to stimulate the production of macro and micro globulin antibodies (Rathore *et al.*, 2004).

C-phycocyanin and polysaccharides isolated from *Spirulina* was found to possess high erythropoeitin (EPO) activity. Effect of these compounds on peripheral blood and bone marrow hematopoietic stem and progenitor cell in normal, irradiated and anemic mice showed the unique capacity of C-phycocyanin and polysaccharide to influence the differentiation and proliferation of committed hematopoietic progenitor cell. In addition to this, they also lower the anemic degree of mice combined with gamma-ray irradiation and Benz hydrazine hydrochloric acid and peritoneri injection (Zhang-Chang-Wu, 1994).

Mice fed with *Spirulina* diet showed increased numbers of spleenic antibody producing cells in the primary immune response to sheep red blood cells (SRBC). *Spirulina* enhances the immune response, particularly the primary response by stimulating the macrophages function, phagocytosis and IL-1 production (Hayashi *et al.*, 1994).

Bronchoalveolar lavage macrophages isolated from cats, when cultured on a glass cover slip with water-soluble extracts of *S. platensis* in a concentration range of 0-60 mg ml^{-1} for 2 h could not cause significant macrophage cytotoxicity over untreated controls. The *S. platensis* extract enhances macrophage phagocytic function which implies that *Spirulina* supplementation might improve the disease resistance potential in cats (Qureshi *et al.*, 1996).

The exposure of chicken macrophage to *S. platensis* extract enhances selected effector functions of cells of the chicken immune system. However, the ability of spleenic natural killer cells to kill tumour cell targets was not affected by *Spirulina* treatment. Macrophage cultures exposed to *Spirulina* produced a factor in their culture supernatant with tumoricidal potential, which was similar in reactivity to the factor produced by macrophage following exposure to lipopolysaccharide (Qureshi *et al.*, 1995).

Chicken macrophage exposed to water soluble extracts of *Spirulina* showed enhanced phagocytic activity *in vitro*, suggesting activation of mononuclear phagocytic system function. Furthermore, dietary supplemention of *Spirulina* (1,000 to 10,000 ppm), improved thymic weights, enhanced CBH response, increased tumor cell killing by NK cells and doubled the macrophage phagocytic potential over chickens fed a basal diet. *Spirulina* enhances several immunological end points in chickens both during *in vitro* and in *in vivo* exposures (Qureshi *et al.*, 1995). *Spirulina* consumption lowers the amount of igE in the blood, which in turn normalizes and reduces allergies in the body without any side effect (Evets *et al.*, 1994). *Spirulina* also shows the immunomdulartory and biomodulatory effects (Rathore *et al.*, 2004).

Anticancer/Antitumor/Anti Proliferative and Cytotoxic Compounds

Micro algae in general and cyanobacteria in particular have been identified as the most significant group of organisms from which new anticancer type natural products have been isolated (Moore and Patterson 1988,1991,1993,1994). Many of these bioactive compounds possess unprecedented structures. Structural novelty is combined with exceptionally high incidences of biological activity. Crude extract of 6 per cent cyanobacterial strains studied exhibited antiviral activity to *Herpes simplex* virus (Moore *et al.*, 1988).

The extracts of some cultured cyanobacteria exhibited good anticancer activity (Barchi *et al.*, 1983). Ethanolic extract of cultured *Tolypothrix byssoideae* was found to show significant inhibitory activity *in vitro* against KB and NIH/3T3 cells and *in vivo* against P-388 lymphocytic leukemia in mice (Barchi *et al.*, 1983). Bioassay directed isolation and identification of the active constituent (tubericidin) have also been reported from *T. byssoideae* (Barchi *et al.*, 1983).

Bioassay directed fractionation of extract of *Phormidium tennue* led to the isolation of three classes of glycolipids *viz.* monogalactosyl diacylglycerols (MGDGs) as antitumor factors. Among the 17 tested MGDGs and digalactosyl diacylglycerols (DGDGs), three of the DGDGs showed more potent inhibitory activity towards tumor proliferation than the others (Shirahashi *et al.*, 1993).

The extract of *Lyngbya majuscula* was found to have strong cytotoxic effect against a Vero cell line ATCC CCI 81 (Gerwick *et al.*, 1994). The extract was also found to be toxic to brine shrimp (LC 50=25ng ml^{-1}). Using the brine shrimp assay to guide fractionation, a unique metabolite was isolated (8-10 per cent of the crude extract) which was found to be responsible for the potent brine shrimp toxicity (LC 50=3ng ml^{-1}) and mammalian cell antiproliferative activity (IC 50=6.8ng ml^{-1}) in the Chinese hamster Aux B1 cell line.

Bioassay guided fractionation of the organic extract of *L. majuscula* led to the isolation of a new lipid, curacin A, with exceptional brine shrimp toxic and tumor antiproliferative activities. Pure curacin A is an antimitotic agent (IC 50 values in three cell lines ranging from 7-200 nm) that inhibits microtubule assembly and the binding of colchicines to tubulin (Gerwick *et al.*, 1994).

In discovering new efficient drugs for the treatment of solid tumors in clinical conditions, Corbett *et al.* (1992) developed a rapid inexpensive disk diffusion assay to find agents that exhibit greater cytotoxicity against tumour cells than leukemia cells. Aulosirazole, the major cytotoxin of *Aulosira fertillissima* shows solid tumor selective activity in the Corbett assay. Its structure has been determined to be 5-hydroxy-3 methoxinapthol (2,3-d) 1,2-Thiazole-4,9-dione (Startman *et al.*, 1994).

The recently discovered antiproliferative agents cryptophycin-A, curacin A (Gerwick *et al.*, 1994) and notodion A (Kobayashi *et al.*, 1990) produced by *Nostoc sp., L. majuscula, and Nostoc commune,* respectively, inhibit microtuble assembly function and show promise as potentially useful anticancer agents.

The extracts of *Spirulina* in which β–carotene is a major component inhibited the growth and destroyed oral cancer cells *in vitro.* β–carotene, a nontoxic carotenoid had a cytostatic and cytotoxic dose-dependent effect on hamster and human oral squamous cell carcinoma cell lines (Schwartz and Shklar 1987).

The chemo preventative activity of *Spirulina* (1g day^{-1} for 12 months) in reversing oral leukoplakia in pan tobacco chewers in Kerala, India showed complete regression of lesions in evaluable subjects supplemented with *Spirulina* (Mathew *et al.*, 1995).

Polysaccharide of *Spirulina* can inhibit the proliferation of ascetic hepatoma cells of mice at a concentration of 200 mg Kg^{-1}. It can inhibit the incorporation of H-thymidine, H-uridine and H-leucine into DNA, RNA and protein synthesis of sarcoma 180 and ascetic hepatoma cells during the period of 24 h after exposure *in vitro*. The degree of inhibition increases with the increase in incubation time. Polysaccharide of *Spirulina* can inhibit DNA synthesis of Sarcoma 180 and ascetic hepatoma cells. The inhibition is due to the alteration in DNA metabolism (Lisheng *et al.*, 1991).

An extract of *Spirulina-Dunaliella* was shown to prevent tumor development in hamster buccal pouch when a solution was applied topically three times weekly for 28 weeks (Schwartz *et al.*, 1988). Anti-cancer properties and antineoplastic activity have also been demonstrated using compounds from marine cyanobacteria (Murray and Mitsui, 1982).

Antitoxin

Mercury induced kidney toxicity in rats was suppressed by feeding *Spirulina*. Para aminophenol and cisplatin induced renal toxicity in rats was significantly reduced by phycocyanin extract of *Spirulina*. (Fukino *et al.*, 1990). Thus, phycocyanin plays a major role in the protective effect of *Spirulina* against mercury and pharmaceutical drugs induced renal failure and that *Spirulina* might be applicable to the reduction of general renal dysfunction (Fukino *et al.*1990).

Radio Protection

The radioprotective effect of *Spirulina* extract was studied using the micronucleus test in polychromatic erythrocytes of bone marrow in mice. The extract caused a significant reduction of the micronucleus frequencies induced by gamma radiation (Qishen *et al.*, 1989).

Spirulina reduces urine radioactivity levels by 50 per cent within 20 days. Use of *Spirulina* also decreases radiation load received from food contaminated with radio nucleotides, Cesium-137 and Stronntium-90. *Spirulina* is favourable for normalizing the adaptive potential of children's bodies in conditions of long lived low dose radiation (Loseva and Dardynskaya, 1993).The intake of easily digestible micro and macro elements of *Spirulina* also has a positive influence on many functions of the immune system (Loseva, 1999).

Reproductive Modulators

Carotenoids are used to improve the health and enhance the fertility of cattle (Jackson, 1981). Crude ethanolic extract of *Oscillatoria willei* caused abnormal changes in the sperm of mouse (Thirunalasundari and Subramanian, 2000). An immune potentiating property coupled with murine antifertility activity, without being toxic to other systems in mice models was exhibited by the extracts of *O. willei* (Krishnaprema, 1996, Jagdeshwari, 1998, Ansarudin, 1999) and *O. late-virens* (Deth, 1999)

Constrains

The scale of utilization of any microbial biomass products is mainly determined by the economic parameters of its production. Efficiency and elegance of cyanobacterial product production depends on many factors which include nutrients availability and its cost, extraction and processing efficiency and acceptance from food industries and consumers. At present, in the free market economics, prices of conventional protein sources available is much less than the *Spirulina* protein, which makes it economically uncompetitive. The high cost of *Spirulina* protein is due to the high cost of production which is primarily due to the cost of major chemicals used in the conventional medium. The cost effective production of cyanobacterial products relates to the low cost biotechnology for its large-scale biomass production. Production of cyanobacterial biomass should therefore be optimized by the development of screening procedures to select an efficient cost effective strain depending on the availability of the low cost nutrients and scaling up of the mass cultivation processes.

Future Prospects of Cyanobacterial Technology in India

The Cyanobacterial technology is considered to be a boon to the developing nations like India. However, the future of this technology depends on expansion of the range of products to be produced from cyanobacteria under economic budget. Molecular biology and genetics of cyanobacteria hold promise for new products and improved yield. At present, it is very difficult to say whether cyanobacteria would be able to compete *E. coli* or yeast strains in hetrotrophic expression of proteins. Many new ideal concepts are yet to be developed to exploit maximum benefit from cyanobacteria. The future of cyanobacterial technology in the production of metabolites having role in nutraceutics, pharmaceutics, food processing and cosmetics industries seems to be bright and needs further attention.

Conclusion

Cyanobacteria are oxygenic, photosynthetic prokaryotes which have long been recognized as having potential in biotechnology as biofertilizers, amino acid production, fuel production, antibiotic production, single cell protein production and for the production of other secondary metabolites and pigments used in food industries. These interesting features have fascinated the Scientists all over the world to understand and explore in detail the basic mechanisms involved for the biotechnological exploitation of cyanobacteria for the betterment of mankind.

The traditional use of cyanobacteria including food, energy source and biofertilizers flourish today and are beginning to reap the benefits of modern molecular biology to enhance their performance. The first generation of high value added byproducts produced by cyanobacteria, including phycoflours are currently available in the market and is quite popular. However, only a few cyanobacterial strains producing intresting compounds have so far been exploited commercially. Screening programs must therefore identify useful products and the cyanobacterial strains having the ability to produce the products in significant amounts should be cultured in mass. Projection for future research on pharmaceutical potential of cyanobacteria includes discovery of novel bioactive metabolites, investigation of the pathways by which selected metabolites are synthesized and regulated and a precise elucidation of the mechanism by which these novel compounds exert their biological effects.

Acknowledgements

The authors thank Dr. R.K. Saxena and Ms. Sangeeta Raina for their help in prepartion of the manuscript.

References

Annapurna V, Shah N, Bhaskaram P, Bamji MS, Reddy V (1991) Bioavailability of *Spirulina* carotene. *J Clinical Biochem Nutrition* 10: 145–151.

Ansarudin PA (1999) Bioactive compound from a marine cyanobacterium *Oscillatoria willei* BDU 130511, *M.Sc. Thesis*, Bharthidasan University, Tiruchirappalli, India.

Ayehunie S, Belay A, Baba TW, Ruprecht RM (1998) Inhibition of HIV-1 replication by an aqueous extract of *Spirulina platensis* (*Arthrospira platensis*). *J Acquired Immune Deficiency Syndromes and Human Retrovirology*. 18: 7-12.

Barchi JJ, Norton TR, Furusawat E, Patterson GM, Moore RE (1983) Identification of a cyatotoxin from *Tolypothrix byssoideae* as tubericidin. *Phytochem* 22: 2851-2855.

Becker EW (1994) Microalgae. In: *Biotechnology and Microbiology*. Cambridge Univ Press, Cambridge, U.K.

Bhat VB, Madyastha KM (2001) Scavenging of peroxynitrite by phycocyanin and phycocyanobilin from *Spirulina platensis*: protection against oxidative damage to DNA. *Biochem Biophys Res Commun* 285: 262-266.

Bloor S, England R R (1991) Elucidation and optimization of the medium constituents controlling antibiotic production by the cyanobacterium *Nostoc muscorum*. *Enzyme Microb Technol* 13: 76-81.

Bobbili V V, Pardhasaradhi A, Mubarak Ali A, Kumari L, Reddanna P, Khar A (2003) Phycocyanin-mediated apoptosis in AK-5 tumor cells involves down-regulation of Bcl-2 and generation of ROS. *Mol Cancer Ther* 2: 1165–1170.

Bonjouklian R, Smitra TA, Doolin LE, Molly RM, Debono M, Shaffer SA, Moore RE, Stewart JB, Patterson GML (1991) Tjipanazoles, new antifungal agents from the blue-green alga *Tolypothrix tjipanasensis*. *Tetrahedron*. 47:7739-7740.

Borowitzka MA, Borowitzka. LJ (1988) *Microalgal Biotechnology*. Cambridge University Press, Cambridge, U.K.

Boyd MR (1988) Strategies for the identification of new agents for the treatments of AIDS: A national program to facilitate development of new drug candidates for clinical evaluation. In: *AIDS Etiology, Diagnosis, Treatment and Prevention*, (eds) De Vita VJ Jr., Hillman S, Rosenberg SA, Philadelphia, Lippincott, pp. 305.

Burja A M, Banaigs B, Mansour EA, Burgess JG, Wright PC (2001) Marine cyanobacteria: A prolific source of natural products. *Tetrahedron* 57; 9347-9377.

Chandrasekaran M (1997) Industrial enzymes from marine microorganisms: The Indian Scenario. *J Marine Biotechnol* 5: 86-89.

Chauhan V S, Singh B, Singh S, Bisen PS (2004) Scrrening of cyanobacterial strains producing cyanovirin–N and cyanovirin–V, protein capable of inhibiting growth of HIV-I and HIV-II. In: *Proceedings of the International Conference on Recent Advances in Biomedical and Therapeutic Sciences*. Bundelkhand University, Jhansi, India. 247.

Ciferri O, Tiboni O, Sanangelantoni AM (1989) The genetic manipulation of cyanobacteria and its potential uses. In: *Algal and Cyanobacterial Biotechnology*, (eds) Cresswell RC, Rees TAV, Shah N, Longman Group UK, pp. 239-262.

Cohen Z (1986) Products from microalgae In: *CRC Handbook of Microalgal Mass Culture*, (ed.) Richmond A. CRC Press, Boca Raton, Florida, USA, pp. 421-454.

Corbett TH, Valeriote FA, Polin LC, Pugh S, White K, (1992) Models and concepts for discovery and development. In: *Cytotoxic Anticancer Drugs*, (eds) Valeriote FA, Corbett TH, Baker LH, Kluwer Academic Publishers. Norwell, pp. 35.

Dainippon Ink and Chemicals Inc. (1981) Cosmetics containing water soluble phycocyanin. *Japanese Patent* 79:138755.

Dainippon Ink and Chemicals Inc. (1980) Production of highly purified alcoholophilic phycocyanin. *Japanese Patent* 80: 77890.

Deth SK (1999) Antimicrobial compound from marine cyanobacteria with special reference to the bioactivity of a purified compound from *Oscillatoria latevirens*. *Ph.D. Thesis,* Bharthidasan University, Tiruchirappalli, India.

Evets LB, Belookaya T, Lyalikov S, Orehov SD, Shipulin E (1994) Means to normalize the levels of immnoglobulin E. Russian Federation Committee of Patents and Trade. Patent Number (19) RU (11)20005486 C1 (51) 5 A 61K35/80.

Fogg GE (1971) Extracellular products of algae in freshwater. *Arch Hydrobiol* 5: 1-25.

Fogg GE, Stewart WDP, Fay P, Walsby AE (1973) *The Blue Green Algae*. Academic Press, London.

Fukino H, Takagi Y, Yamane Y (1990) Effect of *Spirulina* (*S. platensis*) on the renal toxicity induced by inorganic mercury and cisplatin. *Eisei Kagaku*, 36:5.

Gerwick, WH, Mrozek, C, Moghaddam, MF, Agrawal, SK (1989) Novel cytotoxic peptides from the tropical marine cyanobacterium *Hormothamnion enteromorpoides*. 1. Discovery, isolation and initial chemical and biological characterization of the hormothamnins from wild and cultural material. *Experientia* 45:115-117.

Gerwick, WH, Proteau, PJ, Nagle, DG, Hamel, E, Blokhin, A, Slate DI (1994) Structure of curacin A, a novel antimitotic, antiproferative and brine shrimp toxic natural product from the marine Cyanobacterium *Lyngbya majuscula*. *J Org Chem* 59: 1243-1247.

Giraudet H, Berthon JL, Buisson B (1997) A comparison of daily alkaline phosphatase activity of a cyanobacterium (*Microcystis aeruginosa*) and a diatom (*Synedra capitana*). *Comp Ren Acad Sci Ser III Sci Vie Life Sci*, 320: 451-458.

Guerrero M, Ramos JL, Losada M (1982) Photosynthetic production of ammonia. *Experientia* 38: 53-55.

Gustafson KR, Cardellina JH II, Fuller RW, Weslow OS, Kiser RF, Snader KM, Paterson GM, Boyd MR (1989) AIDS-antiviral sulfolipids from cyanobacteria (blue green algae). *J Natl Cancer Inst.* 81: 1254-1258.

Hayashi K, Hayashi T, Morita N (1993) An extract from *Spirulina platensis* is a selective inhibitor of *Herpes simplex* virus type 1 penetration into HeLa cells. *Phytother Res.* 7:76-80.

Hayashi K, Hayashi T, Kojima I (1996) A natural sulfated polysaccharide, calcium spirulan, isolated from *Spirulina platensis*: *in vitro* and *ex vivo* evaluation of anti-*Herpes simplex* virus and anti-human immunodeficiency virus activities. *AIDS Res Human Retroviruses*. 12: 1463-1471.

Hayashi O, Katoh T, Okuwaki Y (1994) Enhancement of antibody production in mice by dietary *Spirulina platensis*. *J Nutr Sci Vitaminol.* 40: 431-441.

Horrobin DF (1990) *Omega-6 Essential Fatty acids: Pathophysiology and Roles in Clinical Medicine*. Wiley-Liss, New York.

Ishibashi M, Moore R, Patterson, GML, Xu C, Clardy J (1986) Scytophycins: cytotoxic and antimycotic agents from the cyanophyte *Scytonema pseudohojmanni*. *J Org Chem* 51: 5300-5305.

Jackson PSA (1981) Note on a possible association between plasma β–carotene levels and conception rate in a group of winter housed dairy cattle. *Animal Production*. 32; 109-112.

Jassby A, Lembi CA, Waaland JR (1988) *Algae and human affairs*. Cambridge University Press, Cambridge, UK.

Jagdeshwari V (1998) Effect of purified compound from *Oscillatoria willei* BDU 130511 on the testis and accessory reproductive organs of swiss albino mice. *M.Sc. Thesis*, Bharthidasan University, Tiruchirappalli, India.

Kerby NW, Niven GW, Rowell P, Steward WDP (1987) Photoproduction of amino acids by mutant strains of N_2–fixing cyanobacteria. *Appl Microbiol Biotechnol* 25: 547-552.

Kerby N, Niven G, Rowell P, Steward WDP (1988). Ammonia and amino acid production by cyanobacteria. In: Algal *Biotechnology. Proceedings of the 4th International Meeting of the SAA.* (eds) Studler T, Mollion J, Verdus MC, Karamanos Y, Morvan H, Christiasen D. Elsevier Appl Sci, London, New York; pp. 277-283.

Kobayashi J, Ishibashi M, Murayama J, Takamatsu M, Iwamura M, Ohizumi Y, Sasaki T (1990) Amphidinolide-E, a novel antileukemic 19-membered microlide from the cultured symbiotic dinoflagellate *Amphidinium sp. J Org Chem* 55: 3421-3425.

Krinshnaprema K (1996) Effect of the crude ethanolic extract of *Oscillatoria willei* BDU 130511 on male BALB C mice. *M.Sc. Thesis*, Bharthidasan University, Tiruchirappalli, India.

Lisheng L, Baojiang G, Jihong R, Guangquan Q, Botang Wu (1991) Inhibitive effect and mechanism of polysaccharide of *Spirulina platensis* on transplanted tumor in mice. *Marine Sc*, 5:33-38.

Loseva LP, Dardynskaya IV (1993) *Spirulina*–natural sorbent of radionucleides. *6th Int Congress of Applied Algology*, Czech Republic, Belrus.

Loseva LP (1999) *Spirulina patensis* and specialties to support detoxifying pollutants and to strengthen the immune system. *8th Int Congress of Applied Algology*, Italy, Republic, Belrus.

Madhava C, Reddy J, Bhat V B, Kiranmai G M, Reddy S, Reddanna P, Madyastha K M (2000) Selective Inhibition of cyclooxygenase-2 by C-phycocyanin, a biliprotein from *Spirulina platensis. Biochem Biophys Res Commun* 277:599-603.

Mathew B, Sankaranarayanan R, Nair PP, Varghese C, Somanathan T, Amma BP, Amma NS, Nair MK (1995) Evaluation of chemoprevention of oral cancer with *Spirulina. Nutrition and Cancer*. 24 (2): 197-202.

Mohanty P (1996) Cyanobacterial photosynthesis: concepts and applications. *J Sci Ind Res* 55: 553-570.

Moore RE (1981) Constituents of blue green algae. In: *Marine Natural Products: Chemical and Biological Perspectives*, (ed) Scheuer, PJ, Academic Press, New York.

Moore RE, Patterson GML, Carmichael WW (1988) New pharmaceuticals from cultured blue green algae. In: *Biomedical Importance of Marine Organisms*, (ed) Fautin DG, Mem California Acad Sci 13:143-146.

Murray RL Jr, Mitsui A (1982) Growth of hybrid Tilapia fry fed nitrogen fixing marine blue green algae in sea water. *J world Maricult Soc* 13: 198-203.

Musgrave SC, Kerby NW, Codd GA, Stewart WDP (1982) Sustained ammonia production by immobilized filaments of the nitrogen-fixing cyanobacterium *Anabaena* 27893. *Biotechnol Lett* 4: 647-52.

Normal J, Temple I (1988) Does β-carotene prevents cancer: a critical appraisal. *Nutrition research*. 8: 685-690.

Patterson GML, Baker KK, Baldwin CL, Bolis CM, Caplan FR, Larson LK, Levine IA, Moore RE, Nelson CS, Tschappat KD,Tuang. G. (1993) Antiviral activity of cultured blue-green algae (cyanophyta). *J Phycol* 29:125-130.

Patterson G M L, Carmeli S (1992) Biological effects of toly-toxin (6-hydroxy-7-O-methyl-scytophycin B), a potent bioactive metabolite from cyanobacteria. *Arch Microbiol* 157: 406-411.

Patterson G M L (1996) Biotechnological application of cyanobacteria. *J Sc Ind Res* 55: 669-679.

Puniya AK, Singh S, Ganeshkumar C, Singh K (1995) Single cell protein: a promising dietary substitute. *Indian J Exp Biol* 33: 545-551.

Qishen P, Baojiang G, Kolman A (1989) Radioprotective effect of extract from *Spirulina platensis* in mouse bone marrow cells studied by using the micronucleus test. *Toxicol Letters;* 48:165-169.

Qureshi M, Ali R (1996) *Spirulina platensis* exposure enhances macrophage phagocytic function in cats. *Immunopharmacol Immunotoxicol.* 18(3):457-463.

Qureshi MA, Kidd MT, Ali RA. (1995) *Spirulina platensis* extract enhances chicken macrophage functions after *in vitro* exposure. *J Nutr Immunol* 3: 35-44.

Qureshi MA, Garlich J, Kidd M. (1996) Dietary *Spirulina platensis* enhances humoral and cell-mediated immune function in chickens. *Immunopharmacol Immunotoxicol* 18:465-476.

Rai A N (1990) *Handbook of Cyanobacteria.* CRC Press Boca Raton, Florida, USA.

Ramirez D, Gonzalez R, Merino N, Rodriguez S, Ancheta O (2002). Inhibitory effects of *Spirulina* in zymozan-induced arthritis in mice. *Mediators Inflamm.* 11: 75-79.

Ramos J L, Guerrero MG, Losada M (1987) Factors affecting the photoproduction of ammonia from dinitrogen and water by the cyanobacterium *Anabaena sp.* Strain ATCC 33047. *Biotechnol Bioeng* 29: 566-571.

Rathore NK, Singh S, Singh V (2004) *Spirulina* as immuno–enhancer and biomodulator. *J. Immunol. Immunopathol.* 6 (2): 1-10.

Riccardi G, Sora S, Cifferi O (1981) Production of amino acids by analog-resistant mutants of the cyanobacterium *Spirulina platensis. J Bacteriol* 147: 1002-1007.

Rinehart K L Jr, Shaw P D, Shield L G, Gloer J B, Harbour G C, Koker M E S, (1981) Marine natural products as sources of antiviral, antimicrobial and antineoplastic agents. *Pure Appl Chem* 53:795-799.

Reichelt J L, Borowitzka MA (1984) Antibiotics from algae: results of a large scale screening programme. *Hydrobiologia* 116/117:143-147.

Roberts R J (1985) Restriction and modification enzymes and their recognition sequences. *Nucleic Acids Res* 13: 2165-2200.

Schwartz J, Shklar G (1987) Regression of experimental hamster cancer by beta-carotene and algae extracts. *J Oral Maxillofac. Surg* 45: 510-515.

Schwartz J, Shklar G, Reid S, Trickler D (1988) Prevention of experimental oral cancer by extracts of *Spirulina-Dunaliela algae. Nutr Cancer* 11:127-134.

Schlegel I, Thanh Doan N, Nola de C, Smith G D (1999) Antibiotic activity of new cyanobacterial isolates from Australia and Asia against green algae and cyanobaceria. *J Appl Phycol* 10 (5): 471-476.

Shih S R, Tsai K N, Li Y S, Chueh C C, Chan E C (2003) Inhibition of enterovirus 71-induced apoptosis by allophycocyanin isolated from a blue-green alga *Spirulina platensis. J Med Virol.* 70: 119–125.

Shirahashi H, Murakami N, Watanabe M, Nagatsu A, Sakakibara J, Tokuda H, Nishino H, Iwashima A (1993) Isolation and identification of antitumour promoting principle from fresh water cyanobacterium *Phormidium tenue. Chem Pharm Bull* 41: (9) 1664-1667.

Singh S (1992) Photobiological production of ammonia by an L-methionine-DL-sulfoximine-resistant mutant of the cyanobiont *Nostoc* ANTH. *Indian J Exp Biol* 30: 284-287.

Singh S (1993) Immobilization-caused physiological and biochemical changes in the cyanobiont *Nostoc* ANTH. *Indian J Exp Biol* 31: 446-449.

Singh S, Rai AK (1989) Methionine sulphoximine induced changes in the cyanobacteria *Anabaena doliolum* and *Anacystis nidulans. Indian J Exp Biol* 27: 227-229.

Startman K, Belli J, Jensen CM, Moore RE and Patterson GML (1994) Aulosirasole, a novel solid tumor selective cytotoxin from the blue green alga *Aulosira fertilissima J Org Chem* 59: 6279-6285.

Sundararaman M (1992) Intra and extra cellular substances in marine cyanobacteria. *Ph.D. Thesis,* Bharathidasan University, Tiruchirappalli, India.

Thirunalasundari T, Subramanian G (2000) Antifertility compound from marine cyanobacterium. In: *Proc International Conference of Reproductive Biology. 9th French–Czech–Slovak Colloquium.* Slovak Academy of Sciences, Stara Lesna.

Upasani CD, Khera A, Balaraman R (2001) Effect of lead with vitamin E, C or *Spirulina* on malondialdehyde, conjugated dienes and hydroperoxides in rats. *Indian J. Exp. Biol.* 39: 70-74.

Venkataraman L V (1983) *Spirulina* in India. In: *Proc National Seminar on Cyanobacterial Research-Indian Scene,* (ed) Subramanian G, National Facility of Marine Cyanobacteria, Tiruchirappalli, pp. 92.

Vincenzini M, De Philippis R, Ena A, Florenzano G (1986) Ammonia photoproduction by *Cyanospira rippkae* cells "entrapped" in dialysis tube. *Experientia* 42: 1040-43.

Whitehead P.R. Brown N.L. (1985) Three restriction endonucleases from *Anabaena flos-aquae. J Gen Microbiol* 131: 951-958.

Whitton B A, Grainger S L J, Hawley G R W, Simon J W (1991) Cell bound and extracellular phosphatase activities of cyanobacterial isolates. *Microbial Ecol* 21: 85-98.

Wikstrom P, Szwajcer E, Brodelius P, Nilsson K N, Mosbach K (1982) Formation of alpha keto acids from amino acids using immobilized bacteria and algae. *Biotechnol Lett* 4: 153-156.

Zhang-Chang-Wu (1994) Effects of polysaccharides and phycocyanins from *Spirulina* on peripheral blood and haemopoietic system of bone marrow in mice. *Proc. Second Asia Pacific Conference on Algal Biotechnol,* Univ. Malasiya, pp. 58.

Authors Index

A

A. Vaishampayan, 1
A.B. Prasad, 1
A.K. Mishra, 142

B

B.S. Tiwari, 261
B.V. Singh, 196,213

C

C. Dayananda, 56
C.B. Singh, 169

D

D. Vyas, 155
D.W. Dhar, 213

G

G. A. Ravishankar, 56
G. Abraham, 97

H

H. K. Sharma, 1
H.K. Jaiswal, 1

I

Iqbal Habib, 209

J

J. N. Maurya, 104
J. Rath, 42
J.P. Shahi, 1

K

K. Ganesan, 85
K. Suresh Kumar, 85
K.K. Choudhary, 142
Kiran Toppo, 179

M

M.K. Das, 186
M.R. Suseela, 179

N

N. K. Rathore, 275

P

P. Jaiswal, 75
P. K. Singh, 75
P. V. Subba Rao, 85

R

R. Prasanna, 75
R. Raghuvanshi, 275
R. Sarada, 56
R.K. Gupta, 131,155,250
R.K. Singh, 1
R.P. Singh, 1
R.P.N. Mishra, 1
Rajinder K. Gupta, 213

S

S. Sharma, 1
S. Singh, 196,213
Surendra Singh, 275
S.K. Singh, 250
S.N. Tripathi, 261
S.P. Adhikary, 42
S.S. Singh, 142

T

T. Dey, 1

U

U.K. Chaturvedi, 209

V

V. Prasad, 131
V.D. Pandey, 250
Vaibhav A. Mantri, 85

Y

Y. Singh, 115
Y.V. Singh, 229

Subject Index

A

ACE, 254
Acrylic acid, 210
AFLP, 219
Algal primary metabolites, 136
Algal secondary metabolites, 136
Algicidal activity, 76
Algicide, 75
Alginate biofilms, 175
Alginate spheres, 175
Amino acids, 281
Anatoxin-*a*, 202
Antialgal activity, 278
Antibacterial compounds, 45
Antibacterial, 278
Anticancer compounds, 286
Anticancerous agents, 252
Antifungal, 278
Anti-inflammatory agents, 254
Antimicrobial agents, 252
Antimicrobials, 277
Antioxidant activity, 46,93
Antiviral agents, 253
Antivirals, 277
Aquatic macrophytes, 170
ARDRA, 221
Arsenic, 188
Artificial symbioses, 20
Azolla-Anabaena, 10

B

Bidirectional hydrogenase, 160
Bioactive compounds, 251,282
Biodiversity, 237
Bioenergy, 57
Biofertilizer, 98,143
Biofuel production, 63
Biofuel, 59
Bioprospecting, 42,43
Bioremediation, 182
Bioresources, 48
Biosorptions, 66
Botryococcus, 61,115,119

C

Cadmium,188
Calothrixin, 252
Carbolines, 253
Carotenoid, 121
Cellular water, 261,262
Chlorella, 61
Chlorellin, 209
Chlorophyll fluorescence, 104
Chromium, 188
Cyanobacterial blooms, 198
Cyanobacterial DNA, 214
Cyanobacterial mutants, 2
Cyanobacterial symbioses, 8
Cyanobacterial toxins, 196
Cylindrospermopsin, 203

D

Debromoaplysiatoxin, 252
Dessication-tolerant cyanobacteria, 261
Diatoms, 63
Diazotrophs, 2
DNA-DNA hybridization, 216
Dunaliella, 61

E

Energy crops, 70
Environmental management, 174
Enzyme inhibitors, 254
Enzymes, 282
Eustigmatophytes, 62

F

Fatty acids, 282
Feed, 131, 138
Fm, 109
Fo, 109
Folk medicines, 137
Freshwater algal food, 134
Fv, 109

G

Goitre, 210
Golden-brown algae, 62
Green algae, 59

H

Hapalindoles, 253
Heavy metal pollution, 179,188
Heavy metal toxicity, 186
Heavy metals, 66,170,180
Heterocyst frequency, 8
HIV, 253
Homoanatoxin-*a*, 203
Hormothamnins, 253
HSV, 253
Human Diet, 85
Hydogenases, 155
Hydrocarbons, 67, 115
Hydrocracking, 67
Hydrogen metabolism, 155
Hyperaccumulators, 172

I

Immobilization, 126
Immobilized cyanobacteria, 175
Immunomodulators, 285
Indolocarbazoles, 253

L

Lead, 188

M

Majusculamide, 253
Malyngamides, 253
Marine algae, 42,43
Marine algal food, 133
Mercury, 188
Metal uptake, 183
Microalgae, 56
Microcystis, 75
Microsatellite, 220
Muscle relaxants, 254

N

Nitrogenase, 156
Nutraceuticals, 99,131,275,279

O

Organic farming, 229,231
Oxyphotobacteria, 251

P

PC-IGS, 222
Pesticides, 4,5
Pharmaceutical compounds, 251
Pharmaceuticals, 99,275,276
Photosynthetic prokaryotes, 172
Phycobiliproteins, 280
Phytoplanktons, 171
Pigments, 147
Plasmids, 215
Plasmin, 254
Plastoquinone, 105
Polysaccharides, 145
Polyurethane foam, 175
Prymnesiophytes, 62
PS I, 105
PS II, 105
Pyrolysis, 69

R

RAPD, 217
Renewable energy, 56
Restriction enzymes, 284
RFLP, 218
Rice field cyanobacteria, 186

S

Saxitoxins, 203
Schizotrin, 253
Scytophycins, 252
Seaweed amino acids, 92
Seaweed proteins, 92
Seaweeds, 85,86
Secondary metabolites, 251
Soil microorganisms, 236

T

Thermochemical liquefaction, 67

Tolypodiol, 254

Tolytoxin, 252

Toxic cyanobacteria, 197

Transcriptional regulation, 155

U

Uptake hydrogenase, 157,159

UV-B radiation, 126

V

Vitamins, 282

W

Water sorption isotherms, 265